W9-BOL-156

'In this rollercoaster of a book, Ben Wilson describes the 1850s as the most explosive period in history, a decade that gave birth to modernity and trampled those who resisted it ... the scholarship is certainly impressive but the drama is what delights ... *Heyday* is a lot like its subject; it's a big-bearded book of enormous scope and unstoppable momentum'
Gerard DeGroot, *The Times*

'This is a scholarly, intelligent and readable book. This book is an original prism through which to view the mid-nineteenth century and, essentially, about the invention not so much of modernity as of globalisation'
Simon Heffer, *Spectator*

'Wilson's account of the 1850s is a wonderfully engrossing and intelligent read ... He has clever and entertaining things to say about even the most banal topics, tracing the Victorian enthusiasm for beards, for instance, to the impact of the Crimean War ... He even manages to make the history of Minnesota exciting'
Dominic Sandbrook, *Sunday Times*

'This is a book written with great verve and with a sharp eye for the obscure fact or the telling details'
Denis Judd, *TLS*

'With a rip-roaring style to match his subject ... excellent ... His grasp of the interplay between politics, economics and individuals is admirable. This is narrative history of the highest quality'
Andrew Lycett, *Daily Telegraph*

Ben Wilson is the author of four critically acclaimed history books, including *What Price Liberty?* for which he received a Somerset Maugham Award, and *Empire of the Deep: The Rise and Fall of the British Navy*, a *Sunday Times* bestseller. He was educated at Pembroke College, Cambridge, where he graduated with a first-class degree and an MPhil in history. He lives in Suffolk.

By the same author

The Laughter of Triumph: William Hone and the Fight for the Free Press

Decency and Disorder: The Age of Cant, 1789–1837

What Price Liberty?: How Freedom Was Won and Is Being Lost

Empire of the Deep: The Rise and Fall of the British Navy

HEYDAY

The 1850s and the Dawn of the Global Age

BEN WILSON

A W&N PAPERBACK

First published in Great Britain in 2016
by Weidenfeld & Nicolson
This paperback edition published in 2017
by Weidenfeld & Nicolson,
an imprint of the Orion Publishing Group Ltd,
Carmelite House, 50 Victoria Embankment,
London EC4Y 0DZ

An Hachette UK company

1 3 5 7 9 10 8 6 4 2

A CIP catalogue record for this book
is available from the British Library.

ISBN 978 0 753 82921 9

Typeset by Input Data Services Ltd, Somerset

Printed and bound by CPI Group (UK) Ltd, Croydon, CR0 4YY

www.orionbooks.co.uk

For my darling Ariane,
born when the idea for this book was forming in my mind

Of all the decades in history, a wise man would
choose the eighteen-fifties to be young in.

G. M. Young

The need of a constantly expanding market for its products chases
the bourgeoisie over the entire surface of the globe. It must nestle
everywhere, settle everywhere, establish connections everywhere.

Karl Marx and Friedrich Engels

CONTENTS

PART III
NEWS OF THE WORLD: THE AGE OF BRONZE

ILLUSTRATIONS AND MAPS

Illustrations

Page

Plate Section

Maps

PREFACE

When the idea for *Heyday* first formed in my mind, I was drawn to the decade of the nineteenth century that seemed big with world-shattering events. The 1850s saw the Australian Gold Rush, the Crimean War and the Indian Rebellion. Alongside them some of the great technological advances of the century took place in this decade, notably the advent of international telecommunications. There were other upheavals too; perhaps most marked of all was the astonishingly rapid settlement of what the Victorians called the 'wilderness' by tidal waves of European and Chinese migrants. It was a decade of accelerated modernisation and astounding economic boom, of scarcely credible optimism and swaggering self-confidence. With so many momentous events crowded into such a short space of time I was confident it warranted closer exploration.

These things had been written about extensively, but they seemed oddly disjointed. I wanted to understand the human side of the story of this extraordinary acceleration of progress thanks to new technologies and set it in the context of the geopolitical events that were also reshaping the world. The task I set myself was to unravel threads of connections that led from one part of the planet to another – to detach events that appeared local and fit them into a globe-spanning jigsaw.

The more I immersed myself in this period the more I began to see it as distinct and very different from what we understand as the 'Victorian era'. It is a modern habit to parcel up time into decades. But it became clear to me that people in the nineteenth century regarded the 1850s as possessing defining characteristics quite different from the periods that sandwiched it.

If you are American the 1850s is significant because it marks the sharp descent to Civil War. For the Chinese, this was a blood-drenched decade of civil war and foreign intrusion that marked a national humiliation that is only now being overcome. In India it has special significance because it was the time not only of the great Rebellion against British rule, but the advent of the railway and telegraph. The birth of nations such as Japan, Italy, Germany, Australia, New Zealand and Canada are traced back to these years.

But for the British the 1850s does not seem to possess such special significance. There was no great political controversy or domestic upheaval to mark it out as particularly significant; all seems tranquil – business as usual for the Victorians. I hope to alter that perception in the following pages. The period from 1851 to 1862 – the eleven years that *Heyday* charts – represents a unique moment in British history, one that often gets obscured. During that brief time Britain was unquestionably at the height of its power, influencing and shaping the events that left such a deep impression in so many other countries. At no other point in its history did it possess such influence over the destinies of the human race; at no other time did confidence ride so high.

That power was short-lived and hard to discern. One of the reasons that the British find it difficult to appreciate the extent and reality of this power was because it was not concentrated or tangible: parts of the map painted red, great fleets on the sea and military victories tend to affirm the reality of power and rivet the attention of history. Britain's influence was of an entirely different order. It was put best by the civil servant and historian Herman Merivale during a series of lectures at Oxford at this time: 'It is a sort of instinctive feeling to us all [the British], that the destiny of our name and nation is not here, in this narrow island which we occupy; that the spirit of England is volatile, not fixed; that it lives in our language, our commerce, our industry, in all those channels of inter-communication by which we embrace and connect the vast multitude of states, both civilized and uncivilized, throughout the world.'[1]

That sentence of Merivale's helped crystallise in my mind the reason why the 1850s is not seen as a pivotal decade in Britain in quite the same way as in other countries. If the home front was relatively tranquil and uneventful – providing few of the headline events that capture the imagination of later generations – Britons were actively engaged in the wider world. The interests of the country and the rest of the planet were intermeshed to an unprecedented – and perhaps unrepeated – degree. This became apparent to me as I was writing this book; open any newspaper from the time, including provincial papers, and there is a wealth of often surprising detail about faraway lands about which today, I wager, we know comparatively little.

The result has been that *Heyday* has taken me to locations – often exotic and perhaps out of the way – that I could never have imagined playing a part in the book when I started. Places and events that do not usually make it to histories of the nineteenth century or the Empire suddenly became hugely significant – Nicaragua, Minnesota, Newfoundland and the Caucasus spring to mind. I ventured to them because I followed strands teased out of smudged newspaper columns or forgotten travel books. Often we will travel to these locations in the company of Britons who saw themselves primarily as citizens of the world, not as foot soldiers of empire. One of the joys of researching and writing this book has been in discovering connections across the planet, or in making comparisons between places, peoples and events that seem unlikely but which jumped out at me as I followed the trail. Researching the history of beards, for example, helped me put my finger on notions of British national identity and masculinity that emerged in these frenetic years.

Some of the places and events in *Heyday* are not so obscure. Much of the time as I researched it I felt myself being pulled westwards across the Atlantic to the United States. Events that I remembered from a paper on US history during my undergraduate days or from reading about elsewhere suddenly began to take on new, global significance. If the United States looms large in *Heyday* it is because it did so for Britons in the 1850s – far more so than the Empire (with the exception of India in 1857). At no other point between the American Revolution and the Second World War were the two countries so tightly bound together or so interdependent. Political controversies, wars, economic fluctuations, elections, even the weather in one country deeply impacted on the other. No wonder that America watched events in Britain and the British Empire like a hawk, and the British feasted on information from places that had recently been obscure to them, such as Illinois or Kansas. And little wonder also that this intimate relationship was prone to breakdown: even in this short passage of time Britain and the US came to the brink of war three times.

If the Britain of 1851, when my narrative commences, was little different on the surface from that of 1862, when it concludes, the same cannot be said for the rest of the world. In this transformational moment of modern history, Britain was actively involved almost everywhere, for good and ill. By the time her empire was at its greatest geographical extent, at the beginning of the twentieth century, Britain's power had long since peaked. Its true zenith coincided with an earlier decade of global tumult and galloping change when – crucially – it exercised dominance on the world stage largely untroubled by a major rival.

According to an editorial in the *Illustrated London News* in 1851 Britain's global influence was measured in 'our widely-stretching commerce, our incessant activity, our wealth diffused in every part of the world, our steamships on every sea, our lines of communication to the remotest parts of the earth, our capital, our skill, no less than our great mechanical and scientific triumphs'. It was self-congratulatory hype, of course, but the newspaper was spot on about the connection between technology, communications, trade and power. I began *Heyday* as a study of the technological revolution that followed hot on the heels of the Industrial Revolution. But I soon realised that in the increasingly interconnected mid-nineteenth century nothing could be written about in isolation.[2]

Heyday is subtitled 'Britain and the Making of the Modern World'. But it is not a history of Britain, nor is it a history of the British Empire, nor a comprehensively global history. Rather, it is about a different kind of relationship with the world – one revealed in the quotes by Merivale and the editor of the *Illustrated London News* – that I came to see as defining this period. Put simply, it is a history of Britain *in* the world.

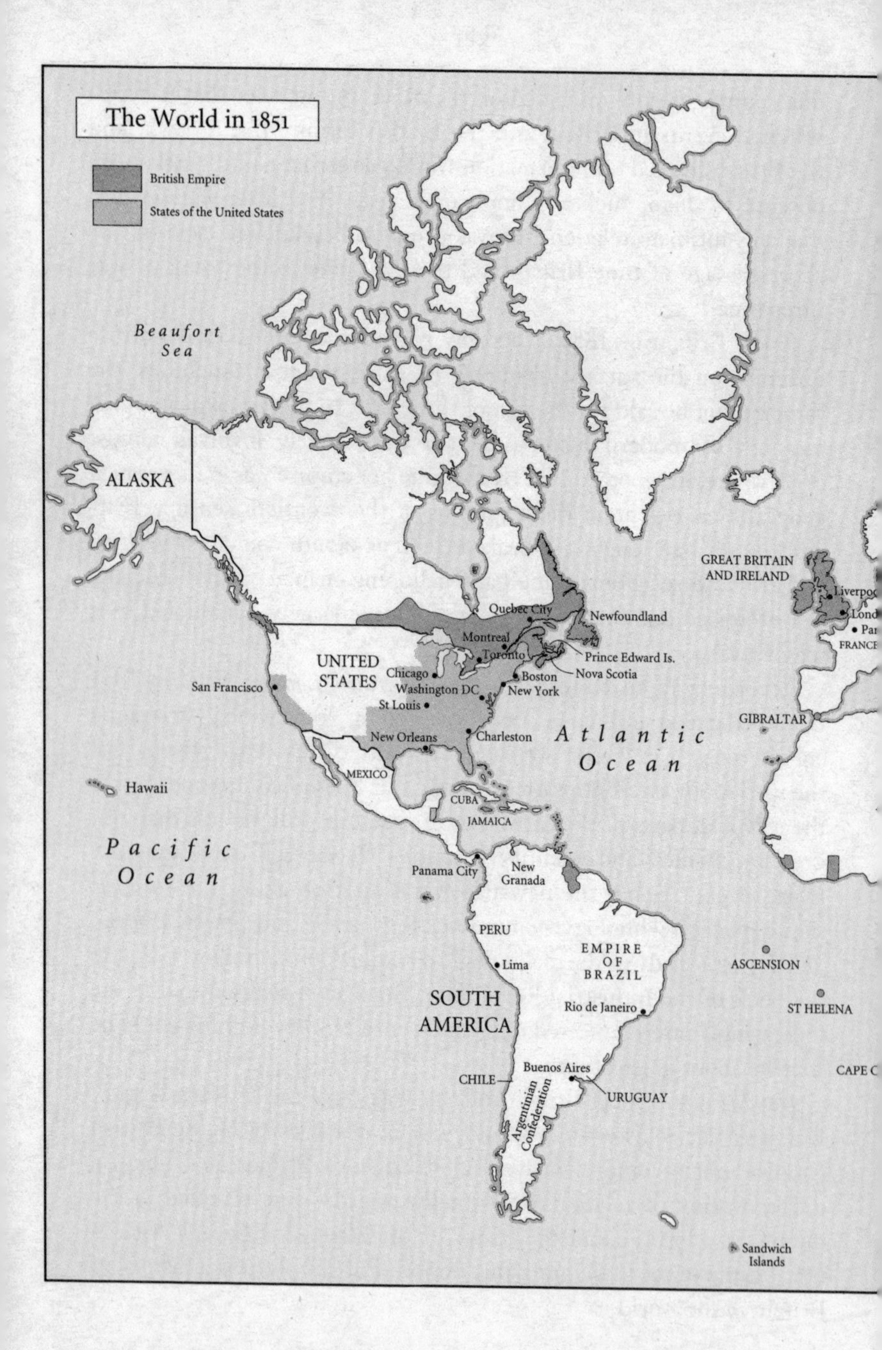

The World in 1851
British Empire
States of the United States
Beaufort Sea
ALASKA
GREAT BRITAIN AND IRELAND
Quebec City
Newfoundland
Montreal
Toronto
Prince Edward Is.
Nova Scotia
UNITED STATES
Chicago
Boston
New York
Washington DC
San Francisco
St Louis
GIBRALTAR
New Orleans
Charleston
Atlantic Ocean
MEXICO
Hawaii
CUBA
JAMAICA
Pacific Ocean
Panama City
New Granada
PERU
EMPIRE OF BRAZIL
ASCENSION
Lima
SOUTH AMERICA
Rio de Janeiro
ST HELENA
Buenos Aires
CHILE
URUGUAY
Argentinian Confederation
Sandwich Islands

Arctic Ocean
RUSSIAN EMPIRE
St Petersburg
Moscow
RUSSIA
USTRO-
NGARIAN
MPIRE
OTTOMAN EMPIRE
PERSIAN
EMPIRE
Khiva
Kokand
Bukhara
AFGHANISTAN
Bering
Sea
Beijing
JAPAN
Edo
CHINA
Nanking
Shanghai
Nagasaki
Pacific
Ocean
Delhi
Calcutta
INDIA
BURMA
Canton
Bombay
HONG
KONG
RICA
BAHRAIN
ADEN
SIAM
PHILIPPINES
MALDIVES
SUMATRA
SINGAPORE
SEYCHELLES
Indian
Ocean
MAURITIUS
AUSTRALIA
NATAL
Adelaide
Sydney
Melbourne
Auckland
Wellington
NEW ZEALAND
Antarctic Ocean

INTRODUCTION

1851: Precipice in Time

. . . change is coming upon us so rapidly, that only
the young can fully comprehend it.
Jesup W. Scott[1]

Drip. Drip. Drip. Agonisingly slowly the sap oozed out of incisions drilled into the trunk of a palaquium fruit tree. The teams of Malay woodsmen waited as the resin steadily filled their bamboo bowls or coconut shells. Braving the insects, tigers or pit vipers that inhabited the jungle, they had ventured deep into the rainforest, armed with *billiongs* (axes) and *parangs* (machetes) to hunt for the mighty palaquiums, which rose sixty feet or more into the sky.[2]

Upon exposure to the air, the sap, called gutta-percha, solidified. This hard, pliable latex was washed and folded into blocks. Loaded into the holds of cargo ships at Singapore, the gutta-percha began its journey to a high-tech factory in London, at Wharf Road, Islington.

'We enter a modest-looking doorway beside a pair of folding gates, on which the words "Gutta Percha Company" are printed,' wrote a visiting journalist in 1851, 'and we become speedily aware that a branch of manufacture of which we hitherto knew next to nothing is being carried on within.'[3] The visitor's journalistic instincts were right to carry him to this anonymous-looking factory. If not many people knew what gutta-percha was in 1851, they soon would. Few other materials have had such a revolutionary impact on the world. And few others have been forgotten so quickly.

The journalist's nose was assailed first; the factory smelt like a tannery, with odours of cheese, drying laundry and gas tar mixed in.

Then he saw the strange blocks of natural rubber that had come from the rainforests of the Malayan Archipelago. Our visitor then observed the purification process: the gutta-percha was boiled, shaved by a state-of-the-art cutting machine with two vertical discs spinning at 200 revolutions per minute, boiled again, then kneaded at high temperature in a masticator. After being cooled in another machine and rolled into sheets, the purified mass of plastic was ready to be sold to manufacturers around the industrialised world.

Gutta-percha had first come to the attention of an English traveller two centuries before. But this hard, rubber-like substance seemed to have no practical use. It was only in 1832 that Dr William Montgomery, assistant surgeon to the Presidency of Singapore, perceived its extraordinary properties. A Malayan labourer showed him how gutta-percha became pliable when hot and hardened again into whatever shape you desired. Malayans had for centuries moulded the hot latex into such things as whips, vases and knife handles. Modern industry found that it had almost endless possibilities.

Soon gutta-percha became ubiquitous in daily life. You could have gutta-percha raincoats, shoelaces, boot soles and walking sticks. Fashion houses used it as a cheap alternative to whalebone for the hoops in dresses. A search on eBay for 'gutta-percha' reveals a world of Victorian domestic appliances: brooches, earrings, lockets, buttons, photograph frames, inkstands, snuff boxes, doorknobs. The material was everywhere, and anyone who has handled a Victorian knick-knack has undoubtedly and probably unwittingly touched gutta-percha. Resistant to acids, saline and chemicals, it was invaluable to industry. The use of the material in golf balls transformed the sport: the 'gutty' ball was a cheap and durable alternative to its leather-and-feather predecessor, and it helped to popularise the game.[4]

But it was not gutta-percha's use as ear trumpets, golf balls, dress hoops or cheap jewellery that transformed the world. In September 1851 its full potential became clear.

Not only is gutta-percha waterproof, but it shows no signs of deterioration when submerged in salt water, making it the perfect insulator for electrical wire. In the summer of 1851 the Gutta Percha Company supplied the entrepreneur John Watkins Brett with 100 miles of copper telegraph wire encased in a tube of the Malayan latex. Brett took it to a company that specialised in producing wire rope for use in mining, R. S. Newall. Newall cut Brett's wire into four equal lengths and used its giant steam-powered machine to twine them together with several strands of hemp. This twenty-five-mile-long cable was taken to another machine, which wove ten thick galvanised iron wires round the gutta-percha-insulated copper core.

Newall's men worked twenty-four hours a day for twenty days until the 200-ton iron rope lay ready, coiled neatly in the factory yard. Out it came into the world, wound by a series of wheels positioned on roofs fifty feet above Wapping High Street. The 'Great Cable' snaked up and over the buildings until it reached the Thames-side wharf. There it was wound into the hold of HMS *Blazer* on 24 September and the warship was towed out of the Thames Estuary to Dover.

While she chugged through stormy seas from Dover to Sangatte, *Blazer* unwound the wire onto the seabed. As the ship pitched and rolled, constant adjustments had to be made to the speed at which *Blazer* was towed and the cable paid out. On one occasion the cable ship was towed too fast and the cable was badly damaged because it got caught in the rudimentary paying-out equipment. Five days after the cable had left the factory in Wapping a cannon on the gates of Calais boomed in celebration; it had been triggered by a pulse of electricity sent down that cable from England a few seconds before. On 13 November the line was opened to the public. Now every city in Britain enjoyed instantaneous communication with scores of cities across Europe.

Looked at from the distance of over a century and a half, from a world wrapped in fibre-optic wire and surrounded by satellites, it is hard to appreciate how monumental the Channel Cable was. Playing with technology imperfectly understood, Brett had plunged into the unknown. Cables had been tested under rivers and in bays, but no one knew how a cable submerged deep into a rough and unpredictable sea would perform. No one knew what sort of cable was suitable or how to lay it correctly from a moving ship. In other words, the Channel Cable was a costly experiment. *The Times* of London commented that 'this conquest gained by science over the waves must ever remain recorded

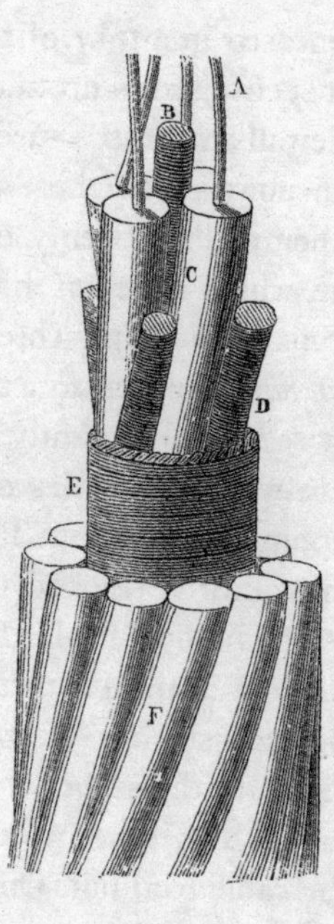

The 'Great Cable' of 1851: the four wires poking out in the centre are the copper telegraph wires, extending out of the gutta-percha insulation. The next layer is the protective hemp; on the outside are the thick galvanised iron ropes that armoured the wire and gave it the weight to sink to the seabed.

as amid the greatest of human achievements since record has existed of the mighty feats accomplished by man'.[5]

The cable was a milestone in human history, *The Times* and others were saying, because if land separated by the twenty-odd miles of the choppy Channel could be connected by electricity, then why not the other gulfs that separated the human race? Having taken one great leap forward, modern technology could advance at limitless speed, connecting ever-greater distances. As one writer prophesied, 'the whole earth will be belted with electric current, palpitating with human thoughts and emotions . . . It shows that nothing is impossible to man.'[6]

*

Thomas Hardy remembered the year 1851 marking 'an extraordinary chronological frontier or transit line, at which there occurred what one might call a precipice in time. As in a geological "fault", we had presented to us a sudden bringing of ancient and modern into absolute contact.' Still later, the British historian G. M. Young (1882–1959) declared that 'of all the decades in history, a wise man would choose the eighteen-fifties to be young in'.[7]

According to Hardy, it was the suddenness of change that burst upon the world in the early 1850s that was so startling. In his single sentence Young encapsulated the excitement and hope of a decade that in his youth many older people looked back upon with nostalgia and regret as a fleeting golden age. Again and again in the pages that follow we will encounter the swaggering self-confidence and extraordinary utopianism that intoxicated millions in the West. Writing in a private letter in 1852, Karl Marx appears breathless at the possibilities: 'Good luck to the new world citizen! There is no more splendid time to enter the world than the present.'[8]

Heyday recreates the exhilaration of the rollercoaster ride of the 1850s and early 1860s, a few years of tumultuous upheaval and accelerated change that saw the birth of the modern world. That is a large claim, but look at some of the great events of these years. The United States expanded into the lands that lie between the Atlantic and Pacific oceans and remade itself on the anvil of civil war. Across the ocean, the Crimean War sent fissures through the old European order, leading to the creation of two powerful new nations, Italy and Germany. The decade was no less convulsive for Asia. In 1854 the US navy intruded into Japan, secluded from the world for centuries, setting in train social upheaval, political revolution and meteoric modernisation. The trajectory of modern Indian history was set by the ferocious cyclone of the Mutiny-Rebellion and the draconian reassertion of British imperial power. For China, recently the most powerful nation on earth, these were traumatic years beginning with the blood-drenched Taiping Rebellion (1851) and culminating in British and French flags flying over Beijing (1860).

Behind the clash of these momentous geopolitical events, however, the world changed in profounder, even more radical ways. It is the inventions and recently discovered raw materials, patterns of trade and settlement, the interactions between widely dispersed societies that made this time a 'chronological frontier' and which are at the heart of *Heyday*.

The hallmark of modernity is its remorseless assault on time. Our foot-stamping intolerance at fractional delays in the transmission of data to our mobile devices is surely a salient feature of our age. Accounts of the fortunes made by shaving milliseconds off the exchange of financial data are astounding. Faster, faster, faster – we vanquish tyrant time over and again, and every victory only makes us indecently impatient for more. Truly, 'we are living at a period of the most wonderful transition . . . distances which separate the different nations and parts of the globe are rapidly vanishing . . . Thought is communicated with the rapidity and even the power of lightning . . . knowledge now acquired becomes at once the property of the community at large.'[9]

As you will have guessed, these are words from the 1850s – uttered by Prince Albert in fact, not by a twenty-first-century techno-utopian. If his words are familiar to us now – wearisomely commonplace, even – it is because of the inextricable connection between instantaneous communication and modernity. Electricity freed the transmission of information from the constraints of the physical world. And, thus liberated, it revolutionised everyday existence.

As a material that changed the world, gutta-percha is sadly neglected. The Channel Cable of 1851 heralded a future of instantaneous global communication, making Albert's prophecy a realisable attainment. One journalist foresaw the 'never-ceasing interchange of news' in real-time across a planetary nervous system. The world seemed to have shifted gear. That was the hallmark of the decade under consideration, what one newspaper in 1851 called the 'practical annihilation of space and time'.[10]

If electricity annihilated time, physical distance was being assaulted by a battery of weapons. The world was being opened up and simultaneously shrunk by steam and electricity. The speeding-up of life was highly significant in the history of the world. But what made it so explosive was its convergence and interaction with three other great forces in the early 1850s: the discovery of huge deposits of gold, unprecedented mass migration, and free trade. The result was boom on a colossal scale, with global trade expanding by an incredible 500 per cent. The bonanza of these years acted as an accelerant on processes that were already in train. As one writer had it, living in the 1850s was like 'railway motion . . . [after] the wearisome progress of a caravan of camels'. No wonder that these were breathless, exhilarating years of breakneck growth and expanding possibilities.[11]

The idea that this was a 'precipice in time' or a heyday would have made perfect sense to people living through the dizzying 1850s. Stand still too long and you would miss it. The American businessman and globetrotter George Train wrote of the few months he spent in Melbourne that they represented 'a lifetime . . . in this startling age'.[12]

Train resided in Melbourne during a period of frantic growth, when the city took in upwards of 60,000 people in a single year and became, almost overnight, one of the wealthiest places on earth. And yet, and yet – exciting as it was to experience, many footloose fortune-seekers who ranged the world had the nagging suspicion that something bigger and better was occurring elsewhere. The astonishingly rapid settlement of California, the American Midwest, Canada, Australia and New Zealand in the 1850s and 1860s was regarded then and now as one of the miracles of the century. California had 15,000 non-native people in 1849 and 380,000 eleven years later. The Australian colony Victoria was founded in 1851 with a population of 76,000; within a decade it had rocketed to 540,000. Cities that benefited from the vertiginous upswing in global trade – such as Chicago, Liverpool, Bombay, Hong Kong, Shanghai and Yokohama – grew with no less impressive speed. In addition, hundreds – perhaps thousands – of towns and cities mushroomed across the globe, erupting out of what Victorians called the wilderness. Human history offered no other example of such demographic explosion.[13]

Here, in the streets, markets, building sites and wharves of these rapidly expanding urban centres, we can savour the exuberance of the 1850s and the unshakeable confidence of the millions who gambled on exponential progress. The travel writer, man of letters and poet William Howitt could have been commenting on hundreds of boom towns the world over when he wrote of 'Marvellous Melbourne' that in the ceaseless hubbub of the new streets, the myriad building sites and the boisterous egalitarianism, 'the future was so palpable and positive that none could miss seeing it'.[14]

Appreciating the sincerity of the widespread belief that humankind stood on the cusp of epochal change is vital to understanding this period. Confidence propelled stupendous economic growth at least as much as tangible things like steam engines, telegraphs and gold. It is striking that so many historians have reached automatically for the word 'heyday' to describe aspects of the 1850s that fall within their specialism. I have used 'heyday' to refer to this short but crowded period because it also embraces a state of mind that pervaded Western societies.

The 1840s were a miserable decade of economic depression, food shortages, failed harvests, famines and suppressed revolutions. By 1851 a boom was unquestionably in full throttle. But it was not merely rising wages, new opportunities and easy money that made this time feel so good for people living through it. As Hardy said, they believed they were transiting a 'chronological frontier': a new civilisation was being built on instantaneous communication and unrestricted trade. Prince Albert said that the annihilation of time and space by modern invention 'tends rapidly to accomplish that great end, to which, indeed, all history points – the realisation of the unity of mankind'.[15]

Such faith, naïve as it sounds in retrospect, exerted a powerful and significant influence. It was what gave people the confidence to gamble on the future by upping sticks and migrating to what they saw as the wilderness or to invest massive sums in visionary engineering projects.

But a perceived heyday can as readily addle minds as inspire great deeds, beckoning the unwary or gullible into a mire of delusions. The high confidence of the 1850s led many to believe that the problems that bedevilled other eras had simply evaporated in the searing white heat of modernity. While the good times lasted, and the laws of economic gravity were held in abeyance, credit sloshed around and the boom-time hotspots of the new world experienced one of the most extravagant property bubbles in history. Believing in exponential growth and limitless returns, people of all classes speculated on increasingly risky ventures with borrowed money, intoxicated by the evanescent mood of the decade.

Credit crunches, busted businesses and foreclosed mortgages – when they came – were slight, however, compared with the permanent scars left by the mid-century boom. Stratospheric economic growth resulted from glittering and showy technological advances, to be sure, as well as the injection of massive quantities of gold and the creation of new markets by mass migration. But there was another ingredient, and it was a very ancient one. Industrial expansion in Britain, Western Europe and the north-eastern United States increased the demand for cotton, the white gold of the mid-nineteenth century. Cotton production soared in the 1850s, more than doubling in those years. This feat was achieved through better fertilisation, the exploitation of rich lands, huge capital investment and by a web of railroads. But most important in ramping up production were the forced migration of hundreds of thousands of slaves to the south-west of America and enhanced (and inhumane) techniques for exploiting their muscle-power.[16]

Slave labour helped sustain the boom, and so too did the appropriation of vast chunks of land from their owners: forests were felled, land fenced off, European crops, vines and grass varieties sown, and millions of domesticated grazing animals displaced game. And on top of all that, railways scythed through the landscape, telegraph wires stretched to the horizon and towns sprang to life. The London-based Aborigines Protection Society declared that there was an urgent need 'to bring protection to the feeble families of mankind against the homicidal process which, in every instance of modern colonization, has accompanied the diffusion of our Anglo-Saxon race'.[17]

Many indigenous peoples living in the pathways of empire had been able to resist Western encroachment by their skills as warriors or by natural ramparts such as deserts, mountains, diseases or, simply, their remoteness. Significant numbers of these native tribes were swamped by resource- and land-hungry white settlers brandishing new technologies and weapons, protected from the diseases that had once checked their ravenous progress. One writer commented on the indignity suffered by a group of Ioway Indians 'exhibited' in Britain in 1851: 'They are slaves chained to the car of the conqueror; they are the shadow of the old races that the victorious and implacable civilization of the West crushes in its progress.' As the 1850s and 1860s wore on, more names would be added to the list of subjected, dispossessed and vanished peoples swept aside by the tsunami of settlement.[18]

It was not just warrior tribes that found their resistance to the 'implacable civilization of the West' blunted. Telegraphs, steamships, railways and the like laid siege to the world's physical barriers. There remained, however, no less tenacious human brakes to the fulfilment of unfettered global exchange.[19]

Societies in India, Burma, Siam, China and Japan came under intense pressure from the West. The effects of their encounter with Europe and America resonate to this day; these decades in the mid-nineteenth century dominate their historical memory and sense of national identity. Most clearly, the advance of the West, under the banner of trade, threatened national independence and traditional ways of life. As Karl Marx and Friedrich Engels wrote, industries and customs were being swept away by mass production and speedy communication. 'In the place of the old wants, satisfied by the production of the country, we find new wants, requiring for their satisfaction the products of distant lands and climes. In place of the old local and national seclusion and self-sufficiency, we have intercourse in every direction, universal interdependence of nations . . . The cheap prices of commodities are the heavy artillery . . . [that] batters down all Chinese walls.'[20]

Early in 1851 the apocalyptic Taiping Rebellion broke out in southern China. It came about as the direct result of Western intrusion, and would leave well over 20 million corpses in its wake. Effects of the rebellion in China reverberated around the planet. It was the leading edge of the hurricane gathering in Asia, a foretaste of the civil wars that would engulf Japan and India. More than that, divisions and distractions in China gave Britain, France, Russia and the United States the opportunity to penetrate deep into Asia and compete for the spoils. In the maelstrom of the 1850s, efforts to prise open China and Japan and integrate them into the world economy, whether they liked it or not, intensified.

Here, in Asia, the quasi-religious faith in the transformative and regenerating effects of free trade and modern technology that characterised the 1850s reached its zenith. The US in Japan and the British in China and India believed that unrestrained exchange of goods and services would work a miracle: ancient societies would be radically remade in an incredibly short space of time by the power of modernity. Napoleon III put this clearly when he proclaimed his intention for France to be part of 'the movement of progress, civilization and commercial expansion of which China was going to be the theatre'. In Asia heyday would reach its climax.[21]

For the Far East, 1851 marked a fault line in time as surely as it did for the West. It was not just the start of the Taiping Rebellion. In that year the telegraph came to Asia, eighty-two miles of line connecting Calcutta and Kedgeree. The first railway in the continent, twenty-one miles of it, opened in Bombay two years later with enormous fanfare. Once Prometheus was unbound his revolution proceeded at velocity: by the end of the decade the telegraph and rail networks extended to thousands of miles each, their skeins reaching into regions that were till recently remote from the main currents of international trade.

Telegraphs and railways underlined, above all else, the West's belief in its superiority and right to rule others. Of all Asian countries, India experienced the full heat of the technological revolution, used as it was as a vast laboratory to showcase the regenerative power of Western science. But here, as elsewhere in Asia, brand-new technology was regarded by many not as an instrument of liberty but a tool of oppression. In India, and in China and Japan, efforts to radically and rapidly reshape societies led to anarchy, rebellion and massacre as people sought to defend their religions and ways of life from the encroachment of the West.

Heyday follows the twists and turns of a decade that *The Times* of London described as being like 'a serial romance, in which each successive chapter is distinguished by some strange event or unexpected turn of fortune'. In order to get to the marrow of this time I ventured to a diverse range of locations, both familiar and unfamiliar. I draw connections from one end of the world to the other, juxtaposing seemingly unrelated events, ideas and people.

The geographical span of *Heyday* is global; the timespan is brief, lasting from 1851 to 1862. These eleven years form a distinct era, sandwiched between darker, more uncertain times. The utopianism and millennial expectations of this short-lived heyday – the sense that it marked a moment of radical change in the history of humanity – provoked flurries of activity that left deep impressions, for good and ill, across the planet and which shaped modern history.

The first stop on the journey into the world of the 1850s is London. For it was there, in 1851, that the hopes and dreams awoken at the beginning of the decade became solid.

PART I

Boom: The Age of Gold

1

1851: Annus Mirabilis

LONDON

Its grandeur does not consist in *one* thing, but in the unique assemblage of *all* things. Whatever human industry has created you find there . . . It seems as if only magic could have gathered this mass of wealth from all the ends of the earth.

Charlotte Brontë[1]

When the Pyramids of Egypt crumble to dust, people will still remember the Great Exhibition. That was one of the least hyperbolic predictions floating around the British press in the summer of 1851.

The Pyramids have existed for millennia; the prefabricated glass-and-iron Crystal Palace housed the Great Exhibition of the Works of Industry of All Nations in Hyde Park for less than five months. One newspaper called the Palace a monument to human progress and the Exhibition 'the most remarkable event in the modern history of mankind'; it represented not merely the dawn of a new era but the 'realisation of Utopia' itself. *The Times* wrote of the opening ceremony on 1 May, in suitably millennial language, that it was 'the first morning since the creation that all peoples have assembled from all parts of the world and done a common act'.[2]

The Crystal Palace resembled a mighty cathedral: the great nave, more than half a kilometre in length, was intersected by a transept where the immense barrel-vaulted glass roof dwarfed the elm trees enclosed by it. This was the heart of the Exhibition. The elms extenuated the height of the building for first-time visitors when they entered through the south entrance. As they progressed through the transept their senses were assaulted. An enormous fountain played in the centre, and its murmur was joined by numerous other fountains. One dispensed eau de cologne. Another provided tourists with an endless stream of Schweppes mineral water (a representation of the Exhibition fountain can still be seen on bottles of their tonic water). Among the fountains and trees were tropical plants and flowers from every part of the globe; the effect was enhanced by statues interspersed in the transept and the nave.

The Exhibition was arranged so that the visitor could make his or her progress along the central nave, the aisles and courts and up to the first-floor galleries and not tread the same ground twice. A single day-long visit would only give you the barest impression of the whole. The effect was deliberately overwhelming. There were over 100,000 exhibits provided by 14,000 individuals and organisations from almost all countries on the planet. As a guidebook put it, 'One of the distinguishing characteristics of the Great Exhibition was its vast comprehensiveness. Nothing was too stupendous, too rare, too costly for its acquisition; nothing too minute or apparently too insignificant for its consideration.'[3]

It was inevitable that the Exhibition itself became a blur. How could it not, with so much stuff on display and so many people to gawp at? A day at the Exhibition meant sensory overload. This reflects the gargantuan ambitions of the organisers. The Great Exhibition was intended to educate the public, ignite ingenuity and competition by comparing British and foreign manufactures, stimulate domestic industry, promote world peace, showcase colonial goods, develop international markets, and much more besides.

The exhibits were divided into raw materials, machinery and mechanical inventions, manufactures, and fine art, and these four broad headings were subdivided into thirty categories. The profusion of artefacts and materials made it impossible for the visitor to hold on to anything in particular; it was the totality of the spectacle which made it great, rather than any individual exhibit. There were gorgeous fabrics, furs and silks; intricate clocks and microscopes; lumps of coal, bales of wheat and samples of wool from around the world; exotic Asian artefacts and modern Western sculpture; medicines and surgical instruments; scissors, Sheffield cutlery and porcelain; state-of-the-art agricultural machinery, gardening equipment and futuristic tools; foodstuffs, minerals, chemicals, resins and dyes; fire engines and lifeboats; furniture, household appliances and decorations. The industrial competed for space with the domestic, the showy with the mundane, the luxurious with the practical. Next to the ultra-modern were the handicrafts and costumes of indigenous people from remote parts of the globe. Even a very long list would scarcely scratch the surface. A week in London in 1851, one Frenchman said, was a 'tour of the world': 'The observer is, as it were, carried away by magic from country to country, from east to west, from iron to cotton, from silk to wool, from machines to manufactures, from implements to produce.'[4]

The point of the event was to revel in modernity. And sure enough, one of the most popular parts of the Exhibition was the machinery court, which was 'indicated by its deep and heavy murmur, like the distant roar of the torrent'. There were always great crowds around the 700-horsepower engines, the steam hammers, hydraulic presses, pile-drivers, Crampton's locomotive (capable of a top speed of seventy-two miles an hour) and other wonders of the age.[5]

The large display of telegraphic apparatus was one of the first things that visitors saw when they walked through the enormous entrance. It told the short history of modern telegraphs, from Cooke and Wheatstone's pioneering telegraph of 1838 to the more sophisticated versions of recent years. Well-heeled visitors were able to summon their carriages to the exit of their choosing by electric means. The venue had a private line direct to Scotland Yard in case of disturbance, and it was connected to the entire national network, allowing visitors to telegraph any of the hundreds of offices in the country. There were maps of the networks and daily weather reports from around the kingdom. 'We went to the Exhibition and had the electric telegraph show explained and demonstrated before us,' Queen Victoria noted in her diary. 'It is the most wonderful thing and the boy who works it does so with great ease and rapidity. Messages were sent out to

Manchester, Edinburgh etc and answers received in a few seconds – truly marvellous!'[6]

Here were the machines, precision instruments, raw materials and technologies that would revolutionise the world. Here modern science was on full display; the Exhibition hummed and purred to the sound of steam power and electricity. Prefabricated in factories around Britain and assembled by mechanical means, the iron-and-glass Crystal Palace was a physical monument to modern mass production and a wonder of the age. The great American journalist Horace Greeley said that the building was more impressive than any one thing it housed: 'it is really a fairy wonder, and it is a work of inestimable value as a suggestion for future architecture . . . Depend on it, stone and timber will have to stand back for iron and glass.' Greeley advocated a Crystal Palace for New York, which he imagined presciently as a colossal shopping mall.[7]

The Great Exhibition has never been rivalled as a world fair. The event caught imaginations all over the world and set the euphoric mood that characterised the 1850s. Under the immense barrel-vaulted glass ceiling the modern world was imagined and arranged in microcosm: the raw materials, manufactures and inventions of all nations were placed side by side for comparison and education. The event was promoted as the first ever international 'peace congress'. In a world united by trade and commerce and unfettered information – exemplified by the juxtaposing displays at the Crystal Palace – 'gain is gain to all . . . the achievements of human intellect are common property'. Commenting upon the opening ceremony, *The Times* said that Queen Victoria sat 'enthroned . . . amid the spoils of the world' like a mighty conqueror, except that, unlike other conquerors in human history, the booty was there by the uncompelled good graces of its owners.[8]

In the past, international peace congresses represented the triumphalism of the victors of bloody wars. But in this case it was a celebration of peace through progress. The Exhibition had, according to a French writer, 'inaugurated a new era' in the history of the world: the 'peaceful struggle of the nations'.[9]

Throughout the world, people read about or saw pictures of the event. As one American journalist put it, 'The accounts published from day to day in the newspapers, the woodcuts in the illustrated papers, the pictures and panoramas, and the endless list of articles on the subject in the magazines, have made most people familiar even to weariness with all the details.' According to Thomas Hardy, 'None of the younger generation can realise the sense of novelty it produced in us who were then in our prime. A noun substantive went so far as to become an adjective in honour of the occasion. It was "exhibition" hat, "exhibition" razor-strop, "exhibition" watch; nay, even "exhibition" weather, "exhibition" spirits, sweethearts, babies, wives.'[10]

By an accident of history the Great Exhibition coincided with the beginning of the long mid-century boom. It became the physical manifestation of the optimism that was beginning to build in the West, giving birth to thousands of articles and books that proclaimed the dawn of a new era of peace, prosperity and unlimited growth. This burst of euphoria was all the more remarkable because it erupted so suddenly after a long period of gloom.

In 1849 the Swedish novelist and feminist Fredrika Bremer visited England. 'A thick heavy atmosphere rested like a storm-charged cloud over the towns,' she remembered; 'hearses rolled through the streets; cities were deserted, for the cholera ruled there . . . I had never seen human misery in such a shape, as I now saw it both in Hull and London.' Bremer visited towards the close of one of the most miserable decades in British history. It was remembered for ever after as the 'hungry forties', a time of unemployment and cholera, harvest failure in Britain and mass starvation in Ireland, economic depression and political volatility.[11]

She contrasted this apocalyptic scene with another tour two years later when 'the spirit of spring' permeated the land. Britain had, for Bremer, been reborn in 1851. Everywhere she saw a prosperous, happy nation, brimming with energy. Fredrika Bremer might have overly dramatised the contrasts in two years. But she was absolutely accurate in reflecting the transformation of mood. She explained the change between 1849 and 1851: 'Free trade had borne its fruits; beneath its flag Commerce and Industry had blossomed with fresh life; the price of corn had fallen, bread was cheap.'[12]

In the 1840s Britain had taken a great leap into the dark. The repeal of the Corn Laws is hardly a subject to set pulses racing. But it had a monumental impact not only in Britain, but all over the world. The Corn Laws stood at the centre of a thicket of protective measures that defended Britain's agriculture, industry, overseas trade and shipping, and which underpinned the imperial system. The Corn Laws, enacted in 1815, kept the price of imported grain high enough to shield British producers from overseas competition. Over a thousand other duties similarly made it expensive to import foreign raw materials, foodstuffs and manufactures. The Navigation Acts restricted trade between Britain and her colonies to British-built, owned and crewed ships. The measures were designed not only to protect agriculture and industry, but glue the Empire together by giving Caribbean sugar producers and Canadian timber companies, among others, exclusive access to the British market.

The first big shock to the old system came in 1842 when Robert Peel, the Conservative Prime Minister, abolished or reduced 1,640 tariffs. In the next four years tariffs on wool, linen, flax and cotton were torn down, so that, in the words of Peel, 'there hardly remains any raw material imported from other countries, on which the duty has not been reduced'. Thus emboldened, he undertook the biggest gamble of all and repealed the tariff on the most important of raw materials – the Corn Laws.[13]

It came as the culmination of years of fierce political battle. For the repealers, not only did the Laws keep the price of bread high, to the misery of the working classes, but they also stifled manufacturing. Industry was hungry for workers, and it needed them from the land. If cheap food came from abroad, and grain prices fell, many more workers would be freed up to enter the mills and factories. Their labour would be fuelled by bread and sugar from the world market.

But there was an even more important motive than simply feeding the growing working class. American, Russian and European merchants who sold their primary produce in the British market would be paid in pounds. They would then use this cash to buy British manufactures. By giving American wheat farmers unrestricted access to their market, British capitalists would gain flourishing American

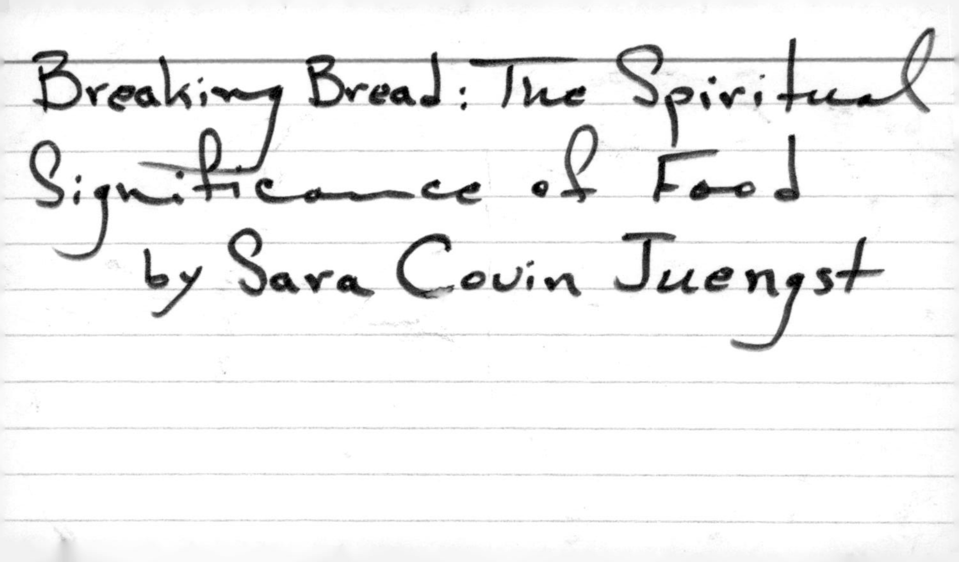
Breaking Bread: The Spiritual
Significance of Food
by Sara Covin Juengst

markets. Better still, with cheaper imported raw materials the cost of mass-produced British manufactures would plummet, allowing British exporters to flood foreign countries and undercut and ruin their industrial competitors. In this vision of the future, Britain would be the workshop and financier of the world, its megacity, leaving the rest of the planet more or less as its farm. Advocating repeal, Thomas Babington Macaulay said that while 'other nations were raising abundant provisions for us on the banks of the Mississippi and the Vistula', the British would 'supply the whole world with manufactures, and have almost a monopoly of the trade of the world'.[14]

The campaign for repeal pitted industrial Britain against landed Britain, commercial power against landed power, the working and middle classes against the aristocracy. Peel passed his Act; but he was turfed out of office by his own party as a result. Nonetheless, the assault on protection carried on. Next to go were the sugar duties. Then, on the first day of the year 1850, the Navigation Acts disappeared from the Statute Book. Now any ship from any country could bid for the trade of the Empire.

The opening-up of the massive British market had a galvanic effect on the world economy. American shipbuilders benefited the earliest from the repeal of the Navigation Acts, selling some of the best vessels in the world to British and colonial shippers. Brazil and Cuba flooded the market with their sugar. An enterprising group of Greek merchants relocated to the City of London and began exporting textiles from Lancashire and importing grain from a region that extended from the Danube Basin through the Black Sea, the Adriatic and the Eastern Mediterranean to Egypt. Seven years after the repeal of the Corn Laws they had netted an eye-watering £30 million. In time, the repeal of the Corn Laws would catalyse the settlement of the American Midwest as the ravenous British market scoured the globe for breadstuffs. People around the world enjoyed the fruits of the liberalising of the British economy.[15]

Exposure to the withering winds of global competition, however, caused economic depression and unemployment in Britain. By the early 1850s the once-mighty shipping industry was in the doldrums, outclassed by its new transatlantic competitors. The tidal wave of free trade engulfed the City and brought down venerable firms and banks. It broke over the Empire as well. Britain's sugar-producing Caribbean colonies were ruined for ever. Canadian timber exporters and wheat farmers had relied upon the special colonial relationship with Britain too; now they were reduced to competitors on the brutal global market and they faced disaster. The bonds that held together the British world had been brutally slashed; the Empire was threatened with violent dissolution.

Not surprisingly, therefore, when the Great Exhibition was conceived in 1849 its leading lights were at pains to avoid making it appear to be a celebration of free trade. It was a contentious issue, one that divided the country. The event was sold as a way of advertising Britain's machinery and manufactures to overseas customers and to promote raw materials from around the Empire.

But fast-forward a few years to 1851 and the gloom had been entirely extinguished. In its place came giddy rejoicing. The Great Exhibition became a triumphalist 'Free Trade Festival'. British agriculture had not collapsed as many had predicted, for one thing. In the early 1850s it was at its peak, enjoying bounteous harvests. Even so, millions of Britons were being fed on imported foodstuffs: prices were down but demand was apparently inexhaustible.[16]

Farmers were doing well, but it was nothing compared to the upsurge in British industry. After the initial dislocation to trade, the effect of liberalising the economy had been positive. The cost of foreign food and raw materials had tumbled, good for the working class and industry alike. Countries that made money exporting their staples to open-door Britain spent their profits importing British manufactures. By the time of the Exhibition, Britain produced 66 per cent of the world's coal, 70 per cent of its steel, 50 per cent of its iron and, most importantly, 50 per cent of its textiles.

'Manchester appeared to my eye like a colossal spider,' wrote Fredrika Bremer, 'in the midst of its factories, towns and suburbs and villages, in which everything seemed to be spinning – spinning – spinning cloths for the people of the world.' The value of British exports had expanded by 51 per cent since Peel had taken his leap into the dark, and that of imports by 34 per cent. *The Times*, in its description of the British section of the Great Exhibition, noted items marked for export to the Continent, China, India, Russia, the Ottoman Empire, South America and the United States: 'manifest proofs that we aspire to world-wide demand and have the mass of the population in every country for our customers'. Britain, with its system of mass production of cheap consumer goods and its open trading policies, was able 'to command, by the element of price, the markets of the world'.[17]

What the British called friendly competition, much of the rest of the world called 'commercial despotism'. All that talk about controlling the markets of the world while the country stifled the development of foreign industry sounded like a declaration of economic warfare. Britain had defended its growing industry behind the fortress of protection. Now it was preaching free trade in order to strangle foreign industry in its cradle by dumping cheap textiles and inexpensive manufactures in ports around the world. And by inflating commodity prices, it used economic power to force other countries to gear their economies to producing foodstuffs and raw materials for export to Liverpool rather than building up their own domestic industries.

The sudden onset of a boom and the lavish display of their machinery and inventions meant that the British walked with a swagger in 1851, on top of the world. Writers sniffed at the contributions of other nations. As Horace Greeley said, 'I apprehend John Bull, whatever else he may learn, will not be taught meekness by this Exhibition.' *The Times*, sneering, called the space allotted to American exhibits a 'desolate prairie' because the transatlantic guests had failed to fill it up. It was blithely assumed that the US was rich in natural resources but poor in technical skill. The American display 'invites emigration' – literally so: manufacturers from Britain and France in need of more room for their wares began to colonise the prairie ground.[18]

But this condescending tone did not last long. Before 1850 the British had held the field in industrial inventions. Their very own self-aggrandising Exhibition, however, gave them a rude shock. When they began to examine the German, French, Belgian, Dutch and American stands they saw manufactures and industrial processes as good as or superior to their own. Many Asian artefacts exceeded cheap

British manufactures in their quality and popularity. And, worse still, as one American journal pointed out, the empty space in the US section was no accident: what self-respecting entrepreneur would give away the secrets of a new invention gratis to rivals at an international exhibition? As the journal said, 'as soon as an Englishman is taught, he is at perfect liberty to use his knowledge, and it is certain he will use it, immediately, to undersell his teachers and ruin their business'. The nagging question, then, was not what *was* at the Exhibition, but what *wasn't*?[19]

Even so, there were enough examples to take the edge off the joyous mood in Britain. American prowess in shipbuilding was confirmed when the schooner *America* won the Royal Yacht Squadron's annual race round the Isle of Wight that year. The silver winner's trophy was renamed 'The America's Cup' and became the prize in the world's most famous yacht race and the oldest international sporting trophy. Among the American exhibits were some of the Exhibition's show-stealers, including Cyrus McCormick's reaping machine, which heralded the advent of mechanical farming and a new agricultural revolution. American skill was underscored with the exhibition of items made from rubber, the results of vulcanisation – a relatively new process invented by Charles Goodyear.

An American entrepreneur was responsible for one of the most popular, admired and influential exhibits – one that was destined to have a big impact on the world in the 1850s and that cut across the grain of a self-proclaimed Peace Conference.

Samuel Colt's new six-shooter revolver caused a sensation. *The Times'* correspondent, tongue in cheek, compared Colt to Edward Jenner: at Colt's stand at the Exhibition 'you may make yourself acquainted with the new method of vaccination as performed by the practitioners of the Far West upon the rude tribes who yet encumber the wilderness with their presence'. This new firearm immediately superseded all other weapons, especially for mounted soldiers. It could fire six rounds without having to be reloaded, allowing a cavalry trooper to triple his rate of fire. A United States army officer said that 'a dragoon armed with a Colt's repeating pistol . . . or perhaps a Sharp's rifle, would be the most formidable [of combatants] for the frontier service, and particularly when encounters with the savages occur – as they generally do – in prairies, defiles, and mountain gorges . . . A few bold men, well skilled in the use of these weapons, can, under such circumstances, encounter and scatter almost any number of savages.'[20]

It was indeed the perfect weapon for 'frontier service' and would 'work a revolution in war'. American Indians were formidable mounted warriors, whether armed with muskets, spears, bows and arrows, knives or tomahawks. They were capable of overwhelming opponents equipped with single-shot firearms. The killing field was more or less level on some of Britain's frontiers as well. In the New Zealand section of the Exhibition there was a model of a Maori *pa*, a hilltop settlement fortified with terraces and stockades. It was constructed by Lieutenant Henry Balneavis, who wanted to show people in Britain that the Maori were not savages but skilful military engineers and warriors, capable of tying down larger numbers of imperial soldiers and inflicting disproportionate casualties from the *pa*s. Elsewhere, the British faced stiff resistance in southern Africa, on the Niger, in Afghanistan and Burma. *The Times* recommended the Colt revolver for British cavalry currently engaged in the frontier skirmishes in Cape Colony against the Xhosa and Kohekohe, and for lancers on the North-West Frontier of India.[21]

The Colt six-shooter became one of the most recognisable brands in the world in the 1850s, flaunted by white migrants in locations as diverse as the Australian bush, the jungles of Panama and the plains and deserts of the trans-Mississippi West. Colt's 1851 Navy Civilian revolver was the iconic weapon of the Wild West for decades after its invention, touted by, among others, such celebrities as Wild Bill Hickok, Doc Holliday and General Robert E. Lee.

But its significance lay not just in its ubiquity in the West's settler disputes on the frontier. It was the way the revolver was made that heralded a 'revolution in war'. British specialists who pored over the weapons at the Crystal Palace were stunned at the advances in firearms technology displayed by Colt. There had been repeating pistols before, but Colt's breakthrough in 1850–51 was to render them safe and reliable for the shooter and, above all, cheap. The parts were the result of mass production and power-driven machinery, making them interchangeable. This gave the user considerable advantages in the field, where he or she could repair, replace or recycle the working parts with negligible skill. Compare that to the British system, where musket parts made for the military were hand-crafted in Birmingham workshops, sent to the Tower of London for quality control, and then dispatched to another set of craftsmen to assemble each firearm. This complicated method of making a firearm, whereby the parts makers were not involved in the assembly, stifled innovation in the industry. As a result of Colt's display at the Crystal Palace, British experts went to the United States to investigate arms manufacture.[22]

THE COLT BREECH-LOADING REVOLVER.

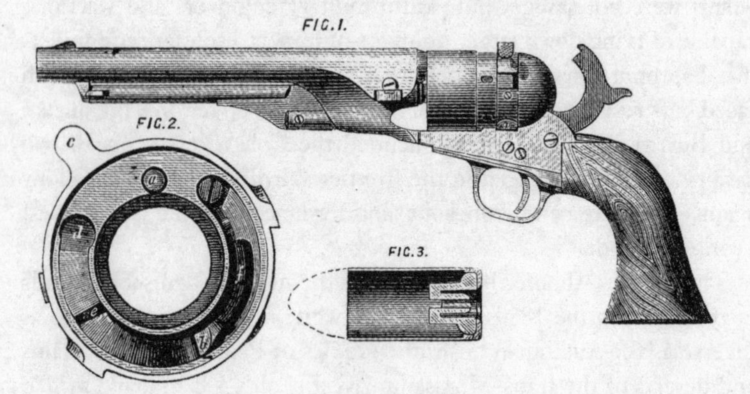

The result of the inquiry was the importation of American machinery and an expert from the United States armoury at Harpers Ferry, Virginia, to oversee the introduction of mass production at the government's small-arms factory at Enfield. The fruit of the new system, inspired by Colt, was the Pattern 1853 Enfield rifle-musket. It became the most popular rifle of the age, produced in the millions for the British, Indian and American armies and for frontiersmen and sportsmen. It was the next phase in the revolution inaugurated by Colt. The

Enfield rifle was reliable, easy to repair (thanks to its interchangeable parts) and incredibly accurate; it gave ordinary infantrymen a good chance of hitting a target at 600 yards. The muskets it superseded in the British army – such as the good old Brown Bess, veteran of Waterloo – allowed a soldier to hit a man-sized target at 100 yards at the very best.

A rifle differs from a musket because the latter has a smooth bore, while the former has a rifled or grooved barrel, which spins the bullet, giving it greater aerodynamic stability. Rifling had been known about for centuries, but it was not practical for military weapons because the cylindrical bullet had to be the same diameter as the barrel so that it engaged the grooves. Forcing such a tight-fitting bullet down the barrel took too long and the detritus left by the bullet fouled the weapon, nullifying the advantages of greater range. A musket was inaccurate and had much lower velocity, but the round ball could be rolled down the barrel with no difficulty. Until a rifle was invented that was as easy to clean and as fast to reload as a musket, it would never be adopted by infantry.

The pointed cylindrical bullet developed by Claude-Étienne Minié in 1849 changed everything. The bullet was carried in a waterproof paper cartridge with the requisite gunpowder; the soldier tore the cartridge with his teeth and rammed the bullet and powder down the barrel. The conical Minié bullet was small enough to be loaded via the barrel like a musket ball, yet suit the purposes of a rifle, because it expanded on firing to engage the grooves of the barrel and spin towards the faraway target.

Up until the 1850s colonial powers were matched against opponents armed with identical muskets. The Sikhs, for example, had sophisticated arms manufacturing. Other peoples adapted Western weapons and learnt techniques which outclassed their opponents; American Indians and the Muslim guerrilla fighters of the Caucasus, for instance, were adept at reloading muskets while riding at speed. The Enfield rifle took the same time to reload as the old technology, but its superior range and accuracy and the velocity of the Minié bullet gave its users deadly advantages. It was the ideal weapon for conventional wars and frontier action alike.

The Colt gave pioneers the edge over indigenous people who stood in the way of settlement. It became the most famous international brand name of the decade. Again and again the Enfield would prove its worth for small European forces pitted against vastly superior armies in China, India and elsewhere. They were joined by other advances – Armstrong's breech-loading artillery gun that lobbed lethal explosive shells over long distances on Chinese soldiers and Maori warriors, for example, or the Gatling gun that sprayed death at unprecedented speed. The irony that one of the most significant legacies of the great 'Festival of Peace' in 1851 was state-of-the-art killing machines was not entirely missed.

Colt's epochal six-shooter, McCormick's reaper and other inventions proved to the discomfort of the hosts of the Great Exhibition that the United States was emerging as a serious industrial power. According to the *Economist*, not naturally inclined to enthusiasm for that country, the 'rapid progress and the greatness of the Americans, like their clipper ships, are, in the main their own and chiefly due to themselves . . . They seem certain to become the greatest and most powerful nation that ever existed.'[23]

With the inventiveness of the United States on display, along with the evident potential of the German states, came the painful realisation that Britain's industrial dominion had peaked in 1851. But the dawning of this fact was to have important implications.

He came 13,000 miles to see the Great Exhibition. It was fun, but failed to excite him. London's physical appearance – the dark, sombre, sooty buildings and the incessant crowds of people – contrasted unfavourably with the bright, cheerful towns he had recently passed through on the Continent. George Russell, a prosperous, dour Australian sheep farmer, was hard to impress. His pen, however, was roused to eulogy by the docks crowded with shipping and 'the large roomy warehouses . . . filled with merchandise from all parts of the world . . . the great business activity and bustle which is here met with . . . gives one an idea of the great commercial importance of England, quite surpassing anything of the kind I have hitherto seen'.[24]

That was why another foreigner came to London in 1851. Julius

Reuter's skill was in keeping ahead of modern communications technology. Before he arrived in London his homing pigeons raced between Brussels and Aachen. His flocks of racing birds brought with them the latest prices from the Paris Bourse, telegraphed to Brussels where the French line terminated. Before the mail train chuffed into town bringing with it the financial news, Reuter's pigeons had reached their loft in Aachen, the furthest extent of the German telegraph network. Without a second's delay Reuter zapped the prices and exchange rates on the French markets down the wire to his subscribers in Berlin and Vienna. For those subscribers time really was money: getting the news first meant making a killing. In 1851 the gap in the telegraph network was closed, Reuter's pigeons took early retirement and the entrepreneur relocated to London, in time for the opening of the Channel Cable. Reuter's Submarine Telegraph Office at the Royal Exchange Buildings offered his clients in the City and throughout the Continent as far as Russia up-to-the minute information from the European financial and commodity markets. They also got American prices as soon as they were cabled from the mail ship at Liverpool.

It is not hard to see why George Russell and Julius Reuter were attracted to the City of London in 1851. By the time of the Great Exhibition the global economy was clearly in a vertiginous upswing and the City was discovering a new role at the centre of the action. The excitement was palpable. On one occasion during the summer of the Exhibition the normally genteel Stock Exchange descended into chaos when 200 young traders took part in an impromptu and unruly game of football. The staid bankers with their pince-nez, black clothing and plodding ways were still there to be sure; but alongside them 'you find yourself in the company of sprightly young gentlemen, who talk about the opera and other amusements of the town with all the ease of connoisseurs in high life; and whose chief study is to give effect to coloured neckerchiefs, showy chains and mogul tie pins'. To the astonishment of the old hands who had relied on punctilious correspondence with brokers in foreign cities, hundreds of international deals – even small ones – were being made over the wires every day. The stuffy old banks, which projected old-fashioned sobriety, were giving way to gaudy new private banks shimmering with enormous plate-glass windows and blazing chandeliers. The old proprieties were dying with the heady, bullish atmosphere and a tempo of financial business dictated by rhythmic pulses of electricity.[25]

The opening of the Channel Cable linked money markets across Europe in real-time at a period when the world economy was beginning to take off. Capital was pouring out of the City, seeking investment overseas in massive infrastructure projects, government bonds, farming operations and businesses. Only a few years before investors had been badly burnt by the defaults on their loans, first of several Latin American republics and then of a number of states of the United States. But in the first years of the 1850s confidence was returning. A loan of £5.5 million had recently been floated for the Russian government; in that year and the ones that followed similarly gargantuan sums were subscribed for, among others, Piedmont, Austria, Brazil, Belgium, Canada and Victoria. The year 1851 also saw the incorporation of the Illinois Central Railroad, which bid fair to be one of the most important iron roads anywhere, opening up the prairies to the world market. It coincided with the comeback of the US as an attractive destination for foreign investment after long-lasting doubts about America's creditworthiness. The injection of Californian gold kindled confidence. US railroad securities were being traded directly in London for the first time, and they became a mainstay of the City's business in the early 1850s. Capital raised throughout Europe funnelled through London and flowed across the Atlantic to help fund the great rail boom. By the 1870s the British overseas investment portfolio totalled a mammoth £1 billion ($5 billion).*

Halcyon days for the young brokers and bankers of the Square Mile, with their gaudy jewellery and brash ways. The City of London was fast becoming the financial hub of the soaraway global economy, providing commodity exchanges, long-term loans, legal services and insurance. Most important of all were the bills of exchange backed by the City's financial houses – and traded on the money markets – which provided the short-term credit that allowed complex multinational

* To put this figure into context, the entire annual expenditure of the British state was £70 million.

trades to take place from Shanghai to San Francisco via Hong Kong, Singapore, Melbourne, Cape Town, Buenos Aires, London, New York and points between. As a parliamentary committee put it without exaggeration or false national pride: 'The trade of the world could hardly be carried on without the intervention of British credit.'[26]

If British industry was facing relative decline in the face of ambitious, resource-rich rivals, its financial sector was revelling in a bonanza. This was, wise observers calculated, the future of Britain's power in the world. It should not – and it could not – try and maintain its industrial dominance by pricing everyone else out of the market. Instead it should do everything it could to stimulate economic activity in foreign countries and encourage uninhibited trade across the world.[27]

By facilitating international trade and the movement of labour, Britain could take a slice of the pie at every stage. The country was well placed to do so. It possessed ports and coaling stations along the international highways west to east, from Halifax to Hong Kong. Its early industrial take-off equipped it with surplus capital to pour into railway construction everywhere from the Pampas to the Russian steppe. In turn, it supplied these ventures with high-tech equipment such as power tools, telegraphs and engines. These railways and telegraphs meant that more foodstuffs and raw materials could reach the global market. The British Merchant Marine was ideally suited to carry the produce of the world and warehouse it, duty-free, in the ports of the Empire. As capital from Europe funnelled out of London, goods from the Americas, Africa and Asia flowed into Liverpool to be distributed for the markets of the Continent.

If you sell cloth and scissors, the politics of your customers does not matter so much. But if you are exporting your cash it matters a great deal. It was in Britain's interests, therefore, to make the world safe for investors and open for business. Throughout the 1850s Britain would use a mixture of hard and soft power to do just that – to tear down the political barriers to trade and integrate hitherto closed markets into the global economy. A country that aspired to marshal and direct global commerce needed to maximise exchange by creating a system of free trade. Later parts of *Heyday* deal with the hard power as Britain used its navy – the most powerful the world had seen – and its military muscle and financial leverage accomplish those ends. The Great Exhibition, however, stood as the most powerful monument to Britain's attempt to persuade the world to follow its example.

Conceived as a way of showcasing British industry, the Great Exhibition became a living representation of Britain's vision of a new world. Above all else, it had to advertise the moral and material benefits of free trade.

What did the rest of the world have to gain from free trade? First there was the possibility of universal peace. Countries and empires that hunkered down behind the barriers of protection were said to be more aggressive and warlike. They used the revenue raised from taxing trade to fund gigantic armies. As *The Times* had it, in the middle of the nineteenth century 'we fight our battles far less with fire and sword, with gabion and howitzer, than with calico and scissors, with coal and with iron . . . Trade is now much more than the acquisition of lucre; it is a great political engine.' When Hardy conjured up the metaphor of 'a precipice in time' opening up in 1851, he was referring explicitly to the Great Exhibition. People sincerely believed that that year marked a new epoch in world history, a 'chronological frontier'. Under the glass roof of the Crystal Palace and surrounded by the produce of the world, some burst into tears; many others were profoundly moved at this visual representation of peace through trade.[28]

The successful completion of the Channel Cable right at the end of the Exhibition intensified the feeling that the human race was joining together in peace and harmony. Many believed that the uninhibited exchange of goods and ideas by the means of modern technology would break down the barriers that separated human societies, engendering mutual understanding and making war impossible. 'The most characteristic fact of our civilization,' said a French economist in a discussion about the Great Exhibition, 'is the growth of that mutual dependence of nations which is the soundest guarantee of peace.'[29]

Then there was the appeal to ordinary men and women around the globe. Competitiveness would release dynamism and enterprise, pulling up living standards and driving down prices. The diffusion of knowledge and prosperity by communication and trade, it was believed, would make oppressed people long to kick away the traditions, prejudices and tyrannies that inhibited their creativity. The Hungarian freedom fighter and international hero Lajos Kossuth told a meeting at Southampton in 1851 that 'All despots fear free trade, because the liberty of commerce is the great vehicle of political liberty.'[30]

The greatest exhibit at the Crystal Palace was used to illustrate these points. That exhibit was not a new technology, work of art or powerful engine; it was the millions of people who coursed through the building over the summer. Before the event, the thought of the population disgorging itself on Hyde Park sent shivers down the spines of the great and the good. Gatherings of working people had always been regarded as a recipe for anarchy and drunkenness; they had been prevented or broken up at the first opportunity. There were plenty of warnings that, when presented with priceless jewellery and invaluable treasures in flimsy cases, the *Lumpenproletariat* would not be able to restrain themselves. The crowned heads of Europe feared that the Exhibition would be a magnet for the revolutionaries and socialists of the Continent, who would unite in lightly policed London; violent insurrection would spill out of Hyde Park and surge back across the Channel.

When Fredrika Bremer travelled down from Manchester to London, so great was the congestion on the line that trains were backed up nose to tail. The train on which she travelled 'had to stop several times and wait, warned by red lights, like fiery eyes; these had been warned by other red lights, attached to the posts of the electric telegraph'. Electric signals pulsing down the wires managed the weight of traffic. It was this sophisticated transportation network that meant that by the time the Exhibition closed in October there had been 6 million paid entrances; when repeat visits and numbers of foreign tourists are taken into account this means that between a fifth and a quarter of the population of the United Kingdom visited the Crystal Palace. According to a guidebook of the event, 'The exceeding popularity of the exhibition eventually became its greatest wonder.'[31]

At no other time in history had so many civilians been transported to a single event. Working-class families were able to get to London thanks to subscription clubs, working men's associations, mechanics' institutes, savings banks and friendly societies, all of which arranged travel and lodgings. Railway companies offered package deals. The tour operator Thomas Cook brought his experience organising excursions for the Temperance Society to see 165,000 people to the Great Exhibition.

Impressive feats of transportation helped cement the image of Britain as a hyper-modern country. But it was the behaviour of the millions when they got to Hyde Park which really excited commentators. In the five months of the Exhibition there were a mere twenty-five criminal offences – nine for pickpocketing, six attempts to do so and ten instances of pilfering from the stalls. Queen Victoria, guarded by half a dozen unarmed policemen, walked amid 50,000 people (the amount of people who daily filled the Crystal Palace) 'as quietly as if she was in her own drawing-room', according to *The Times*. For one writer it was pleasant 'to observe how completely all social distinctions were for the moment merged in the general feeling of pride and admiration at the wondrous result of science and labour exhibited in the Palace of Glass'.[32]

The results of industrialisation had not, the British said, been immiseration and revolution. Repeal of the Corn Laws meant that the institutions of the country were ready to listen and respond to the demands of the people. While Europe was convulsed with revolution in 1848, Britain remained peaceful. The cost of living there was cheaper thanks to liberal policies. Indeed, the Crystal Palace came in under budget because of the recent abolition of taxes on glass. The Exhibition itself stood as an immense advertisement for laissez-faire economics: it was the people, not the government, that had brought the spectacular show into being. The event was a business, run for profit not national aggrandisement; it did not beg for taxpayers' money. According to Henry Cole, one of the leading lights of the Exhibition, 'the English people do much better for themselves than any Government can do for them'. The retreat of the state from the sphere of economics, and the opening-up of the country to worldwide competition, had unleashed the entrepreneurial zeal of the people and brought into being triumphs like the Crystal Palace.[33]

As one journalist put it, the Exhibition showed that in Britain all sections of society 'combine the most indomitable energy with the most ingrained love of order and respect for their own institutions' – the opposite, it was said, of Continental Europe, which was bedevilled by industrial underdevelopment, un-competitiveness, class tensions and arbitrary power.[34]

The great glass dome, the mass-producing machines and super-fast locomotives, the high-tech machines, the scientific discoveries and the peaceful throngs of contented people: here was a new civilisation exhibited to the world in Hyde Park, a prototype that others would adopt. According to the census of 1851 Britain had become the most urbanised and industrial society in history: the majority of people now worked in towns and cities rather than on farms. If you wanted to see the future, in other words, come to Britain.

The layout of the Exhibition reflected Britain's vision of a new world order. The groupings of the exhibits represented human societies open to each other, bonded by mutual exchange, free communication and peaceful competition. One half of the floor space, the entire western side of the Crystal Palace's idealised world, was given to British manufactures and inventions. On the other side of the nave, facing the hosts' lavish displays, were the contributions of Britain's industrial rivals, dominated spatially by France, the German Confederation and the United States. In the transept stood India and Britain's other Asian colonies. Facing them were China, Brazil, Turkey, Persia and Egypt, 'grouped . . . like a kind of torrid zone'. If we count Brazil as representing Latin America, these were countries regarded by Britain as the lucrative markets of the future, societies which could be shaped and regenerated by liberal values, but whose potential was being stifled by corrupt and rickety elites and by restrictive practices. 'Let us try and improve all these countries by the general influence of our commerce,' said Lord Palmerston, Foreign Secretary at the time of the Exhibition, 'but let us abstain from the crusade of conquest.'[35]

The rejuvenating breath of free trade would shortly gust through the world, it was believed, toppling tyrants and replacing closed empires with free exchange. Visitors to the planet recreated under the glass roof of the Crystal Palace were invited to compare and contrast its various countries. Here were the old and tired societies of Asia, their vitality suffocated by despotism and superstition. Here was Europe, the majority of its people crushed under the jackboots of the soldiers of tyrants, their dreams of nationhood and modernisation stifled by the military power of Russia, the Austro-Hungarian Empire and the Ottomans. At the end of the year the President of France, Louis Napoléon (nephew of Bonaparte), faced with having to leave office at the expiration of his second term, seized power in a brutal coup. France came under the authoritarian rule of the newly styled Napoleon III and Europe faced the prospect of war after the decades of peace since the final defeat of the first Napoleon.

Across the planet, dreams of republicanism, democracy, socialism and nationalism were dead or dying. There was surely one glaring exception. Venture into the stands of the United States and you could see an alternative pathway for humanity in the nineteenth century. Horace Greeley said that the British and European ruling classes '*dread the contagion of our example*; and this dread must increase and be diffused as the rapidly increasing power, population and wealth of our country commend it more and more to the attention of the world'. When news received in Continental Europe from the interior of the US – mainly concerning 'Indian fights and massacres, fatal steamboat explosions or insolvent banks' – was sixty days old, the 'contagion' of republican egalitarianism was quarantined by distance. But now, thanks to a new generation of fast steam packets and land telegraphs, that gap was slashed to a mere ten days and there was every indication it would be further truncated as technology advanced. America stood poised to influence world affairs.[36]

Revelling in the Californian gold bonanza, fuelled by dreams of 'Manifest Destiny', and on the cusp of a railroad revolution, the dynamism of the United States was given fair representation at the Exhibition by its inventions. But, crucially for a country bidding to stamp its mark on civilisation, it showed off its cultural prowess as well. One of the most discussed pieces of art in the Crystal Palace was Hiram Powers' marble statue depicting a naked and shackled Greek slave.

Universally admired the statue might have been, but it was a magnet for barbed comments. The statue was continually revolved so that it was presented at all angles to the encircling throngs. The man operating it reminded some of a dealer at a Virginian slave market, exhibiting human flesh for sale. The satirical magazine *Punch* produced a cartoon in which Powers' statue was rendered as a black slave shackled to the Stars and Stripes. One day the former slave Henry 'Box' Brown (so-called because he escaped bondage by mailing himself in a crate to a free state) deposited the *Punch* cartoon next to the statue, loudly declaring, 'As an American fugitive slave, I place this "Virginian Slave" by the side of the *Greek Slave*, as its most fitting companion.'[37]

While the Exhibition was under way the American periodical *The National Era* was serialising Harriet Beecher Stowe's visceral anti-slavery novel *Uncle Tom's Cabin*. When it was published as a book it sold 300,000 copies in the States but more than a million in Britain in its first year. Here were the realities of slavery exposed in all their horror. The British could be smugly satisfied that they had spent years and a fortune fighting the slave trade on the high seas, a long, slow campaign that was reaching its climax in 1851 with British naval patrols operating off the coast of Brazil, the last major importer of African slaves. Thus confident in their moral superiority, in excess of 500,000 British women put their signatures to a petition urging their sisters across the Atlantic to mobilise against slavery.

For the British, the world in 1851 seemed to be dominated by slavery and serfdom, tyranny and superstition, poverty and backwardness. A year prior to the Exhibition the admission of California to the American Union as a free state had sparked political crisis. For the first time there were more free states than slave states; the American South had to be placated with concessions to prevent the break-up of the federal union. American unity was exposed as inherently fragile. In Europe and America, as far as the British were concerned, you could see the sorry wreckage of doomed experiments in republicanism and democracy. The Crystal Palace was more, much more than an exposition of hardware and high-tech: it was an exhibition of British values, and that is what gave the event its lasting potency.

For a few months that summer the hyperbole went into overdrive, the press and much of the public intoxicated with the idea that

(left) The Greek Slave *by Hiram Powers and* (right) Punch's *response: 'The Virginian Slave'*

Britain was the torchbearer of a new civilisation. Only liberalism held out the possibility of worldwide renewal. Free-market capitalism was the most radical idea going – for through the medium of the uninhibited exchange of information and goods it would regenerate the human race by silently and invisibly subverting the ancient structures of power everywhere. 'Our duty – our vocation – is not to enslave, but to set free,' declared Lord Palmerston, 'and I may say without any vainglorious boast, or without great offence to anyone, that we stand at the head of moral, social and political civilisation. Our task is to lead the way and direct the march of nations.'[38]

In retrospect such language was a tempting target for satire. Fyodor Dostoyevsky wrote sardonically of such attitudes that 'new economic relations will be established, all ready-made and worked out with mathematical exactitude, so that every possible question will vanish in the twinkling of an eye'. In similar vein, the *New York Daily Tribune* mockingly said that the British supposed that when Manchester textiles and Birmingham manufactures flooded the ports of the world, 'the toughest old social wrongs will disappear as incontinently as harlequin through a stage-trap'. But it was exactly this kind of magical thinking that made the Great Exhibition so important. The notion that the age-old problems facing humanity would be blasted away by the unstoppable force of progress exerted a powerful hold in 1851 and well beyond. The world it conjured up of interlinked telegraphs and trade routes, of countries open for trade and bonded in peace by friendly exchange, was a long way off. The Crystal Palace was selling a dream, not a reality.[39]

And it sold this vision aggressively, not only in Britain but throughout the world. The sense that people stood at the dawn of a new civilisation was one of the most important outcomes of the Great Exhibition. It is an attitude we will encounter again and again in this narrative. Utopianism was one of the major engines of rampant economic growth, and it found its clearest expression in Hyde Park in 1851.

Around the world – and particularly in the Anglo-American part of it – optimism inspired people to brave the rigours of migrating thousands of miles, to settle in inhospitable terrain, to invest their money in visionary schemes such as long-distance underwater telegraph lines, gravity-defying mountain railways or the Suez Canal. Settlers in far-off lands carried with them the arrogant assumptions that Western capitalism and science ruled supreme. This belief – manifested in the American and the British worlds – that they stood in the vanguard of a new civilisation imbued them with the notion that they had the moral right to appropriate land from indigenous people.

Much of the stratospheric growth of the 1850s was based on excitement and speculation; often that expectation of profit leapt far ahead of economic realities. The Great Exhibition did not create such confidence; but it was the physical embodiment of faith in the progressive forces of the age and the progenitor of a prodigious international literature on the blessings of unchained capitalism.

Utopianism and hype can change the world to a greater extent than tangible things. From Sweden, Fredrika Bremer reflected on Britain's 'mission in the history of nations' and believed that the country was 'destined to extend her civilisation over the greatest part of the world'. One of the effects of the Great Exhibition was to implant a quasi-evangelical belief in Britain's moral crusade to remake the world through the free market. As one writer put it, the British should be thankful to God 'that it has been vouchsafed to us in our generation to lead the peoples onward in the march of peaceful enterprises and industrial triumphs'.[40]

This mixture of utopianism and opportunism determined Britain's relationship with the rest of the world in the 1850s, not to say its own domestic health. The country became deeply involved in international affairs as it sought to extend the frontiers of free trade by persuasion and force. Britain turned its attention and energies to the wider world; this period is remarkable for its lack of domestic social or political reforms. In Charles Dickens' *Bleak House* (1852) Mrs Jellyby famously neglects her children as she invests all her energies in philanthropic projects in Africa. And, like Mrs Jellyby, Britain seemed far more interested in rejuvenating the world than in sorting out its own problems, particularly those of the working poor. Another work that belied the roseate visions of the Great Exhibition was *London Labour and the London Poor* (1851), in which Henry Mayhew presented the underbelly of the capital city of world trade as a sink of prostitutes and drug addicts, sweatshop workers and beggars, pickpockets and quacks, mudlarks and professional dog-shit scrapers (among others), all negotiating a miserably precarious existence. Such works published at the same time as the Exhibition struck a discordant note, but not loud enough to drown the cacophony of celebration.

One paper called London 'the real capital, not alone of the world's material civilization, but of the spiritual idea of progress'. Britain in the 1850s aspired to be the midwife of a new world order based on unrestrained free-market capitalism. The Great Exhibition was instrumental in turning the attention of the public outwards, towards the world. It gave the British the opportunity to articulate their ambitions and imbued them with the confidence to back them up with action. As the planet's premier military and trading superpower, the future seemed theirs.[41]

The *Illustrated London News* of 10 May 1851 was a strange issue. In one article it hailed the Great Exhibition as 'the most remarkable event in the modern history of mankind' and a 'moral and religious work'. In another, entitled 'The Depopulation of Ireland', it described the misery of Irish emigrants to the United States and Canada, packed onto 'coffin ships' at Liverpool and Cork to escape the 'ravages of famine and pestilence' that Britain – for all its supposed wealth and liberality – had allowed to engulf their homeland. Not for these migrants the wonders of modern transport – the 250,000 who fled across the Atlantic every year might spend weeks in a rickety ship huddled together 'like pigs at a fair'. This edition of the paper was the spirit of 1851 in microcosm, with all its hypocrisy, contradictions and faith in the miraculous properties of capitalism as a regenerative force.

In the same issue the dark side of emigration was counterbalanced by a much more positive story. The *ILN* reported that a group of men and women about to embark upon the 'rugged work of colonisation' in Australia were given a private tour of the Great Exhibition. These emigrants (families 'of a very superior class', not famished Irish) were exporting their 'country's spirit of improvement' to the distant shores of the world, and it was deemed fitting that they should attest to 'the crowning act of her civilisation' when they arrived at Melbourne.

Whether they found anyone willing to listen to their tales when they arrived is doubtful. For if the Great Exhibition offered a vision of the future, Australians in 1851 – as yet unbeknown to the rest of the world – were experiencing the reality of breakneck progress and tumultuous upheaval.

2

The Hairystocracy

MELBOURNE

Among the accidents which have influenced the fortune of the world will hereafter be ranked the erection of a saw-mill on a branch of the Sacramento.

D. T. Coulton[1]

John Hunter Kerr's sheep station had been one of the most remote spots on earth; he had been attracted by the 'indescribable sense of the vastness and freedom of nature' found near the River Loddon, some 160 miles north of Melbourne. The route to the station was by way of a scarcely defined track through great tracts of forest, over hill, dale and countless unbridged streams. It was a region 'truly Australian in all its features': vast plains extended for miles in every direction to the horizon. In winter they presented 'one radiant field of brilliant verdure'; the scene turned pink, blue and gold in spring when millions of tiny flowers carpeted the plains; pitiless summer scorched everything 'to a great sea of yellowish brown'. The few settlers were thinly scattered over this huge area; 'the mode of living was primitive, and almost idyllic in its simplicity'.[2]

When Kerr ventured to Melbourne after the shearing season of 1851 he witnessed the first tremors of one of the greatest upheavals of the century. That year was eventful by anyone's reckoning. Summer had been intensely hot; the country frazzled into an enormous hay-field. On 6 February – Black Thursday – one of the fiercest and most destructive of all bush fires rampaged through the land, turning it into a 'smoking wilderness'. A few months later, on 1 July, this singed land was given a name and a political identity. At its founding in 1834 it was christened the Port Phillip district of New South Wales. Since then its European population had grown from a handful to 76,000 and it was emerging as one of the most prosperous and promising places in the world, its economy thriving on exports of wool, tallow and whale oil. It was now called Victoria, 'a colony formed by colonists' eager to differentiate itself from the penal colonies that stained Australia with a black reputation.[3]

The lonely path through the forest had been widened and furrowed by a weight of traffic. Many of the ancient trees had been felled, and 'low black stumps remained as solitary relics of what had been the forest's pride'. In their place were mounds of earth and thousands of tents. On the 'wild rough roads' Kerr encountered caravans of fifty or more drays at a time negotiating the swampy tracks. In their wake tramped brawny, bearded men pushing handcarts and shouldering guns; at their heels trotted dogs of all breeds.

The inn Kerr stopped at was once the genteel resort of the colonial farming gentry; now it resembled an enormous impromptu fair, with drays and wagons drawn up, tents pitched, fires blazing and a raucous mob in holiday spirits. The air was thick with drunken song, filthy oaths and swirling dust. Men, their faces illuminated by the glare of fires, stood in rings around bare-knuckled boxers; one of them lolled insensible against a log, his arm embracing a case of champagne. There was not enough food to feed this multitude which had burst upon the wilderness. Kerr had to make do with a tin of herrings, which he flambéed with brandy, and two bottles of champagne. The primeval stillness of the bush and the Arcadian simplicity of colonial life were shattered for good.[4]

The colony was in uproar. The man responsible was named Edward Hargraves. In January 1848 James Marshall discovered gold in the tailrace of a lumber mill he was building for John Sutter, a Swiss pioneer, on the bank of the American River, California. When US President James Polk announced the news that December, immigrants from all over the Pacific region – from Oregon, the Sandwich Islands, Mexico, Peru, Chile, New Zealand and Australia – rushed by sea to

San Francisco and onwards to the Sierra Nevada Mountains. One of these was Hargraves.

The Australian Forty-niner detected similarities between the geography of his homeland of New South Wales and the goldfields of the Sierra Nevada. This prompted Hargraves' return to Australia and his discovery of gold near Bathurst. The news sparked similar searches in other colonies. The first great discovery of gold in Victoria was made at Buninyong in August 1851, then at Ballarat. The bacchanalian scene that Kerr encountered was one manifestation of the gold madness that was gripping Victoria as it had recently done California. News of fantastic discoveries of gold began to dominate the Australian press. Cities and farms were deserted, and areas where gold had been discovered inundated with fortune seekers.

Stories of fabulous riches spread around the globe. One digger described the scene at Bendigo in December 1851: 'You could see the gold shining in the heaps of dirt, and every man sat on his heap all night with a pistol or some weapon in his hand. I thought it would come to our making picks and shovels of the gold, it was so plentiful.' Small claims were said to yield gold worth between £18,000 and £55,200, enough money to live like a lord till your dying days. The papers were stuffed full of stories of sudden and amazing accumulations of wealth.[5]

News of such wonders reached Britain at the end of 1851 and the beginning of 1852. There was a tantalising two-month gap in the mails. Then, in April, came news of the staggering riches of Mount Alexander. A few weeks later eight tons of Victorian gold was landed in England, confirming the fevered rumours of limitless riches. With it came accounts of diggers eating sandwiches filled with £10 notes and swilling them down with champagne. They were also said to ride horses shod with golden shoes.[6]

It was, of course, a joke. But few other rumours did more to stimulate emigration to the ends of the earth. Charles Dickens described the scene at shipping offices in London: 'Legions of bankers' clerks, merchants' lads, embryo secretaries, and incipient cashiers; all going with the rush, and all possessing but faint and confused ideas of where they *are* going, or what they are going to do.' The press was full of miracles: 'The marvellous tales of boundless wealth, of unceasing prosperity, of unexampled progress read almost like some story of enchantment from the *Arabian Nights*. Yet they are all true.'[7]

Kerr threaded his way through this pandemonium in his long journey from the bush to Melbourne. When he got there he found it all but deserted. Then at Christmas thousands of successful gold diggers flowed back into the city to celebrate their sudden accumulation of wealth. These were rough, bearded men with unwashed faces, dressed in the finest broadcloth and their fingers heavy with rings. They were with barefooted, straggly-haired women, bedecked in the latest fashions and weighed down with costly jewellery. There was a rage among the diggers to celebrate their bonanzas with lavish weddings, and they were a remarkable sight in 1851 and 1852 as they paraded through the streets on carriages pulled by horses sporting huge plumes and attended by liveried servants. 'There they go,' wrote Ellen Clacy, an English immigrant, 'the bridegroom with one arm round his lady's waist, the other raising a champagne-bottle to his lips; the gay vehicles that follow contain company even more unrestrained, and from them noisier demonstrations of merriment may be heard.' They taunted the colonial gentry as they whirled past: 'It is our turn to be masters now; you will be our servants yet!'[8]

The importance of discoveries of gold in California and Victoria, and later in Colorado and New Zealand, in accelerating the pace of change in the mid-nineteenth century cannot be overstated. By 1851 the wealth of California had exploded on the global economy. Fresh discoveries in Australia almost doubled global production of the precious metal, injecting more energy into the world economy. The 1851 Victorian gold rush is less well known than the Californian version, but in many ways it is as significant, if not more so, in the history of the mid-century boom and the changes that assailed the world at this time. For one thing, the second great discovery of gold in the Antipodes closed a circuit in the Pacific. California and Australia were both remote places in the world before this time; their troves focused attention on the entire Pacific, including its islands, the Asian littoral and the American coastline from Vancouver to Tierra del Fuego. They created busy new trade routes that criss-crossed the planet, connecting hitherto isolated places into a global system.

The impact of gold on the world is the subject of the chapter that follows. This chapter concerns itself with the effects of sudden and unprecedented wealth on new communities. The Australian experience followed, almost exactly, the trajectory of the Californian rush. Both lured hundreds of thousands of migrants over distances of thousands of miles. In both cases the individualistic experience of early pioneers quickly gave way to capitalist organisation. The west coast of America and Victoria alike witnessed the phenomena of boom towns and lawlessness, ecological devastation and rapid development that so astounded the age. If anything, Victoria offered a more exaggerated version of what had recently happened in California.

It was as if a tornado had ripped through the colony. Port Phillip's growth in its first two decades had been rapid enough. When Kerr arrived in Melbourne in 1839 it was a tiny settlement of wattle-and-daub one-storey houses amid 'a silent wilderness, where tea-tree scrub grew close and thick amid the tall gum-forest'. In 1851 it was flourishing, with a population of 23,000 out of the 76,000 who had settled in Victoria. When gold was discovered in California it had only just become a territory of the US; its population and economic development were consequently less advanced than its Australian counterpart, and it was separated from its nearest American state by thousands of miles of inhospitable terrain. So when news of gold reached Sydney, Adelaide and Melbourne, there were more people available to join the rush. As a consequence things moved faster there; the early rush was more topsy-turvy as it impacted on a relatively prosperous and populous community.[9]

According to William à Beckett, Chief Justice of Victoria, 'a moral and thoroughly British population, which was every day expanding and strengthening, and giving promise of similitude of their institutions, manners and tastes, to those of our glorious mother-country' was debauched by gold. To men such as à Beckett their colony, which had only recently been granted independence, bid fair to be an Antipodean Britain. Instead Melbourne was virtually depopulated and the scum – as they saw it – of Australia swarmed to their lands. The colony entered a state of anarchy. One observer described the rush to Forest Creek: 'Slim shopmen, stout-calved butlers, government clerks, doctors, lawyers, runaway sailors, deserting soldiers, self-ordained divines, and strong-minded females in ultra-bloomer costume, flocked to Forest Creek like flies round a treacle-butt.'[10]

The women in 'ultra-bloomer costume' were wearing scandalously loose trousers gathered at the ankle. The flowing bloomer suits originated in New England in 1851 as part of the 'rational dress' movement, an early feminist attempt to liberate women from cramping fashions. The sight of women working in practical clothing alongside civil servants playing hooky and rowdy sailors symbolised the sudden and disturbing upheaval of a gold bonanza. Towns were unprotected as policemen resigned en masse; warders of lunatic asylums walked out, leaving their inmates in control. The landowning elite found their stations deserted by their shepherds and hutkeepers. Elsewhere in Australia economic life ground to a halt as thousands of labourers and clerks rushed to Victoria.

The early rush was as nothing compared to what followed when news spread around the world. Australian colonies had long been in need of emigrants of all classes in order to exploit the land and establish fully functional political societies. They invested in promoters and advertisements. Would-be emigrants were promised a land 'resembling England in all but climate and poverty'. Melbourne was a fine city and thoroughly English, with cricket clubs, societies, newspapers and schools. The city and its suburbs, said an Australian promoter, 'have all the quiet, subdued interest of an old English domain'. But, unlike Britain, the labourer would find independence and spectacularly high wages in booming Victoria. All were welcome: 'if all England were to pour in one vast torrent upon our shores, we should still have room and employment for all!'.[11]

Before 1851 Australians fretted (with good reason) that people saw their continent as 'a vast jail in the wilderness, a criminal lazar-house at the antipodes, a voyage to which was as much dreaded, as would have been a trip to Siberia or Russian Tartary'. A feature of the age was the 'booster' – a self-motivated or hired pen who depicted an emerging British colony or American territory in the most flattering colours to attract settlers and capital. Now gold did the work of a thousand boosters, bathing Australia in a bright, alluring glare. Australia became the subject of intense interest in the international press. Victoria had precious metals as well as cricket pitches; in other words, it was a place of wealth and recognisably British civilisation, unlike California, which was often stigmatised as a lawless hellhole. Emigrants were enticed with 'an abundance of the precious metal, which only waits to be gathered'. Tales were told of gold lying near the surface which could be dug out with a penknife. Diggers were making over £2,000 per month, some in excess of £5,000, it was reported. Even labourers far removed from the diggings were commanding £1,200 a year.[12]

Fired by stories of high wages, abundant gold and wonderful opportunities, thousands in Britain, Ireland, the United States, European countries and China made preparations to leave. One of Britain's favourite paintings is Ford Madox Brown's *The Last of England*. Inspired by the artist's friend, the sculptor Thomas Woolner, who left for Australia at the height of the gold fever intent on making a fortune, it depicts a couple leaving England in 1852. Grim-faced and determined, hand clasped in hand, they face resolutely away from the White Cliffs of Dover and towards the future.

As thousands of eager gold seekers sailed to Australia, Victoria entered an agony of adjustment to its new situation. Economic and political crises loomed. The relationships between masters and servants, employers and employees broke down because the price of labour went through the roof and everyone expected to be rich immediately. The Lieutenant-Governor, Charles La Trobe, imposed on the diggers the obligation of purchasing a licence at a cost of £1 per month. But the measure was fiercely resisted and the small colony did not have enough police or troops to enforce the order. By the summer of 1851–2 Victoria was becoming ungovernable. 'This used to be a pleasant, quiet country,' opined one Victorian woman; 'but it is all over now!'[13]

'The wharves, till lately so silent and rural in their appearance,' wrote Kerr, 'now presented a scene of the most bustling activity and confusion.' One of the new immigrants was the English writer William Howitt, who arrived on a ship full of eager gold hunters. They had paid £3 per ton for their luggage to be carried 13,000 miles to the other end of the world. When they arrived at Melbourne it cost them exactly the same amount of money to move their goods from ship to city, a distance of seven miles. Many had spent all their savings on travelling to Australia, or had wasted what they had in gambling on board; they were not prepared for the costs of getting ashore.[14]

For years the beaches around Melbourne would be littered with mattresses jettisoned from emigrant ships by people who simply could not afford to get their possessions ashore. Those who paid for their luggage to be taken to the wharf found that Melbourne was bursting with migrants; people were lucky to rent a space in a crowded room in a lodging house. Most slept on their luggage in the open air for the first few nights, then simply abandoned it or sold their possessions on the wharves for a pittance when they discovered the ludicrous prices to convey items to the diggings. Then they had to sleep in Tent Town, a vast canvas suburb outside Melbourne, and contemplate how they would ever escape to El Dorado and make their fortunes.

They discovered that Australia was not the Eden of the imagination. Howitt described a wild landscape of lagoons, gum trees and scrub. '"And is THIS the beautiful scenery of Australia?" was my first melancholy reflection,' wrote Ellen Clacy of her arrival. 'Mud and swamp – swamp and mud – relieved here and there by some few trees which looked as starved and miserable as ourselves.' Melbourne itself was a let-down. Howitt found it 'straggling and unfinished', grids of muddy, unpaved streets chock-a-block with roaming goats, chickens and dogs, and flanked with half-finished wooden one-storey houses. The suburbs were something to behold, 'a wilderness of wooden huts of Lilliputian dimensions' that stretched for miles. What was most galling was the price of land in these endless shanty towns, which was six times greater than in London or New York.[15]

Welcome to boom town. Melbourne was a gigantic, daily-expanding campsite. Recently arrived migrants found it perplexing and dangerous. William Howitt described Bourke Street in the centre of Melbourne, where successful diggers raced up and down on newly purchased horses: 'The whole street swarmed with diggers and diggeresses. Men in slouching wide-awakes, with their long untrimmed hair and beards, and like navvies in their costume. Some had heavy horsewhips in their hands, and were looking at the exploits of other diggers on horseback with a knowing air. Others were swearing, about the doors of pot-houses, where others, again, were drinking and smoking.' He dubbed the swaggering, bearded diggers the 'hairystocracy', for they seemed to rule Melbourne.[16]

Many migrants were men, women and families of respectable, middle-class or even noble backgrounds but embarrassed situations who left Britain with high hopes of making a quick buck and returning to their social niche. They had to adapt to a new world. Charles Stretton, a man of gentry background and empty pockets, arrived in Melbourne as if he were promenading on Regent Street, complete with a top hat: 'I have no hesitation in saying that I heard more oaths sworn on that first day . . . than I had heard in the whole course of my life.' They were directed at him. He and his companions sold their swallowtail coats as quickly as they could and purchased the accepted outfit of the digger community: a battered wideawake hat, hard-wearing moleskin or corduroy trousers, a thick shirt coloured blue or red and a pair of stout boots. The costume was complemented with a revolver (preferably a Colt) and a bowie knife prominently displayed in the belt, and swag – bedding, cooking utensils and foodstuffs – slung over the shoulder in a blanket. It was not complete without a bristling beard. Travel writers recounted stories of Oxbridge graduates marking billiard tables or shining boots for rowdy diggers, barristers toiling on the roads, aristocrats taking up carpentry or young gentlemen whiling away a forlorn existence as shepherds in the middle of nowhere.[17]

Such stories amazed people back in Britain, who lapped up tales of the world turned on its head, of the upper classes reduced to working for the 'new aristocracy' of diggers. Stretton failed at the goldfields and spent six precarious years in Victoria keeping body and soul together variously as pedlar, barber, storekeeper, fisherman, brickmaker, shepherd, actor and prison warder. Victoria was full of Mr Micawbers, waiting for something to turn up, and the name of Dickens' character was given to a whole class of broken-down, disorientated gentlefolk.

Yet there were plenty who embraced colonial life. Howitt encountered a gentleman who purchased a wagon and made £1,000 a year carting people and goods to the diggings. When asked what friends at home would think of his new, reduced circumstances he replied, 'Everyone does as he likes here, you know.' When Stretton made his first attempt at digging for gold he met a rough, tall man who looked, spoke and drank like a 'brigand'. He had as his mates 'two of the most villainous-looking scoundrels that I have ever seen'. This alarming man turned out to be the scion of a noble family. He had little trouble fitting in with the diggers and toiling with his bare hands. And he had no problems slipping back into his upper-class lifestyle: when Stretton was reacquainted with him in England years later, he was every bit the refined gentleman.[18]

'To those of my own sex who desire to emigrate to Australia,' wrote Ellen Clacy, 'I say do so by all means, if you can go under suitable protection, possess good health, are not fastidious or "fine-lady-like," can milk cows, churn butter, cook a good damper, and mix a pudding.' Those who left for Australia prepared for a life of adventure, rather than of easy gain, were able to adapt and survive, if not prosper. One such, a young lady called Harriet, 'brought up in the lap of luxury' in Dublin, emigrated with her brother and their £300 inheritance. She was happy to give up her previous life of waltzes, composition and fancy fashions, cut her hair short and don rough masculine clothes. Harriet made a good income selling puddings and mending clothes – valuable accomplishments in the overwhelmingly male digger camps. 'Wild the life is, certainly,' she wrote home, 'but full of excitement and hope; and strange as it is, I cannot fear to tell you, that I do not wish it to end!'[19]

The romance of what Clacy called 'the gipsy sort of life' and the exhilaration of roughing it in the open air were appealing to those tough enough to overcome disappointment and privation. William Howitt was in his sixties when he worked his way through Victoria. He described the scene as he sat alone enjoying the verdant beauty of Australia: 'Beside me a pile of goods under a tarpaulin, my steak cooking in the frying pan, or a slice of bacon frizzling on a pointed stick, and tea boiling in a quart-pot . . . always looking very formidable, in case of any visit from bushrangers,* by a revolver lying on a camp-stool at hand, and a bowie knife in my belt.'[20]

The thousands of emigrants who poured into Melbourne had to possess this kind of resilience to survive. Such new worlds were rough, tough and often lawless; they were egalitarian and proudly so. Above all, these places of boundless opportunity favoured those with the physical and mental perseverance to subdue them.

Making enough money to get out of Tent Town was the first hurdle. At any one time there were about 20,000 people employed by the colony to build roads; every one of them hoped to save enough to get under way to the diggings. The roads outside Melbourne scarcely deserved the name. Men, women and children trudged through the quagmire and ruts, taking what they could carry. Those who could afford it loaded their digging gear and tents into drays and wagons pulled by eight or ten bullocks and walked behind the vehicles, carrying swag and shouldering guns. Howitt noted a well-dressed woman wearing a sun bonnet; on one shoulder rested her rifle, on the other a basket. Her male companion carried their baby.[21]

The ways to the diggings were congested with backed-up drays. Often they had to be hauled out of the bogs, then surmount steep hills or ford rivers. Every step of the way – which took months – involved continual hard labour along roads lined with broken-down vehicles and littered with the skeletons of horses and bullocks. They camped for the night, terrified of the bushrangers who plagued the roads and held up parties of travellers at gunpoint.

No one ever forgot the first sight of the diggings. 'Every tree is felled; every feature of Nature is annihilated,' wrote Howitt of the Ovens diggings near Beechworth, 250 miles north-east of Melbourne. The forlorn stumps stood in a wilderness pockmarked with holes and piles of gravel: 'it looked like a sandy plain,' wrote Ellen Clacy, 'or one vast unbroken succession of countless gravel pits – the earth was everywhere turned up'.[22]

* A bushranger was an escaped convict who survived in the Australian bush and preyed on gold-rushers.

Howitt noted how the creek had been diverted and the bed of the old river was being dug up. The diggers sank holes here, sometimes as deep as thirty feet and packed as closely together as possible. Some of each gold-digging party were employed bailing or pumping out the fetid black water that filled these holes. Others removed the clay and gravel by hand, or with buckets or pulleys. 'In the midst of all these holes, these heaps of clay and gravel, and this stench, the diggers were working away, thick as ants in an ant-hill.' The lumps of mud were broken down and the gravel washed in puddling-tubs, cradles and tin dishes. Further away from the watercourse were the 'dry-diggings'. Here, every inch of ground was dug up from ten to forty feet. The dry digger was spared flooding, but he had to dig through hard ground in very hot temperatures.

Tents made of rags and pieces of bark, perched on the ground that had not yet been turned up, stretched for miles. People who had not brought tents slept wrapped up in rugs. Among the tents and shops were heaps of rubbish, discarded bottles, the flyblown, rotting heads and feet of sheep and bullocks. Rheumatism, dysentery and scurvy were endemic in these camps.

'The diggers seem to have two especial propensities, those of firing guns and felling trees,' wrote Howitt. The camps were noisome and dangerous places. During the day the diggings resounded to the noise of picks and shovels and the rattle of cradles that separated gravel and gold dust. In the evening the diggers wielded their axes with a kind of fury against the remaining trees. 'But night at the diggings is the characteristic time,' Clacy wrote. 'Imagine some hundreds of revolvers almost instantaneously fired – the sound reverberating through the mighty forests, and echoed far and near – again and again till the last faint echo died away in the distance. Then a hundred blazing fires burst upon the sight – around them gathered the rough miners themselves – their sun-burnt, hair-covered faces illuminated by the ruddy glare. Wild songs, and still wilder bursts of laughter are heard; gradually the flames sink and disappear, and an oppressive stillness follows . . . broken only by some midnight carouser, as he vainly endeavours to find his tent.'[23]

Although vast quantities of gold were dug, only a few individuals made staggering amounts of money. Because prices were unreal, the majority just about covered their costs with the gold they unearthed. To the question of why anyone in their right mind would put up with the privations, hard labour and uncertain returns of gold-digging, the English migrant businessman Henry Brown answered: 'The feeling of being one's own master, to work just when it pleases, and to leave off when it suits, has a great attraction.' Many men and women who had been brought up in overcrowded slums or in rural poverty in the Old World and who felt stifled in a highly stratified, class-obsessed society found release in the New World. Whatever else it was, life in gold-rush Australia or California was one of independence and equality. 'There are no gentlemen in the colonies now,' one guidebook writer said. 'All barriers are broken down. There are only rich men and poor men; and as the latter may be rich in a week one is "hail fellow, well met" with everyone else.'[24]

The rowdiness of the camp, then, was the celebration of personal freedom. 'They enjoy . . . a new liberty, of which they never dreamed before,' recalled Howitt of the working class in the Victoria of 1853: 'how different to the state of the silk-stocking, crimson-plushed, livery-servants at home, who stand touching their hats at every word of their employers. There, on the contrary, released from whatever species or degree of control they were accustomed to, these men run into the rudest and most impertinent license.'[25]

Even if the overwhelming majority of diggers never made a fortune, most could afford meat, grog, horses, dogs and revolvers with the dust they recovered – things which were beyond their means in the Old World. In Britain the pleasures of rural sports were closed to most people by well-guarded private estates and the notoriously draconian Game Laws. No wonder diggers loved nothing more than blasting wildlife and firing their guns into the night sky.

Any manifestation of authority which interrupted the fun was deeply resented. And there was plenty of friction between diggers and government officials. Alcohol was banned on the diggings, but 'sly-groggers' did a roaring trade smuggling in illicit spirits to the camps concealed in such things as horse collars or saddles. The illegal trade in Indian rum gave the authorities the excuse to ransack tents

Diggers search for gold in Victoria

and stores. But the chief resentment was the system by which diggers had to buy licences from the government. It was not just the money involved, although that was a serious grievance. It was the nature of the policing encouraged by the licensing system. Diggers were constantly being hassled by goldfield commissioners to show their licences. If they had left them in their tents while they were labouring down a hole, or for whatever reason, they were fined or imprisoned.

William Howitt said that the goldfield management was like living under martial law. The magistrates acted as judge and jury, and they had armed police to back them up. These officials took a very firm line with people they considered the scum of the earth. After the initial wildness at the beginning of a rush, they imposed order as rigorously as they could. Howitt commented that the diggers had real and serious grievances, and no means of justice or redress, 'such as no Englishman should, or will long, tolerate'.

But however strong the authority, it could not entirely contain the lawlessness of Victoria. Thievery was prolific in the camps. Diggers slept with their gold under their pillows. Often crooks would cut slits in the tents and remove gold from under the heads of sleeping men and women. Henry Brown said that keeping hold of his money during a rush was 'a game of skill, between myself and the thieves'. Carrying around a loaded revolver – and even learning to sleep with one in his hand – 'produced a fierceness, and a disregard of human life, that those who have lived in security can hardly understand, and which I can now look back upon with shame and dread'.[26]

Victoria was not for the faint-hearted. Charles Ferguson, an Ohioan who had experienced the Californian gold rush in all its goriness, said of a roadside inn on the way to the diggings in Victoria: 'We did not stop inside. We had stopped at many hard places, but this was a little too tough.' The British and Victorians prided themselves on the belief that their colony was more law-abiding and civilised than its American counterpart. The opposite was the case. Bendigo, Ferguson said, 'was certainly one of the worst places on earth in 1852–3. One was not safe in going outside of his tent after dark, as he was liable to be either shot or sand-bagged and robbed . . . They would steal wash-dirt, rob a claim, or kill a man without compunction.'[27]

It was no wonder that the roads back to Melbourne were thick with former clerks and shopkeepers fleeing the diggings. These migrants had crossed the planet enticed by stories of people tripping over nuggets. The reality was very different. This was navvies' work – and it was done in the main at navvies' wages. As Howitt remarked, if a young man told you he wanted to venture forth to make his fortune at the diggings, you should make him work in a coal mine and then a stone quarry; next ask him to sink a well in the wettest spot you can find and then have him clear a space in a bog sixteen feet square and twenty deep. 'If, after that, he still has a fancy for the goldfields, let him go.'[28]

William Howitt and his party decided very quickly to escape the rowdiness and poor prospects of the Ovens goldfield and lit out for the unexplored mountain streams above the goldfields. For the first time they left the beaten track. 'Onward we went, up hills and down valleys, and over creeks, wade, wade, wade, through the deep scrub. The country appeared as if no mortal had ever traversed it.' Of course, mortals had traversed it, for millennia. Howitt and his mates entered this remote part of Australia just after the exodus of the Aborigines and immediately before the onslaught of squatters and gold hunters. When the first British convicts arrived at Botany Bay the indigenous population of what would become Victoria stood at 60,000. As one popular guidebook to the Great Exhibition put it: 'In our forty years' possession of [Australia] we have utterly destroyed them, by as atrocious a series of oppressions as ever were perpetrated by the unscrupulous strong upon the defenceless feeble.'[29]

Ravaged by smallpox, menaced or killed by the intruders and displaced by the 6 million sheep that had overrun the district's grasslands in sixteen years, just 2,000 Aborigines remained by the time of the gold rush. Gold-digging devastated their remaining ancestral lands and poisoned the rivers. The tidal wave of settlement in the 1850s and its attendant roads, inns, camps, boom towns and competitiveness forced the survivors deeper into the wilderness and demoralised those who remained. One gold seeker put it this way: 'strong drink effectively civilised them off the face of the earth'.[30]

So, it was an empty and apparently virgin landscape that Howitt and his party entered in the early part of 1853. After much searching they found a promising place beside the Yackandandah stream. For the moment it seemed like one of the most solitary spots on earth, and one of the most beautiful. The morning after they pitched their tents they discovered abundant gold near the surface of the creek – so near in fact that when they pulled up shrubs they found gold in their roots.

'We had quietness and greenness,' remembered Howitt, 'and the most deliciously cool water, sweet and clear.' It was the Arcadian fantasy that lured thousands to Australia. 'But,' he continued, 'this quietness and greenness could not last.'

First Howitt's party destroyed everything that gave lustre to the scene. The creek ceased to run clear. Trees were cleared away and the site was littered with heaps of gravel and mud. Then the water became muddied and smoke hung in the air. Diggers prowling for successful claims learnt to look for these telltale signs; they started arriving at the camp, group by group. The tea trees and wattles were systematically stripped away and the valley became studded with tents. The path Howitt's party had cleared through virgin grass, scrub and forest became a road busy with bullock drays passing in both directions. Within days of being the first Europeans to arrive at this secluded creek, Howitt and his men now shared the location with stores, butchers' huts, a doctor's and a government camp. 'These revolutions,' he mused, 'are about as rapid as the shifting scenes in a theatre.'[31]

If you wanted to witness the future being made, then Australia or California were the places to be in the early 1850s. With the ecological devastation, displacement of native peoples, overnight cities and the mingling of diverse classes and nationalities, they were harbingers of the change that was about to rip through other parts of the world on scarcely less impressive scales.

The pace of change in Victoria, as in California, became its chief attraction and main talking point after gold fever had abated. A rush in full spate was something to behold. As the numbers of gold seekers increased in 1852–3 and opportunities for massive accumulation decreased, people became desperate and stampeded new goldfields on the strength of mere rumours. Storekeepers, horse dealers, auctioneers, gold buyers, barbers, doctors, prostitutes and anyone else with goods or services raced with them. The wilderness was cleared of trees and a township containing 15,000 or 20,000 people started into life with astonishing speed. Within ten days of the rush to Ararat in south-west Victoria, for example, a township sprang up 'like the work of enchantment': a street in the bush a mile long lined with stores, saloons, auctioneers, theatres, music halls, boxing booths, brothels and banks.[32]

The gold rush saw the creation of many a town that exploded into 'a gay Babel of joyous confusion and extravagance' full of life, entertainment and danger and then faded to a ghost town. But camp towns in regions where gold continued to be mined extensively underwent astonishing transformations. Bendigo Creek was named after its original hut-dwelling inhabitant, a pugilistically inclined sailor turned shepherd nicknamed 'Bendigo' after William Abednego Thompson, a famous bare-fist fighter. Ballarat, further south, was a sheep run. Both places stood in the centre of districts which were once lush but were now 'a desert expanse of grey bleached soil'.[33]

The towns swelled in size, both peaking at 60,000. That represented an unprecedented population explosion in the wilderness; Melbourne only a couple of years before had numbered 23,000, and San Francisco contained fewer souls than Bendigo and Ballarat in their pomp. The early populations were shifting ones; diggers and their families came and went in the course of their odyssey. They were shanty towns consisting of thousands of tents, huts and log cabins with bullock-hide roofs. Every time rumours came of a new strike, the men would swarm to it. The women and children remained behind. They were better dressed than their menfolk: at Bendigo in 1853 the fashion was for white wideawake hats with broad ribbons, riding jackets and handsome dresses. It was in this attire that they could be seen every morning outside their tents and cabins, washing, cooking and swinging axes.[34]

Within a few short years Ballarat and Bendigo had made the transition to become major cities. The former was built upon a grid system and in the centre of the town the thoroughfares were wide and flanked with grand stone buildings, including banks, first-class hotels, stores, a post office, churches and other public edifices. The streets of Ballarat, and those leading to the city, were macadamised with quartz and lined with trees which were 'carefully enclosed and lovingly tended' by the residents. Neat cottages with ornamental verandas and pretty gardens made up the suburbs of these elegant, bustling cities.

By the end of the 1850s such goldfield cities still depended on gold-mining. But it was to a far lesser extent than in their feverish early years. Farms and vineyards reclaimed land made barren by gold-digging. The gold rush cities were connected to Melbourne and Sydney by telegraph in 1854 and by rail to Melbourne in 1862. The progress from shanty towns to flourishing inland cities was a truly remarkable feature of the age.

And then there was Melbourne.

A correspondent for *The Sydney Empire* wrote that he had escaped the 'horrible discomforts' of the city in 1852. Two years later he returned and was astounded by the wide streets, handsome shops, abundance of luxurious hotels, 'the torrent of population' and the splendour of the buildings. 'Melbourne at this moment presents the most extraordinary instance of the power and enterprise of the human race that the world ever saw.'[35]

No matter at what point people arrived in 'Marvellous Melbourne' in the 1850s, they were astonished at the rapidity of change when they returned a few months later. The city increased from 23,000 to 80,000 within a year. There were plenty of people who had made fortunes from the rush – and very few of them were diggers. The squatters did good business exchanging mutton for gold with hungry diggers. Added to that, the export value of wool doubled in the decade after 1851. Merchants and storekeepers became inordinately wealthy importing goods to sell at the diggings. Innkeepers on the roads made considerable fortunes from the thousands of thirsty diggers; some retired on £20,000 after a short time. During boom time entrepreneurs found ways to turn the rush to good account: one man gathered the skeletons of the thousands of horses and bullocks that had died on the road to Bendigo, started a bone mill and became wealthy. People who set up in the haulage business did very well conveying humans and goods to the diggings. As ever, lawyers, bankers and property speculators cashed in.

'The Melbournians . . . have shown themselves . . . a most mercurial race – the maddest speculators in the world,' wrote William Howitt. 'At the slightest touch of prosperity, up they go beyond the clouds, and if they met the man-in-the-moon in their flight, he would never convince them that they must descend again – till they actually fell.' This feeling of wealth and confidence made Melbourne the 'Paradise of Labour' in the early 1850s. The inadequate wharves and docking facilities of 1851 were greatly improved. Imported goods and immigrants no longer had to make the tortuous journey from port to city after the opening of a railway in 1854. The suburbs of Sandridge and Emerald Hill were once covered with tents; by 1857 they boasted stone-and-brick houses, excellent inns and shops.

Melbourne itself swelled in grandeur. Stone replaced wood. There were shops with plate-glass windows, fine banks, mansions, theatres,

Collins Street, Melbourne, in 1857

parks, and inns and hotels to rival Vienna. Antonio Gabrielli, a merchant and financier of London and Bombay, raised a loan in the City of London of £1 million for public works such as street paving and drainage in Melbourne and Geelong. The capital was lit by gas for the first time in 1856. There was a botanic gardens, a Mechanics' Institution, Philosophical Society and new schools. The gold rush bequeathed Melbourne some of the finest public buildings of the Victorian period, including the State Library, the University and Parliament House, among others. Internal improvements increased prosperity, with almost £10 million invested in railways and roads.

The progress of Melbourne, said one American, would astound even those who had seen San Francisco. The coarseness of 1851–2 had disappeared by the middle of the decade. Victoria purchased a third of all books exported from Britain, the same as the United States, and it had a respectable periodical press of its own. The streets still bustled as they did at the height of the rush. But gone were the chaos and the diggers' impromptu horse races and bling weddings. The hairystocracy was extinct. The scene was much like London at rush hour, with respectable clerks taking omnibuses or walking home to their charming suburban villas.[36] 'Ah! where are the jovial, rowdy, reckless fellows of the days when diggers were diggers?' mused a Victoria newspaper in nostalgic mood towards the end of the decade. The colony had experienced a rapid process of gentrification, familiar enough to post-industrial districts of Western cities in the twenty-first century.[37]

The Melbourne of the early 1850s had appeared to be headed towards gold-induced disaster as easy riches addled the population and diverted productive labour into speculative mining. The Melbourne of the middle decade, however, accorded with a certain idealised vision of empire. The aim of colonisation, it was said in Britain, was to replicate 'so many happy Englands' in faraway oceans and continents, independent states united by 'bonds of freedom' that germinated British institutions and values in alien soil. 'The greatness of England,' declared one politician, 'exists not in the geographical extent of her Empire, but in the spirit which animates those who inhabit it.' Out on the frontier manliness and rugged individualism flourished, perpetuating supposedly ancient characteristics belonging to their motherlands. The frontier was where diverse peoples could be shaped into nations; where Western civilisation would be born anew.[38]

As one Melbourne clergyman put it, the aim of life was to cultivate a 'high style of character'; the end result of 'energy, self-denial and enterprise' would make a man the 'architect of his own fortune'. Success in the colonies, the idealists held, required doggedness, steady accumulation and frugality. As George Earp, a promoter of Australian emigration, put it, the Antipodean colonies were perfect for building men and forming character. They presented a level playing field, unlike Britain, with its entrenched and unfairly distributed privileges. A poor man who worked diligently, saved money and built up a reputation was 'creating capital out of character', a situation impossible in the Old World. It meant independence and freedom.[39]

The hairystocracy aspired to independence and freedom. But it was a rough kind of libertarianism. In part it was freedom from employers and obligations of deference. But they were also prepared to resist the unfairness of the Miner's Licence and the high-handedness of the goldfield police. In the spring and early summer of 1854 some 12,000 disaffected miners formed the Ballarat Reform League. The movement called for no taxation without representation and for Australian independence from Britain if their demands were not met. The miners held enormous outdoor meetings and they swore allegiance to their flag, which depicted the Southern Cross, the symbol of nascent Australian nationalism. Protest turned into rebellion: the miners burnt their licences and built a defensive stockade at the Eureka Lead, eastward of Ballarat.

It was inevitable that the colonial authorities would intervene to restore order. At 3 a.m. on 3 December the mounted police and redcoated soldiers of the Suffolk and Somerset infantry regiments stormed the stockade. It was all over in ten minutes. At least twenty-seven people died and many more were wounded. Order was restored to the goldfields. But the system of administration was discredited in the process.

The rebellion represented a clash between different visions of colonial life. Individualism and manliness were venerated in particular contexts, only if they accorded with the dominant assumptions. The rough, egalitarian, anti-authoritarian miners were not the kind of colonists envisaged as replicating new Englands in the wilderness. They could not be allowed to upset the imperial dream. With or without the Eureka Rebellion, however, the golden age of individualistic mining, and the lifestyle that went with it, was becoming extinct. By the middle of the 1850s most of the surface alluvial gold had been taken. As in California, mining now required deep sinkings, horse-powered puddling machines and quartz-crushing machines; in short, it needed long-term investment and business organisation.

A man might mine, but he now had to do so for wages with a boss overseeing him. The original diggers did not like it. One of them said that 'he had come from England, where labourers were slaves, and capital had made them so'. Now the Victorian goldfields were coming to resemble European capitalism. Another pioneer declared: 'I would . . . require to be reduced to starvation point before I would work for a capitalist.' Victoria was becoming more like her parent country, with all its vices and virtues, less like an Edenic blank slate. For all that, the gold-rush pioneers shaped their societies. Nostalgia for the rough, manly and gloriously independent ways of the original miners still exerts a powerful influence today in regions that owe their fortunes to gold.[40]

3

Bonanza

NEWFOUNDLAND

Good luck to the new world citizen! There is no more splendid time to enter the world than the present . . . Australia and California and the Pacific Ocean! The new world citizens won't be able to comprehend how small our world was.

Karl Marx, private letter, 25 March 1852[1]

When the clipper *Marco Polo* arrived at Port Phillip Heads, Melbourne, in 1852 her master, James Nicol Forbes, could find no one prepared to believe his claim that he had departed Britain a mere seventy-four days ago. A fast sailing time just two years before had been 120 days. But the evidence of British newspapers dating back only eleven weeks confirmed the diminutive Scottish captain's mad boast.

Marco Polo carried 930 emigrants to Victoria during the height of gold fever. They arrived in Melbourne dizzy from the voyage. Young clipper captains such as 'Bully' Forbes were superstars of the 1850s. They were prepared to sacrifice everything to break world records. As a result of the rushes the ocean-going sailing ship reached its technological apogee in the 1850s, the fleeting golden age of the circumnavigating American clipper. It was a sleek, narrow vessel designed to convey high-value, low-volume freight at the fastest possible speeds, sometimes over 400 miles a day. The clipper was built for speed, not comfort. In the 1840s clippers raced around the world on established routes – opium to China and Chinese tea to ravenous markets in Europe and the United States. The clipper which won the race and brought the first of the new season's tea to markets in New York and London made staggering profits and the crew received a handsome bounty. The discovery of gold gave them new destinations and new cargoes.

Captains such as Forbes combined the sailing qualities of the clipper with recent navigational knowledge. Look at a map of the world and your eye tells you that the quickest route from Cape Town to Melbourne is a straight line. But the one-dimensional world map is deceptive. By sailing in an enormous arc curving first towards the north-east tip of Brazil and then Antarctica – the Great Circle – a ship could shave 1,000 miles off the voyage to Australia. This navigational information was augmented by the work of Matthew Maury of the United States navy. Maury and his team at the Naval Observatory in Washington analysed thousands of ships' logs and built up a mass of data on the currents and winds of the world's oceans. A seafarer who digested his *Charts* and *Sailing Directions* could extend his victory over time.

This new knowledge was particularly important for clipper captains bound for Port Phillip Heads. When Bully Forbes took the *Marco Polo* on the Great Circle and went below forty degrees parallel he met some of the fastest wind systems on the planet: the 'Roaring Forties' and the 'Furious Fifties'. It was at this point that the passengers became aware that it was a voyage from hell: wind shrieked through the rigging; waves towered above the masts and crashed over the decks; the clipper bucked and strained under the pressure.

There were stories that Bully Forbes padlocked his sails to prevent nervous crews from shortening sails in storms; that he loured on his quarterdeck, revolver in each hand, to enforce his madcap orders. These were probably apocryphal: identical stories were told of other lionised British and American clipper captains, who were magnets for legends.

But it was certainly true that Forbes endangered passengers and crew in his fervour to smash world records, that his overbearing personality silenced dissent, and that he closed his ears to the terror of those who were being hurtled through the unknown. Record-breaking speed was only achievable with almost total disregard for safety. The further south a clipper raced in search of powerful winds, the more likely she was to smash into icebergs obscured by the storms Forbes was desperate to meet. 'Hell or Melbourne!' was Forbes' terrifying and apt battle cry.[2]

'The diggings have bridged the ocean,' it was said by one writer. Over a million people left for the goldfields, and the inhabitants of New York and London became accustomed to the regular arrival of ships from San Francisco and Melbourne carrying £1 million worth of gold at a time. By 1856 £60 million ($300 million) of California's trove and a similar amount from Australia had been injected into the world market. Bank vaults were brimful of bullion. 'The discovery of gold in immense quantities,' wrote an economist, '. . . is the most important event of modern times.'[3]

Its importance was not measured in the wealth it brought to the world in terms of money. Gold had an effect far beyond its intrinsic value. According to one contemporary economist, it was impossible 'to overestimate the *indirect* effects of these discoveries'. They set in train a sequence of events that detonated around the world. As another writer put it, gold was so powerful because it 'has led to the development of new branches of enterprise, to new discoveries, and to the establishment in remote regions of populations carrying with them energy, intelligence, and the rudiments of a great society'.[4]

'Because of California,' wrote Marx and Engels in the *Neue Rheinische Zeitung Revue*, 'completely new international routes have become necessary, routes which will inevitably soon surpass all others in importance.' Add Victoria to the mix, and the whole pattern of international commerce was transformed. Beginning with the California gold rush, and given an even mightier impulse by the Australian version, the world became interconnected as never before.[5]

Clippers are so called because they 'clipped' time off journey times. *Marco Polo* belonged to the Black Ball shipping line, founded by James Baines of Liverpool. Bully Forbes astounded that city, as he had done Melbourne, when he brought his ship back to her home port via Cape Horn in a mere seventy-six days carrying £100,000 in gold dust and nuggets. The company achieved instant celebrity with Forbes' record-smashing round-the-world voyage, and the *Marco Polo* bore the legend THE FASTEST SHIP IN THE WORLD.

This was a young man's game: Baines and Forbes were both thirty years old when gold was struck. Risk-taking was essential for achieving high velocity, whether at sea or from behind a desk in Liverpool. James Baines procured ships that were so enormous that most old hands believed they would fail in their endeavours. Described 'as square as a brick fore and aft, with a bow like a savage bulldog', the *Marco Polo* began life as a common-or-garden vessel, built in New Brunswick for the timber trade. Baines obviously had a keen eye for a fast ship, *Marco Polo* belying her looks. His lust for speed took him to Donald McKay's yard in Boston. McKay provided the Black Ball Line (and its rival the White Star Line) with some of the largest and speediest 'extreme' clippers of all time. The *Sovereign of the Seas* recorded the fastest speed ever for a sailing ship in 1854. That same year Black Ball's *Champion of the Seas* set the record, which stood for 130 years, of travelling 465 nautical miles in twenty-four hours, and its *Lightning* and *James Baines* both made the run from the Mersey to Melbourne in sixty-five days. Two years later and the records were rewritten – again – with the arrival of the *Royal Charter*, a luxurious clipper that had a screw propeller to augment her sails.[6]

Gold had indeed bridged the oceans. People in Melbourne were astounded by the regular arrival of sleek, record-breaking clippers in their port, bringing settlers and such things as foodstuffs, machinery, wine, manufactured goods and tobacco. Most importantly, from a psychological point of view at least, these great ships brought letters and newspapers. Fresh news destroyed for ever the imprisoning sense of isolation that Australians had felt since their lands were first settled and which deterred many a migrant from venturing to the ends of the earth.

Wheels grinding endlessly through rough ground wore down and broke or got stuck in mud or snow. Ships skimming over the waves with minimal friction were many times more efficient – and cheap. The revolution in shipping had been matched on dry land only in the few places where there were railways and canals. Elsewhere, people

The Star of Peace *clipper, built by the White Star Line of Aberdeen in 1855*

and goods moved as they had always done – on foot, on mules and horses or hauled by oxen and bullocks – that is to say, slowly and expensively. Great distances over oceans were being drastically shortened; but leave the water and it was the same old story.

This has a distorting effect on the perception of time and distance. We need to recalibrate how we see the world map. We need to look at it through the eyes of a person in 1850.

Thirteen thousand miles of sea swirl between Liverpool and Melbourne. With an express journey time of between sixty and eighty days it was a smoother, pleasanter, quicker and cheaper experience than the overland trek made by William Howitt from Melbourne to the diggings, about 200 miles, in the early months of the gold rush. New York is separated from San Francisco by 2,900 miles as the crow flies; but in 1850 there were no graded roads, let alone railroads, that traversed the prairies, deserts, canyons and mountain ranges west of the Missouri River. Freight could only go by ship, a journey of 16,000 miles that involved beating through the hostile waters off Cape Horn and typically took 110 days by clipper and a lot longer by other forms of ship. It was quicker to sail from Hong Kong to New York, or New

York to Melbourne, than New York to San Francisco. Cornwall was closer in point of time to California than was Michigan: news of the gold rush reached tin miners (enthusiastic participants in the rush) in that English county a long time before it was heard in some inland parts of the US.[7]

These facts of relative distance were of paramount importance when it came to the reordering of world trade in the wake of the gold strikes. Deluged with crowds of settlers, California and Victoria simply did not have the resources to feed themselves; they had to be supplied by sea. As soon as news of gold rushes and population explosions came from California or Victoria, advertisements appeared in the commercial press the world over: 'great scarcity!', 'high prices!', 'enormous consumption!', 'extraordinary demand!'.

Miners and diggers needed pickaxes and shovels, hammers and nails, rifles and revolvers, to be sure, but also mouthwash, musical instruments, books, baby linen, boots, tobacco. And not just basic necessities; there was demand for flashy jewellery, the latest fashions, silver tableware, playing cards, perfumes, lace, walking canes and everything else an entrepreneur could possibly conceive of. Ships from around the world – from Liverpool, London, Cherbourg, Hamburg, Valparaiso, Lima, Canton, Shanghai, Hong Kong, Singapore, Sydney, Auckland, Honolulu, Boston, New York, New Orleans – converged on San Francisco carrying high-paying migrants and valuable merchandise. Britain secured a multi-million-dollar export trade to California very quickly, selling manufactured goods and corrugated-iron prefabs. French wine merchants loaded ships as fast as they could. The first stone building in San Francisco was made from material imported from China; granite was supplied from Hong Kong; bricks and lumber came from New Zealand and Van Diemen's Land. Hawaii's economy boomed as it exported its sugar, potatoes, coffee and fruits. Indeed, most of the major countries in the Pacific relished the bonanza as places such as Oregon, Mexico, Chile, China, Peru, New Zealand and Australia supplied California with such things as foodstuffs, horses and coal.[8]

East-coast American merchants and speculators, therefore, had to contend with the global market for the spoils of California. In order to compete they needed to enlist speed as a weapon in their armoury. News travelled faster than freight; it went eastwards via the Isthmus of Panama and arrived at New Orleans or New York in weeks rather than months. When word got back that barrels of pork, beef and flour were selling for $60, and pairs of shoes for $45, it became clear that the real treasure of California was to be found in its shops rather than its mountains. The profit on lumber alone was 1,000 per cent. Shippers paid off the costs of constructing their vessels before they were even fully loaded. The most dazzling profits, however, went to those who could capitalise on time-sensitive market information and sell their goods at the height of a boom. Goods languishing in the holds of ships lost value as the clock ticked.

When the clipper *Flying Cloud*, built by Donald McKay, set the record for a voyage from New York to San Francisco in 1851 at eighty-nine days, she did so at least twenty days faster than most other clippers and over three months quicker than a conventional ship. By annihilating time in this way, east-coast merchants could respond to demand and set their prices much more advantageously than rivals who used slower vessels. They could also supply the cash-rich gold-rush market with perishable luxury foods as yet unavailable in the new state, such as canned oysters, hams, eggs and butter. Little wonder that clippers intended for the China trade quickly transferred to California, and that the yards of Boston resounded day and night with clattering hammers and the rasp of saws making new ones to keep up with the intense demand for clippers. So high was the profit made by those that won the race that a single voyage could more than make back the dauntingly high cost of constructing a clipper.

Very often these ships would be met by a man in a little boat, waving at them with his wide-brimmed straw hat. He was Collis P. Huntington, the canniest of gold-rush entrepreneurs. He purchased prime goods at the best prices from the ships he waylaid, beating his competitors waiting onshore at Long Wharf in San Francisco, and shipped them to his store in Sacramento as fast as he could. Those who got rich in the 1850s did so because they appreciated better than anyone else the truth of the cliché that time is money. Huntington was born in Connecticut in 1821; when he was sixteen he set out as an itinerant pedlar and then became a partner in his brother's store at Oneonta, New York. The gold rush brought him to Sacramento, where he set up a store selling mining equipment, provisions and luxuries. A teetotaller and abstainer from all other vice and frivolity, Collis Huntington dedicated his life to riding the rollercoaster of the market.

He found his analogue in 'Uncle' Mark Hopkins, an austere workaholic vegetarian blessed with a mind capable of scrutinising accounts and squeezing '106 cents out of every dollar'. The two went into partnership. Information was key for a man like Huntington: his skill was anticipating demand and keeping up a constant stream of letters back to his brother in Oneonta giving detailed requests for merchandise to be sent west immediately.

The skills and determination of Huntington and Hopkins provided the foundations of a mighty fortune accumulated in the 1850s. It was one thing to sell picks and shovels to miners, quite another to make mind-blowing profits. That required doggedness and the ability to co-ordinate markets in the east and west. The golden west was where fortunes were made, and made with lightning speed. Companies which are instantly recognisable today owed their origins to the boom of the 1850s: Levi Strauss, Wells Fargo, Spreckels Sugar, Folgers Coffee, Studebaker and Ghirardelli chocolate.

The export bonanza associated with California was swiftly succeeded by the speculative fever engendered by the discovery of gold in Victoria. If anything, the frenzy surrounding the Australian gold rush was more intense. Imports to Victoria stood at £875,828 in 1851. Two years later they were £16 million, making the Victoria market virtually overnight more valuable than some of Britain's principal trading partners, such as Spain or Russia, and accounting for one-seventh of British overseas trade. Even so, Britain's share of the total imports to her golden colony fell from 70 per cent to 52 per cent. The rest of the world latched on to the dream of exponential profits in the southern El Dorado. India, China and Chile raced to meet the demand.

But it was merchants in the United States who seized upon the opportunities offered by the Victoria gold rush with the greatest alacrity. The export trade from America had been worth a paltry $3,670 in 1851; two years later it was $8.5 million. California had once been supplied from Australia; now California was providing Victoria with basic foodstuffs, sold at immense profit. There was high demand for lumber during the peak of the building bonanza, and much of it came from the US. The Pacific coast of the Americas was closest in point of time to Australia. But Boston and New York were nearer to Victoria than to California in terms of journey time: an American trading house could make higher profits selling in Australia than Sacramento.

When George Train left New York for Melbourne he described the scene on the East River, from where no fewer than 134 cargo ships departed for Victoria in a single season: 'Drays and trucks were early and late carting down the merchandise, and piers and wharves were full of packages . . . filled with every description of merchandise.'[9] And in Melbourne the hubbub was just as intense. A forest of masts in the harbour indicated an international armada. 'All down near the wharves, it was a scene of dust, drays and carts, hurrying to and fro, and heaps of boxes, trunks, bundles and digging tools.' These goods were bound for the 80,000 or more gold diggers out in the fields, who exchanged their hard-won gold dust for basic necessities: 'It required a constant train of ponderous loads of goods to supply them. You saw huge piles of goods on heavy drays, dragged by ten to sixteen bullocks, labouring and ploughing their way along the deep mire and clay.'[10]

American entrepreneurs appeared in every gold-rush town running stores, and auctioneers selling, at inflated prices, hams, bacon, preserved foodstuffs, butter, cheese, furniture, mining equipment, hard-wearing clothing, firearms and furniture imported from the United States. American tools, stoves and carriages were reckoned the best, and Colt's revolvers were, of course, de rigueur. Fast, robust Yankee coaches were better suited to the roughness of Australia's roads than those of their rivals. One company of Californians operated an express passenger line to Bendigo with sprung coaches made in Concord, New Hampshire. Foreign merchants also provided the luxuries craved in a booming economy. George Train made money from, among other things, importing ice from Boston. Americans opened the first decent restaurants in Melbourne, offering French cuisine.[11]

Hundreds, perhaps thousands, of ships entered Port Phillip Heads or passed the Golden Gate; many never left. Crews deserted en masse during both rushes, attracted by the possibility of instant fortunes at the goldfields over regular wages and months at sea. Ellen Clacy wrote of the 'excellent view of the innumerable vessels . . . lying useless and half deserted in . . . [Port Phillip] Bay . . . It was a sad though pretty sight. There were fine East Indiamen, emigrant ships, American clippers, steamers, traders – foreign and English – whalers, etc, waiting there only through want of seamen.' When Bully Forbes arrived at Melbourne he found the bay clogged with forty or fifty ships abandoned by their crews (Forbes solved the problem by having the men of the *Marco Polo* clapped in irons on charges of insubordination so that they were ready to leave for Liverpool within three weeks). The scene had been similar at San Francisco as the discovery of gold temporarily stalled global commerce.[12]

Even these massive subtractions from the merchant marine had important consequences. The dislocation of trade increased the demand for swift new clippers to fill the gap and make up for lost time. In turn, this massive investment in state-of-the-art vessels that cost a minimum of $65,000 meant that each voyage had to maximise its profits. That was easier said than done. Ships that entered San Francisco or Melbourne laden with emigrants and merchandise had little to take back. Few people wanted to leave, and gold is very low-volume: even if all the gold dug in Victoria in the 1850s – about 1,000 tons – were transported in one go to the Bank of England it could have been carried in the hold of a single ship. Even wool, Victoria's staple export, does not take up much hold space.

But the ships still needed a return cargo. Many clippers that rounded Cape Horn and took migrants and merchandise to San Francisco crossed the Pacific for the Chinese treaty ports. In Hong Kong, Shanghai, Canton and Macao these sleek ships beat every other vessel out of the race at first sight. British merchants clamoured for hold space; clipper captains could charge three times more than a British ship to export tea and silk. They did not disappoint. The clippers astonished the world with their record-breaking dashes to London, leaving behind British East Indiamen in Chinese ports desperate for cargoes. Others ventured to the Philippines or India, where freight charges could be as high, sometimes higher, as for Chinese tea. In 1851 McKay built the enormous *Stag Hound* in just 100 days at a cost of about $100,000 in response to the upsurge in global trade. On her maiden voyage she raced from New York to San Francisco, sold her cargo, and headed to Canton and Manila. On arrival in New York eleven months after she set sail, *Stag Hound* had not only paid back the cost of construction and her running costs, but made a profit of $80,000.[13]

Similarly, the extreme clippers that delivered migrants to Melbourne returned via one or more of the booming ports that made up the great chain of global trade. Some went to Liverpool by way of Calcutta, Bombay and/or Cape Town; many headed to China for tea or silk, which were then carried to Britain or back to Australia (which became the world's fourth-biggest consumer of tea in the 1850s).

The great age of the clipper came about as a result of the convergence of several factors. The initial impetus was the opening-up of five Chinese ports to international trade as a result of Britain's war with China in 1840. Then the expiration of the Navigation Acts on the first day of 1850 gave American shippers the right to trade between Hong Kong and London. So when the gold rushes massively increased demand for fast vessels, shippers were able to benefit from freer global trade. The profit won by the owners of the *Stag Hound* encouraged many more to join the worldwide bonanza.

Gold quickened the pace of international trade – literally so in the case of extreme clippers and clippers with auxiliary steam engines that sped round the circumference of the planet. It also created new routes and opened up markets far away from the goldfields. Gold, as much as opium, was a means of enticing long-secluded parts of Asia where precious metals were highly prized to enter world trade; according to the French economist Léon Faucher, 'it acts like roads, canals, or other modes of transport, which, by opening up the means of reaching markets, extends the radius of operations'.[14] Ports along these routes grew rich on the volume of trade and showers of gold dust. The queen city of these ports was Liverpool. Transformed by the gold rushes, the city became the epicentre of global exchange during the heyday.

Liverpool could never be described as prepossessing in its beauty or grandeur. 'I believe there is sunshine overhead; but a sea-cloud, composed of fog and coal-smoke envelops Liverpool,' wrote the American novelist Nathaniel Hawthorne. 'The Mersey has the color of a mud-puddle, and no atmospheric effect, as far as I have seen, ever gives it a more agreeable tinge.' Hawthorne arrived in Liverpool to take up the post of US consul in 1852, at the very onset of the port's spectacular rise and the clipper revolution. From the bay window of his hotel parlour he enjoyed a ringside seat overlooking the busiest port in the world. The river itself was strangely empty of shipping. Instead an 'immense multitude of ships' from all corners of the planet were 'ensconced in the docks, where their masts make an intricate forest for miles up and down the Liverpool shore'. The port possessed the most sophisticated docking system in the world, allowing ships to lay up alongside, and be loaded and unloaded directly from, the warehouses.[15]

A Liverpool newspaper spelt out the reasons for the city's sudden surge as she ran ahead of London and Glasgow. Unlike her rivals she had 'all the essentials of trade at her finger's end'. Behind lay the world's industrial heartland, the great manufacturing districts of Lancashire and Yorkshire; in front was the Atlantic, from where the trade of all the world came on its way to markets in Europe. She stood at the 'converging point for the population of the three kingdoms [England, Scotland and Ireland], where people as well as produce find their way in the course of transit to the remotest parts of the earth'.[16]

At all hours there was an incessant parade of 'little black steamers puffing unquietly along' in front of Hawthorne's window. They plied between Britain's industrial heartlands (via river and canal) and Liverpool's docks towing strings of narrow boats that contained merchandise either brought from or being brought to the warehouses. Occasionally Hawthorne would witness the arrival of a battered ship, her masts reduced to stumps, a survivor of an oceanic storm. Once a week the scheduled Cunard steamer, 'its red funnel pipe whitened by the salt spray', came in from New York, heralded by cannon. As soon as she was moored to a large iron buoy 100 yards from Hawthorne's parlour window a little steamer was puffing alongside her, impatient to bring ashore her most valuable cargo, the sacks of mail that contained the latest information from the financial and commodity markets of New York and New Orleans. Within minutes this news would be cabled around Europe to newspapers and financial houses.[17]

Liverpool's fortunes had been bound up with trade between Britain and America for generations. But in the 1850s Liverpool became the port of destination for the whole world's commodities, not just America's. By the end of the decade fully one-third of total global exports entered British ports, with Liverpool taking the lion's share. Cargo vessels from all over the world brought their commodities to Liverpool, where they were warehoused awaiting distribution to the European market.

Liverpool's capture of the world market, its second great age, was heralded by the big bang of the Australian trade from 1851. The most lucrative export of all was people. New shipping lines, such as the Black Ball and the White Star, were founded to carry hundreds of thousands of migrants and valuable cargo to the Antipodean El Dorado. In 1852 alone 370,000 people left the ports of Great Britain and Ireland, bound not just for the Pacific gold countries, but for New Zealand, Cape Town or the American and Canadian prairies. Facilitating these great migrations was one of the biggest businesses going in the 1850s.

Charles Dickens wrote of gold-addled people clamouring for passage on the 'last-advertised, teak-built, poop-decked, copper-bottomed, double-fastened, fast-sailing, surgeon-carrying emigrant ship' at the height of the gold mania. When a rush was on, speed was a valuable commodity; people intoxicated by the millions they would inevitably make were prepared to pay to be whisked to remote and unknown places like the Sierra Nevada or Bendigo and beat their rivals – the thousands of people you tortured yourself imagining scooping up nuggets while you languished at sea for months.[18]

When the fever dampened down, the need for speed might have abated, but the craving for it did not. After the gold bonanza came the population bonanza, which turned boom into sustained growth. El Dorado became the Garden of Eden. Thousands emigrated throughout the 1850s to California, Oregon and Australia, not as gold miners but in search of land to cultivate, or they went seeking high wages as servants, skilled labourers, carpenters, artisans, sawyers, saddlers, tailors, gardeners, chefs, drivers, policemen, mechanics, engineers, clerks, teachers, architects, civil servants, lawyers, bankers, merchants, doctors and anything else rapidly growing societies required.

They did not *need* to get to Melbourne or San Francisco quickly for fear of missing the nuggets; but they *wanted* to get there speedily, comfortably and safely. The enormous Black Ball and White Star clippers, the bounding greyhounds of the oceans in the 1850s, provided those who could pay with luxurious state rooms and lavishly furnished mahogany-panelled cabins. As time wore on the clippers became more comfortable and, thanks to technological advances, faster.

Nathaniel Hawthorne enjoyed a view of the biggest ship in the world, Brunel's iron steamer *Great Britain*, moored a few yards from his hotel. She was about to commence her thirty-year career carrying thousands of emigrants to Australia. The recently refitted ship was a 'floating palace' with a sail plan, a screw propeller and sumptuous saloons. On her maiden voyage to Melbourne in 1852, 630 passengers and an abundance of live animals to provide fresh meat, milk and eggs were carried from Liverpool in eighty days. The human passengers were prepared to pay for luxury and safety to offset the length and dangers of the journey. Gold made the export of human beings the growth industry of the 1850s; one of its results was to make long-distance emigration swift and luxurious, inducing significant numbers of middle- and upper-class people to join the tides of migrants for the first time, altering the character of newly settled lands.[19]

As she departed the Mersey, *Great Britain* was waved off by Liverpudlians lining the shore or in innumerable small steamers. On her arrival in Melbourne she caused a sensation; over 4,000 people paid a shilling each to tour the palatial steamer. As the Australian farmer John Kerr observed, with a regular stream of state-of-the-art clippers and steamers arriving from all over the world Melbournians forgot their sense of physical distance from the 'other great centres of civilisation' in the twinkling of an eye.[20]

Gold sparked chain reactions that reverberated around the planet. They were felt in Liverpool, but also in another great port city on the other side of the world – Canton. In 1852 20,000 Chinese people arrived in San Francisco, the start of a wave of migration. Even more went to Victoria, so that by the end of the decade 8 per cent of the population was Chinese. The newcomers were renowned as diligent workers, unlike the footloose white gold seekers who were prone to down tools and abandon claims at the merest hint of a new discovery. They were also resented and vulnerable to violent abuse and exploitation.

'The importation of Chinese has become part of the mercantile system of this colony,' complained Sir Charles Hotham, Governor of Victoria. He was right: Chinese people were treated and traded with as much consideration as a commodity. The problem stemmed from the shipping frenzy of 1852, when the world's traders rushed to Victoria. By then the spectacular returns made from supplying California had dropped off, as the mad prices for imports decreased and the state began to feed and clothe itself. Shippers had to find other ways to make a voyage profitable. Such was the traffic of ships searching for return cargoes that all the tea in China was exported, leaving many empty clippers in Chinese ports desperate for business. In response American shipping companies began to spread news of the Californian gold rush in Kwangtung Province, near the southern Chinese tea ports. These were the regions devastated by the Taiping Rebellion, and there were thousands eager to escape the anarchy unleashed by the West by fleeing to the West.[21]

As Sir Charles Hotham said, merchants 'send their ships to Hong Kong, or other ports in China, and receive a living cargo with as little scruple as they would ship bales of dry goods'. When the American clippers *Westward Ho!* and *Neptune's Car* raced across the Pacific from San Francisco to Hong Kong the stakes were higher than honour and large personal bets; the winner, the *Car*, secured the prize: a valuable cargo of tea to take to London. The captain of the *Westward Ho!* was left with nothing to load except human beings. Back across the Pacific he went with 800 Chinese. They were not destined for the goldfields of California, but the guano pits on the Chincha Islands off Peru as indentured labourers doomed to a short, backbreaking life mining the toxic avian excrement. *Westward Ho!* loaded up with guano to take back to fertilise cotton fields in the southern US.

A 'magnificently beautiful ship' designed for maximum speed, with gilded carvings, mahogany panelling, swanky state rooms and stained-glass windows, the *Westward Ho!* was not built at huge cost to traffic indentured coolies and have bird shit pumped into her holds. Her story was all too common by the mid-1850s. After their brief but glorious heyday the American clippers faced stiff competition in the Asian trades from a revivified British merchant marine. Shipbuilders in Aberdeen and on the Clyde, emulating their Bostonian rivals, hurriedly built and got to sea even faster and bigger clippers that beat Americans out of the tea trade by 1855. The gold rushes bequeathed the world a magnificent fleet of clippers to carry its trade. But with so many of these glamorous ships criss-crossing the seas, freight charges plummeted and competition for lucrative commodities was intense. Clippers had to scour the world for relatively low-value cargoes: Alaskan ice, sugar from Mauritius, whale oil from Honolulu, wheat grown in California, jute and linseed from India, Brazilian coffee.[22]

The majority of Chinese labourers did not go by clipper but by rickety tubs. Onboard conditions for the living cargo were abysmal, and numerous overcrowded ships sank with huge loss of life. They borrowed money from local middlemen for their passage, the so-called 'credit-ticket' system. In return, they submitted to supervised labour until their debts were cleared. The lucky ones made it to California or Victoria. Many more were shipped to Peru or to Hawaii and Cuba to toil in the sugar plantations. This was capitalism red in tooth and claw, with humans reduced to figures on a balance sheet or goods traded on the global market.

The Pacific became one of the most important highways of international trade in the 1850s, fringed as it was by the lands of gold and the tempting Chinese tea, opium and labour markets. The rapid development of California and the surge in shipping in the Pacific connected the United States to Asia as never before. Asia and Australasia were, after all, much closer in journey time to California than the east coast of the US. The *Daily Alta California* spoke for all Californians in 1851 when it said that 'we stand looking out upon the vast field of waters before us, its green islands and populous nations . . . anxiously desirous to make its varied wealth and business our own . . . We are at . . . the point whence must depart the argosies, whose business it will be to bring back the riches of Oceanica [*sic*], the golden fleece of the Indies, the fruits of its tropics, the spermaceti of the Arctic.'

But of most interest to the editor of the *Alta California* was Japan. That mysterious, isolated empire offered limitless riches to the United States. The breakneck expansion of California and its precocious status in the American economy made it all but inevitable that the attention of the United States would become riveted on Japan. The *Illustrated London News*, commenting on the 'collision' of European and Asian civilisations as a result of the gold rushes, said that 'the determination of the far-seeing Yankees to walk unbidden in the long-sealed fastness of Japan' would benefit the whole world.[23]

The time when Japan and the United States 'collided' was not far off. In the early 1850s, however, it was another long-neglected part of the world that suddenly became the focus of international attention during the Pacific boom. 'The establishment of links across the Isthmus [of Panama] by highways, railways and canals,' noted Marx and Engels, 'is now the most urgent requirement of world trade.'[24]

But at the time, it was uneconomical to transport freight through the jungles, swamps and mountains of Panama. There was another trade, a highly lucrative one, which burst upon Panama. The quickest route bar none to California from the east coast or Europe was over the sixty miles of the Isthmus. The hordes that made the crossing from 1849 would have dearly loved a railway. For, although this was the quickest and most expensive way to get to El Dorado, it was hazardous and unpleasant.

Steamships from the eastern and southern ports of the United States brought myriads to the port of Chagres on the east side of the Isthmus of Panama. The British emigrant J. D. Borthwick said that in 1851 Chagres consisted of 'a few miserable cane-and-mud huts'. It was a lethal spot; life insurance companies had a clause which invalidated the policy if a migrant slept ashore in the port because of 'Chagres fever' – malaria and cholera. The priority was to get out of

Chagres as quickly as possible. That meant contracting with locals for passage upriver to Gorgona by boat.[25]

Those who wanted to move quickly hired dugout canoes, but migrants weighted down with luggage had to take larger boats. The scenery was delightful, even if the passage against rapid currents in alternating scorching sun and thunderous downpours was miserable: 'There was a vast variety of beautiful foliage,' remembered Borthwick, 'and many of the trees were draped in creepers, covered with large flowers of most brilliant colours.' Going upriver were at all times lines of all manner of vessel; coming down at higher speeds on the other side were empty boats making the return journey. During these hard days and nights, migrants broke the monotony by firing their Colts at parrots, monkeys and alligators. Along the way were 'hotels' – in reality log cabins or tents with overcrowded dormitories and meals of ham and beans – owned by seedy Americans who had displaced Panamanian competitors with 'the persuasive eloquence of Colt's revolvers'.[26]

At Gorgona travellers began the overland trek to Panama City. Once again, contracts had to be struck with Panama-hat-wearing, cigar-chomping locals for pack mules and porters for the journey through the jungle and over precipitous gullies. The tracks were like a conveyor belt, with a chain of mules and migrants and porters straining under seemingly impossible weights of luggage trudging south, passing unladen mules and porters climbing north to Gorgona to pick up the ceaseless tide of migrants for passage over the continental divide.

This was the ancient gold and silver trail by which the Spanish had, from the sixteenth to the early nineteenth centuries, transported the pillaged riches of the Andes to the treasuries of Europe. And once again gold was being carried up and down these steep, slippery ravines and through the jungles. The British traveller Frank Marryat encountered a richly decorated mule train going the other way to him towards the Atlantic coast, twenty beasts laden with gold bars uncovered for all the world to see and reverently fondle as they passed. They were guarded by half a dozen shuffling soldiers from the Republic of New Granada (modern-day Colombia, of which Panama was a province) armed with rusty muskets; ambling behind on a jennet was a magnificent-looking Don languidly drawing on a cigarette, wearing a muslin shirt, polished jackboots with oversized silver spurs and richly embellished pistols protruding from holsters. This exotic Don turned out to be an English clerk employed by a British house in Panama City, seeing gold to his bank's private steamer, which would speed the bullion to a bank vault in the Old World.[27]

The mule train led by the nonchalant English clerk in fancy dress was one of many such which crossed the Isthmus bearing the treasure of California. In all, more than $750 million in gold was conveyed east over this route. The peak year was 1856, when $60 million came through Panama. The Isthmus route, in the early 1850s, was suitable for people, precious metals and mail; only high-priority freight was brought along the rivers and over the mountains.

After the rigours of the climb, the Argonauts left the suffocating humidity of the jungle and descended into open country, enjoying the respite of the sea breeze before entering Panama City. 'Never were modern improvements so suddenly and so effectually applied to a dilapidated relic of former grandeur as here,' wrote Marryat. Decaying Spanish colonial buildings had been patched up with whitewash, and the streets were dominated by gigantic signboards and Stars and Stripes advertising stores, hotels, restaurants, drinking saloons and gambling dens owned by Americans. From here the travellers waited for a steamer to complete their journey to San Francisco.[28]

The Isthmus had been propelled from near-irrelevance to one of the most important areas for global trade. Routes connecting the Atlantic and the Pacific, the east coast and California, were the most lucrative in the world. Gold went in one direction and millions of items of mail passed both ways twice-monthly. Between 1849 and 1857 520,109 individual journeys, with each passenger paying between $150 and $400, were made to and from San Francisco by sea. If you wanted to get spectacularly rich in the 1850s, this was the game to join.

And so, bit by bit, barriers to smooth intercourse were removed. Steamboats took over from the canoes and miscellaneous craft on the Chagres River. In 1851 the tycoon Cornelius Vanderbilt short-circuited the Panama route when he opened up a new crossing at

Nicaragua, cutting 400 miles from the journey and bypassing the malignant fevers of Panama. Steamships carried his clients to the San Juan River, where smaller steamers took them upriver and across Lake Nicaragua. The journey to steamships waiting at San Juan del Sur on the Pacific was completed by stagecoach. Over the next four years, as many migrants – 156,000 – took Vanderbilt's route as went via Panama.[29]

But then, in January 1855, the railroad came. It took five years to complete the Panama Railroad, with vast teams of labourers drawn from Jamaica, Ireland, India, China and elsewhere toiling to build roadbeds across seemingly bottomless swamps, contending with alligators, fevers, tropical heat and torrential rain. The first twenty-three miles took over two years to complete at a cost of a million dollars. In all, 300 bridges and culverts were constructed, including a massive iron bridge over the River Chagres, and a deep cut was gouged at the continental divide. It came at a cost of $8 million and thousands of lives.

The enormous costs were gladly paid because the Panama Railroad instantly became one of the most important rivets in the global economy. Freight could now be transported from Atlantic to Pacific by steam power, a saving in money and months compared to the Cape Horn route. It connected east-coast American exporters with California, and it lubricated Britain's highly lucrative trade with Latin America. British manufactured goods went one way; guano, whale oil, Costa Rican coffee and silver coin the other. The transmission of market information was speeded up considerably by the railroad. Above all, it had a major psychological impact on the perception of distance.

Now you could travel between New York and San Francisco powered entirely by steam, first-class all the way if you could afford it. The journey across the Isthmus was now four hours as opposed to a gruelling eight days on canoes and muleback. A journey that had been fraught with danger and adventure became relatively comfortable and routine. The inter-oceanic railroad sped up and encouraged the flow of people both ways between the west and east coasts of the United States. Going to California once seemed like a one-way ticket to the end of the world; after 1855 the well-to-do could zip between coasts in a mere twenty-one days without dirtying their coats or dresses. Another yawning distance had been tamed.

'The world seems very prosperous since the discovery of gold in California and Australia,' commented a partner at Barings Bank, 'and the extension of the railways and navigation by steam are working great changes in the world.'[30] Note the words 'seems' and 'are': gold engendered a feeling of optimism while steam power actually made the difference. Also noteworthy is the twinning of gold and modern technology. People in the early 1850s were better than later historians at realising how the value of gold was augmented by its interaction with the other great changes that were assailing the world.

Gold on its own was not enough to make the world a richer place. Michel Chevalier, the French economist, observed in 1852 'that vast as the whole sum of gold in the world is, it sinks into insignificance when contrasted with the aggregate product of other branches of industry'. According to another commentator, the bounties of California and Victoria were 'infinitesimal' compared to the mass of wealth being created in the 1850s by increasing cotton, opium and grain production, the spread of railroads and telegraphs, and the expansion of industry. The City of London and Wall Street were very good at manufacturing capital through complicated financial instruments: 'the creation of money by banking facilities far surpasses in quantity all the produce of the mines of the world within the same period'.[31]

But gold possessed power far beyond its face value. With its timeless hold on human imaginations and emotions, gold acted as an accelerant on developments already in train. Take the clipper. These ships already existed to cater for the Asian trade. The gold rushes multiplied their number, improved their technology and increased their reach until these thoroughbreds became the packhorses of the oceans. Or take migration. The lust for shiny metal lured people to settle parts of the globe in greater numbers; gold was the bait that led to the true treasure chests of the 1850s – population explosions, new cities, the exploitation of raw materials, railways and proliferating trade routes.

Gold was a force multiplier in an already fast-changing world.

Contemporaries were very clear that the gold bonanzas, technology, migration and trade were intermeshed and inseparable from each other. For them the gold discoveries were at once a consequence and a cause of accelerated modernity. The metal was found because energetic settlers were already exploiting the land and scouring it for new forms of wealth; it was extracted in amazing quantities because sophisticated forms of transportation could deliver huge numbers of people to the goldfields quickly and supply them with food and equipment over daunting distances. Extraction of the precious metal was multiplied by 'the union of science and capital' – machinery and business organisation.[32]

The gold rushes, the increasing productivity of the world and the tech revolution of the 1850s were not coincidences. The gold bonanza would not have been possible without modern techniques. In the same way, the booming economy magnified the value of gold. In the past its increased production had debased currencies, sparked inflation and reduced its price. But international trade and commerce were increasing faster than the supply of gold. Financial houses in London and New York preferred liquid forms of capital with quick returns – such as railroad stock, cotton bonds, government securities, bills of exchange and shares – to large, immobile stocks of bullion. They were more than happy to export gold to where demand was unlimited and where it fuelled growth – to Asia and Africa, East European grain producers and frontier societies. The troves of California and Victoria were quickly absorbed by the world, used to purchase commodities and raw materials and invest in big infrastructure projects.

The gold rushes, then, did not in themselves create the heyday; it was ignited by a diffuse combination of combustible materials that, when they interacted with each other, proved explosive. 'One miracle seems to beget another,' commented *The Times*, reflecting on the intertwined forces that were reshaping the world.[33]

Gold's greatest effect, however, was psychological. As the Barings banker put it, its abundance created a magical aura; it was the preeminent *symbol* of the increasing prosperity of the world, a tangible form of wealth in a liquid age. The idealism as expressed by the Great Exhibition, the annihilation of time by new technologies, the settlement of the wilderness and the gold rushes combined to boost to the ionosphere the mood of exuberant optimism and epochal change that defined the period. It added up to the most important single ingredient for rapid growth – unbridled confidence.

Above all, this expectation of limitless wealth and exponential progress encouraged people to invest their money and effort in ever bigger and grandiose projects. With interest rates at a historic low and banks groaning with bullion, money was cheap to borrow. It flowed out of Wall Street and the City of London seeking high returns in the rapidly developing regions of the world. One of the most salient and unexpected results of the discoveries had been the connection between gold and speed. The annihilation of space and time offered immense returns to entrepreneurs and investors. With clippers on the seas and railroads closing the gaps on the map, the world was getting perceptibly faster and smaller. Nothing seemed impossible. Plans were afoot to connect the Atlantic and Pacific with a shipping canal in Central America and for railroads that would link New York and San Francisco. But the most futuristic scheme of all envisaged cutting the distances that separated the world down to mere seconds.

The son of a clergyman from Stockbridge, Massachusetts, Cyrus West Field started work at the age of fifteen as an errand boy in New York. By the early 1850s he had made a vast fortune in the paper-manufacturing business and, in his early thirties, was living in retirement in his mansion in Gramercy Park, New York City. But men such as Field do not retire; he was bored and in search of adventure.

During this time that frothed with confidence and ambition, Field was introduced to a young British engineer named Frederic Newton Gisborne. When the Channel Cable of 1851 was laid Gisborne was in his early twenties and making his career as a telegraph engineer in Nova Scotia. The success of that venture led him to dream of a cable that would connect Europe and the United States.

The shortest route across the Atlantic was between the western tip of Ireland and St John's on the north-east coast of Newfoundland. Ireland was already connected to the European network, so Gisborne's first challenge was to link St John's with New York. He travelled to London and purchased a submarine cable made by the Gutta Percha Co. and R. S. Newall to lay across the Cabot Strait that separated Newfoundland from the North American mainland. 'I was looked upon as a visionary by my friends, and pronounced a fool by my relatives,' Gisborne remembered.[34]

Two years later he was $50,000 in debt and facing imprisonment on fraud charges. His precious cable lay broken on the seabed and only forty miles of landline had been established across Newfoundland's inhospitable terrain. In desperation Gisborne sought investment.

He found the right man. With his wealth and political connections, the young millionaire Cyrus West Field was in a better position than an obscure engineer to bridge the Atlantic with electric cable. He wasn't in it for the money; like a present-day Silicon Valley entrepreneur Field wanted to change the world. He foresaw a planet girdled in wire and buzzing with human thoughts. In such a brave new world war would become impossible as unrestrained information broke down the barriers that separated humankind.

His first step was to canvass the opinion of Lieutenant Matthew Maury as to the feasibility of the project. Most of the Atlantic was, said Maury, deep and the bed made up of uneven, jagged ridges which would snag and tear a cable. But by great good fortune the US navy had just completed a survey of the ocean. What it found was that the seabed between Newfoundland and Ireland was a plateau 'which,' said Maury, 'seems to have been placed there especially for the purpose of holding the wires of a submarine cable'. Maury, however, warned Field that there were more serious barriers to the fulfilment of the venture: he needed a calm sea, the longest cable the world had ever seen and a ship big enough to lay it. Even if these obstacles were overcome, was it even possible to transmit electrical messages over such a distance? But with the same breath Maury swept these doubts aside. His words encapsulate the faith in the miraculous powers of progress so evident in the 1850s: 'I have no fear but that the enterprise and ingenuity of the age, whenever called on with these problems, will be ready with a satisfactory and practical solution to them.'[35]

The pioneers of the transatlantic link were American citizens proposing to connect two points of the British Empire using British-made

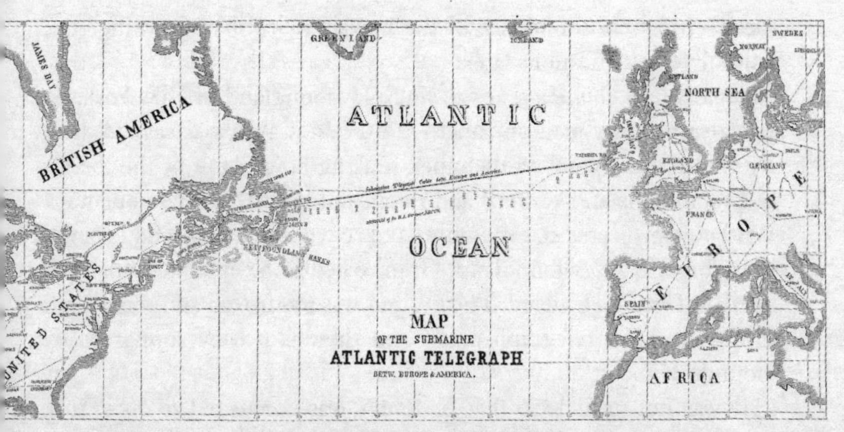

Map of the proposed Atlantic Cable

technology. But the politics did not matter to the five wealthy New Yorkers Field met with in the first months of 1854. Among them was Moses Taylor, a banker who was soon to become president of the City Bank of New York (now known as Citibank) and one of the wealthiest men of his century. He was joined by Robert Marshall, who had made his money from a fleet of steamers that connected New York to Chagres and Panama City to San Francisco, and by Peter Cooper, one of the most successful industrialists and railroad entrepreneurs in the States.

Initially sceptical, these hardnosed businessmen were won over not so much by Cyrus Field's business case as his messianic fervour. 'It seemed to me,' Peter Cooper remembered of that meeting with Field, 'as though it were the consummation of that great prophecy, that "knowledge shall cover the earth, as the waters cover the deep", and with that feeling I joined him in what then appeared to most men a wild and visionary scheme; a scheme that fitted those who engaged in it for an asylum . . . But believing as I did, that it offered the possibility of a mighty power for the good of the world, I embarked on it.'[36]

The project was fuelled by exactly this sort of utopian confidence. It would never have got off the ground, or rather dry land, if it had not been. No one had attempted to lay a cable at such depths or over such a long distance; the science was not even properly understood. There was nothing straightforward about this scheme. It embodied the irrepressible confidence of the heyday more than anything else: build first, ask questions later.

Field's first challenge was to link Newfoundland to New York. 'It was a very pretty plan on paper,' said Field. '. . . It was easy to draw a line from one point to the other, making no account of the forests and mountains and swamps and rivers and gulfs that lay in our way.' What was projected to take months took years of labour and expense in unforgiving Newfoundland. Granite had to be exploded, marshes filled and ravines bridged. The ground was too frozen and rocky even to dig holes for telegraph posts, and they had to be supported by mounds of rocks.[37]

It was not until 1856 that St John's was connected to New York by landlines and two submarine cables at a cost of $1 million. It had been hard enough to lay a sixty-mile submarine cable in rough waters off Newfoundland. Now the challenge was to construct a cable 2,500 miles long and find a ship capable of carrying such a weight and develop paying-out equipment good enough to feed it to the bottom of the stormy Atlantic.

Yet uncertainty was not enough to jeopardise this utopian vision. Cyrus Field arrived in Britain riding a tidal wave of enthusiasm. He told an audience of cotton brokers in the underwriters' room at the Liverpool Exchange that they would soon be able to join their European network to 45,000 miles of telegraph in North America. The single cable that would make this possible could be laid in a mere week at a cost of £350,000. Field was speaking in the epicentre of the global economy. Every year hundreds of millions of pounds of cotton arrived in the high-tech docks of the city, bound for Cottonopolis – Manchester, where it was spun into the textiles that would be exported to the world market. The price of cotton was set in the room in which Field spoke.

He reminded the brokers of Liverpool that with modern communications, it only took forty days to send orders for cotton to New Orleans and then receive the order in Liverpool. When the cable was operational their messages would effectively arrive *before* they were sent, because Liverpool was five hours ahead of the United States. When Field passed round a sample of the cable a broker exclaimed with wonder, 'Now that's the thing to tell the price of cotton!'[38]

Field promised a new world. The 300 shares in the newly formed Atlantic Telegraph Company were sold without advertisement, priced at £1,000 each. The excitement generated in anticipation of this, the greatest engineering project the world had ever seen, pushed all doubts aside. As *The Times* put it: 'Habit has so familiarised us with the marvellous triumphs of physical science that we have ceased to feel or express astonishment at results, which, not so many years ago, would have been dismissed from the consideration of rational men as the visions of a fantastic imagination . . . Already we are beginning to meditate even greater conquests over time and space. India soon, and Australia, will be brought within the current of the electrical stream.'[39]

With gold sloshing around, trade booming and confidence riding high, there was plenty of money begging for employment. It didn't get better than the Atlantic Cable. This futuristic scheme offered unimaginable profits as the global economy was reshaped by instantaneous electric communication.

The fate of Field's great project belongs to another part of this history. We must leave it on the point of climax, and pick up the strands of the cable later. Suffice it to say, the cable was the result of a period when it was believed with utmost conviction that the problems 'which now seem to present insuperable difficulties will . . . be easily solved by the mere force of circumstances and the progress of thought'.[40]

A world connected by shipping lanes and submarine cable sparked significant changes in the way it worked, creating wealth, slashing costs and establishing new societies. But in many ways these developments were refinements of something very old. Most of the great world cities were ports; much of the most productive land lay near water transportation. Even more revolutionary in the 1850s was the way in which the planet's landlocked interiors were opened up. This was accomplished not by dazzling new technology but by one of mankind's oldest forms of locomotion: the humble ox.

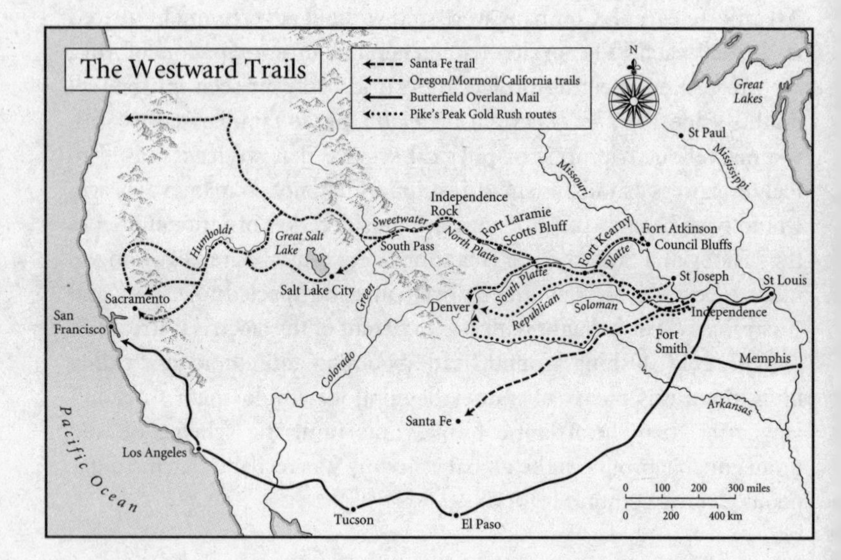

The Westward Trails
Santa Fe trail
Oregon/Mormon/California trails
Butterfield Overland Mail
Pike's Peak Gold Rush routes
N
Great Lakes
St Paul
Mississippi
Missouri
Independence Rock
Fort Laramie
Scotts Bluff
Fort Kearny
Fort Atkinson
Council Bluffs
Sweetwater
North Platte
Platte
South Pass
Great Salt Lake
Humboldt
Salt Lake City
Sacramento
San Francisco
Green
South Platte
St Joseph
St Louis
Denver
Republican
Soloman
Independence
Fort Smith
Memphis
Colorado
Arkansas
Santa Fe
Pacific Ocean
Los Angeles
Tucson
El Paso
0 100 200 300 miles
0 200 400 km

4

On the Road

NEBRASKA

Ho – for California – at last we are on the way . . . and with good luck may some day reach the 'promised land'.
Helen Carpenter, 'A Trip Across the Plains'[1]

Go away: you come to kill our game, to burn our wood, and to destroy our grass.
Left Hand, Arapaho chief[2]

'They came within about four hundred yards of us,' remembered Charles Ferguson; 'then suddenly wheeled their horses and rode around us two or three times, at the same time going through many of their warlike motions, drawing their bows as if to send an arrow.' The Ohioan teenager was part of a wagon train bound for California. The party had just left Chimney Rock, an astonishing pillar spiking out of the prairie in what is now western Nebraska, when a band of eighty mounted Crow braves charged down a hill at full tilt towards them.

Some braves rode straight at the wagons, veering at the last minute; as they did so they turned round in their saddles, feigning to shoot. Ferguson's company became very frightened. 'We expected an attack and closed up our teams as close as possible, but still kept on the move. The men all examined their rifles and pistols.' The Crows continued their hostile manoeuvres for half an hour, pausing only to let their horses blow.

When one young brave came too close to the wagons, Ferguson and a friend suddenly went in pursuit, headed him off and took him prisoner. It had an immediate effect on the Crows. They dismounted and called over the migrants. The two parties shared a pipe of peace, exchanged pleasantries and began to trade. Ferguson swapped brandy and sugar for a fine horse and his woollen shirt for a buckskin suit.[3]

A lonely wagon train in the shadow of Chimney Rock months from home, a swooping band of Indians, the steely courage of grizzled migrants with their Colt revolvers and Sharps rifles – these are the stuff of romance and legend and are still, perhaps, the dominant perception of the overland journey to California in the 1850s that overshadows its catastrophic effects on the indigenous inhabitants and their environment. The great migrations of this time are a key component of the American Dream. We can picture Charles Ferguson's encounter with the Crows because it comes straight out of hundreds of movies.

And in the 1850s the diaries, reminiscences and letters of pioneers trekking across the great interior of America were eagerly consumed around the world. 'The story of thirty thousand souls accomplishing a journey of more than two thousand miles through a savage but partially explored wilderness, crossing on their way two mountain chains equal to the Alps in height and asperity, besides broad tracts of burning desert, and plains of nearly equal desolation, where a few patches of stunted shrubs and springs of brackish water were their only stay,' wrote Bayard Taylor of the Forty-niners, 'has in it so much of heroism, of daring, and of sublime endurance that we may vainly question the records of any age for its equal.'[4]

That was said of the overlanders of 1849–50. Within only a few years the stirring tales of derring-do had become a little humdrum, a bit too compromised by reality. In 1852 the way to California was described by a journal as 'comparatively easy'; it had become a 'great thoroughfare' traversed by thousands of travellers and serviced by merchants. What was recently 'a journey of perpetual doubt, difficulty, toil and danger, can now be merely properly designated as one of weariness and privation'.[5]

People in the 1850s were not amazed so much by the romance of the journey (something that had already passed into history and legend) as the speed by which a trail cut through the wilderness had become a substantial highway. The story had global ramifications. In connecting the Atlantic and Pacific coasts the United States was becoming a gigantic empire with incredible swiftness. The pathway developed by the Americans across their continent created another vital link in the global network of communication and opened up a vast blank on the map to exploitation by settlers. And perhaps most importantly, the manner in which they had achieved it, and the technology and hardware they had developed on the road, offered Western powers lessons in how to make similar inroads into the unknown interior regions of other continents.

Readers of newspapers and books in the 1850s knew the details of almost every mile on the trek, so great was the mass of information. It is worth briefly sketching the route taken by close to 300,000 people. Those who travelled overland utilised the network of rivers, canals and railroads which bound together the US, taking steamboats to the main 'jumping-off' points and outfitting towns on the Missouri River Frontier: St Joseph and Independence in Missouri or Council Bluffs, Iowa. From there they crossed the frontier into a vast territory then known simply as Nebraska.

The trail followed, as much as it could, river valleys – twisting migrant corridors across the continent that afforded water, firewood, game and grass for the overlanders and their draught animals. They followed the valleys of the Platte, North Platte, Sweetwater and Green rivers for hundreds of miles, in the territory that became the states of Nebraska, Wyoming, Idaho and Utah. Monotony of landscape alternated with glorious vistas on the long trail to California. Stunning rock formations, such as Chimney Rock and Independence Rock, erupted out of the level prairie. The trail involved repeated crossings of meandering rivers and occasional rocky, waterless, grassless stretches that tested migrants and animals as they dashed from the lifeline of one river valley to another.

After fording the Sweetwater River the migrants reached the continental divide and crossed the Rockies at South Pass, an important milestone. Anyone expecting dramatic mountain scenery, as Bayard Taylor evidently did in his romantic description of the trail, would have been disappointed: South Pass is an unprepossessing grassy saddle in a ridge of hills.

Those who took the route towards Salt Lake City up steep slopes and through canyons were rewarded with a stunning sight. Charles Ferguson recalled the view from 8,000 feet: the Great Salt Lake 'glistened in the sun . . . Away to the south, as far as the eye could reach, was one broad, beautiful, level plain, covered . . . with a carpet of deepest green. All this loveliness of lake and landscape was bordered and framed by snow-capped mountains whose silver summits seemed to touch the blue vault of heaven.' Ferguson would go on to travel the world; but nothing surpassed this memory of the 'most beautiful prospect on earth'.[6]

Overlanders knew that after the respite of the Salt Lake region and its river valleys lay the most testing routes. The most notorious was the 300-mile road along the Humboldt River, in what became Nevada. The Humboldt was a sickly, shallow, crooked little stream, but it provided a narrow strip of life in the desert, just about enough to support man and beast for the initial 150 miles. The last 150 miles were horribly barren, with little vegetation and scant firewood. This stretch was a trial for exhausted migrants who had battled for months over a distance in excess of 1,500 miles. The Indians who lived in the dismal, desolate valley were known as the 'Digger Indians', who lived off roots and insects. They were hungry for meat, and the overlanders' oxen and cattle were tempting propositions. The Diggers fired arrows at the animals at night or from behind cover and waited until the migrants moved on to butcher the abandoned wounded and fallen stock.

Then, after 300 miles of pain, the muddy little river died in the desert, at Humboldt's Sink. Emigrant trains were faced with a waterless, alkaline wasteland dusted white with salt. Its barrenness pushed horses to the limits of their endurance and to the brink of starvation. 'I thought I had seen barrenness before,' Horace Greeley wrote of the Humboldt valley and the desert; but there 'famine sits enthroned, and waves his scepter over a dominion expressly made for him'. The roads in the salt desert were fringed with graves, abandoned wagons and the sun-bleached skeletons of horses, oxen and cattle.[7]

The end of the desert was a moment to be savoured. 'The sand gradually became a little less deep,' wrote Helen Carpenter in her diary, 'and about the first intimation that we had of the nearness of the Truckee River the cattle began to bawl and those that had the strength to run bolted ahead of the wagons and made for the river, dashing down the steep banks and into the water.' The cool waters of the Truckee and the shade of trees, the first for hundreds of miles, meant a fleeting reprieve; the test was far from over. Ahead lay canyons and precipitous, rock-strewn climbs up and over the passes of the Sierra Nevada mountain range. The hundreds of thousands of migrants knew by heart the tragedy of the Donner Party: in 1846–7 a group of pioneers resorted to cannibalism when early snow left them stranded for the winter in these mountains.[8]

The horrors faced by the Donner Party teased at migrants throughout their journey. No party ever experienced such catastrophe again, but even in fair conditions it was the most gruelling part of the trek. Wagons that had withstood months of river crossings, mountains and desert were frequently abandoned on the climb, along with the emigrants' luggage. But then, as weary, downcast, defeated travellers reached the summit, after at least five months on the road, they saw the Sacramento valley stretched out in front of them. Their journey of over 2,000 miles from the Missouri Frontier was near its end. They had arrived in the golden state.[9]

The number of people going west in 1852 was truly phenomenal, with 70,000 on the road that summer. 'It was a grand spectacle, when we came, for the first time, in view of the vast emigration, slowly winding its way westward over the broad plain,' wrote Margaret Frink. 'The country was so level we could see the long trains of white-topped wagons for many miles . . . I had never seen so many human beings before in all my life.' An average of 2,000 people a year had travelled to California, Oregon and Utah in the decade prior to the gold rushes. In the first four years of the 1850s the yearly average was close to 40,000.[10]

So many wagons headed out of St Joseph in 1852 that they travelled twelve abreast. It was more like a congested road, with wagons backed up all along the way, than a lonely pioneer's trail. In places the road

Emigrants' wagons wait to cross the Missouri River in 1852

was forty-five paces across; the ruts worn by such a volume of traffic are still visible today. There were frequent incidences of road rage in the congestion, and stories of fathers separated from their sons in the crowd and only reunited months later in California. Great drives of cattle and sheep competed for space on the trails with the wagon trains; 300,000 stock animals were trailed overland in 1853 alone. They came in droves of between 500 and 2,000 head of cattle and up to 10,000 sheep at a time. You had to ride five miles or more to overtake some of these epic drives.[11]

Trails once used by mountain men, fur trappers and the occasional migrant train now became some of the busiest and most important roads in the world. No one could expect to enjoy the wonders of the interior in splendid isolation any more. Wise overlanders brought goggles to protect their eyes from the dust thrown up by so much human and animal traffic. So great, indeed, was the press of humans, wagons and animals that Indians on the plains talked of migrating east to the surely by now depopulated lands beyond the Missouri Frontier.

This land bore the scars of mass migration. From the Upper Platte

Ferry in Wyoming onwards (under halfway from the Missouri to California), migrants were faced with the daily stench of death emanating from the rotting carcasses of horses, mules, oxen, cattle and sheep which fringed the trails. Tens of thousands of travellers carved their names onto Independence Rock. From the Missouri River to the Sacramento valley lay a line of rubbish – items which had been jettisoned from wagons as draught animals wearied. The debris included everything imaginable, from pots and pans, stoves and mining tools to, outlandishly, a huge metal safe and a diving bell complete with its equipment. The Forty Mile Desert resembled a battlefield, with thousands of dead draught and stock animals, abandoned wagons, scattered firearms and personal possessions strewn in the alkaline waste. Many who travelled these routes looked wistfully at the plethora of discarded items, realising that they could become millionaires if there was some way of carting them away.

The migration heyday of the 1850s bore little resemblance to the experience of the previous decade. Tens of thousands of travellers on the trails each year proved irresistible for entrepreneurs. They came from both directions, east and west; some who had endured the trek in previous years realised that more gold was to be made from the pockets of hungry travellers than in the goldfields and returned eastwards from California to set up shop. By the monster emigration year of 1852 the 2,000-mile corridor from St Joseph to Sacramento was flanked with services of all kinds. Enterprising people built bridges spanning the rivers, for which they levied a hefty toll. Where there were no bridges there were people offering ferrying services, often at exorbitant prices. Some ferrymen and bridge owners were making over $1,000 a day. Every obstacle overcome by private initiative made the formidable journey that bit easier, quicker and considerably more expensive.

Innumerable blacksmiths operated all along the way by 1852, grog shops were ubiquitous and bored travellers could fritter what they had in gambling dens or brothels. And, unimaginable for the pioneers of the previous decade who endured endless privations, there were trading stations selling provisions. Helen Carpenter, westward bound in 1857, was able to purchase tinned peaches, for example. Canned oysters tickled the palate of jaded travellers. The British explorer Captain Richard Burton, travelling three years later, bought whisky at regular intervals. The spirit was always available on the California trail, one way or another; what made the experience so novel by the end of the decade was that the liquor was served to the accompaniment of chinking ice, a sure sign of civilisation for globetrotters. He could have bought, if he wanted, anything from a needle to champagne at a popular store on the Platte River. Druggists and quacks sold expensive quinine pills, opium and patent medicines to guard against the pestilence that stalked emigrants westwards from the Mississippi River system.[12]

The Mormon 'halfway house' – Salt Lake City – had emerged as a major urban centre by 1852 and offered all manner of comforts, including draught animals, fresh food, baths, haircuts and law courts. It was an oasis prior to venturing into the Great Basin, where lay the Humboldt, the desert and the Sierra Nevada. But even the rigour of these obstacles, which once chilled the spine, had been eased somewhat. Mrs Carpenter was able to purchase such things as coffee, butter, flour, vinegar, molasses and vegetables in the desolate Humboldt valley, albeit at sky-high prices. Frances Sawyer noted in 1852 that the Forty Mile Desert 'is easier to cross this year than it has ever been before. There are seven or eight trading posts on it now, where refreshments and supplies are kept for sale.'

Private enterprise transformed the experience of travel. But so too did the federal government. On 3 March 1853 Congress authorised the Secretary of War 'to ascertain the most practical and economical route for a railroad from the Mississippi River to the Pacific Ocean'. Over the next eight years teams of army engineers surveyed 400,000 square miles. Their findings were produced in twelve volumes – published at a cost of $1.2 million – that contained a wealth of written information on the scenery, ethnography, history, geography, fauna, flora and geology of the west and rich visual imagery in the form of tinted lithographs and engravings. Those twelve volumes were storehouses of information and fulfilled the original orders of Secretary of War Jefferson Davis to the army engineers to 'discover, open up, and make accessible the American West'.

At times during the decade 90 per cent of the entire US army was stationed in the trans-Missouri West, many of the troopers at the string of garrisons guarding the roads. These forts became vitally important staging posts for migrants, offering provisions for sale, blacksmiths and medical services; trading towns sprang up in their protective shadows. The US army was drawn west by the need to guard the emigrant road from Indians and to bring to heel the Mormon state in Utah. By the time Burton went west the overland route had been shortened by hundreds of miles thanks to US government surveys; the military had transformed difficult parts of the trails into graded roads. The desolate and dangerous Humboldt valley and the Forty Mile Desert could be bypassed entirely, with the opening of a military road across southern Utah and Nevada to Carson City.

Then in 1859 a new gold rush was ignited with the discovery of gold at Pike's Peak, Colorado, on the South Platte River. In that year in excess of 100,000 people, along with animals, followed in the footsteps of the Californian Argonauts and ventured forth along the Platte valley and other new roads through the valleys of the Great Plains to the Front Range of the Rocky Mountains. Unlike California, Colorado could not be supplied by sea. All the necessities and luxuries required by gold-rich miners and newly emerging cities such as Denver had to be carried overland. The massive freight trains that sallied forth to the Rockies were a sight to behold. Up close a convoy of twenty-six overloaded wagons – each packed high with 3.5 tons of cargo – resembled a herd of ponderous elephants lumbering through the Plains.[13]

Supplying the new El Dorado was big business. Horace Greeley described the headquarters of the Russell, Majors and Waddell transportation company near Fort Leavenworth with wonder: 'Such acres of wagons! such pyramids of extra axletrees! such herds of oxen! such regiments of drivers! No one who does not see can realize how vast a business this is.' The multi-million-dollar venture employed 6,000 teamsters and owned 45,000 head of oxen to supply the army in the Plains and the growing population in the valleys of the Rocky Mountains.[14]

*

Of all those hundreds of thousands who took the wagon trail west in the middle of the century there could have been few who offered such a breadth of perspective as Captain Richard Burton. He regarded the American interior with the eyes of a seasoned world traveller. His account of the American West stands out from the myriad others, for when he gazed out of the window of the stagecoach or observed the native inhabitants he was not looking solely at America but at the entire world.

Almost every experience is reminiscent of something he had seen elsewhere in the world. Travelling through the prairie, for example, recalled the Arabian desert, the jungles of India or the African bush. He compared the network of rivers in the prairies to the steep-sided *nullahs* of north-west India and the *wadi* of Arabia. The geography of central Wyoming struck Burton as East African; Independence Rock was identical to the Jiwe la Mkoa in Unyamwezi (modern-day Tanzania). Utah 'is like Central Equatorial Africa, a great depression in a mountain land', and its climate was similar to 'the Tartar plains of High Asia'.[15]

This kind of comparative travel writing at once dispelled some of the uniqueness of the west and elevated the global importance of the road. The gap between Missouri and Sacramento had been bridged, Burton was at pains to explain, by important innovations in hardware. The coaches, freighters and wagons that traversed the great distance were far superior in point of comfort and robustness to those used to cross the desert at Suez, in India or in southern Africa. Burton examined the design of these conveyances and the details of their equipment – the bridles, saddles, stirrups, buckles, whips and so on – with the eyes of a connoisseur and explained how such innovations could be used in India or Africa. In a world of high-tech – of clippers, telegraphs and steam engines – this might seem mundane; but they were vitally important in suggesting the way in which the great interiors of the world could be opened up.

Other than the topography, the 'wilds of America' were familiar to Burton. Stepping into the forts along the way, he felt that if he closed his mind he could fancy himself entering a military cantonment in Gujarat, Algeria, southern Africa or Australia. Grouped around a

The British explorer, Captain Richard Burton. With his scarred face, moustache and beard he would have fitted in easily in the American West of the 1850s.

parade ground, the whitewashed veranda-ed bungalows, the low-rise barracks, storehouses and offices were virtually indistinguishable from the outposts of British India or French North Africa. And the duties and operations of the US army in the west and the British on the frontiers of India were almost identical. Both forces were scattered in lonely cantonments thousands of miles from home, required to police roads and launch punitive expeditions against the native populations.[16]

Burton could readily appreciate that Emigration Road was a pathway of empire like so many others in the world. He described the road west as 'a great thoroughfare, broad and well worn as a European turnpike or a Roman military route, and undoubtedly the best and longest natural highway in the world'. The development of the American highway was paralleled by other long-distance routes at the same time. The Georgian Military Road, built at a cost of £4 million by the Russian army and connecting Vladikavkaz and Tiflis (now called Tbilisi) through the Caucasus Mountains, had the dual aim of crushing the lawless Muslim tribes of the region and connecting Russia and Georgia, its distant province. At the same time the British were improving 568 miles of the ancient Grand Trunk Road between Delhi, Lahore and Peshawar following the conquest of the Punjab in 1849. In all these cases the aim was both strategic and commercial, designed for the speedy transit of troops, people and merchandise through hostile lands.

Burton's comparison of the American highway with a Roman road was apt. Looked at on a map, the routes heading west were innocuous dotted lines across a great blank. They represented causeways linking geographically separate parts of the United States, cutting through what was considered uninhabitable wilderness. In the early 1850s few thought of settling the Great Plains or even saw them as habitable. 'All the great interior passed under the general name of Nebraska,' wrote Charles Ferguson; '. . . [it] was but the home of the red man and the range of the buffalo.' Over and again travellers compared travelling through the Plains to being adrift in a mighty ocean. Not for nothing were their wagons called 'Prairie Schooners'.[17]

But roads have a habit of developing a momentum of their own: connecting distant parts of an empire, they end up becoming empires in their own right. Like the sea lanes that bound together maritime empires or long-distance roads that led through bush, mountains or desert elsewhere in the world, the westward trail had to be defended by strings of military outposts and patrols of armed escorts. Rather than being thin threads of communication, roads and shipping routes impacted across entire regions.

Richard Burton offered another important insight as he bounced along the road in the Platte valley, enjoying the wild sunflowers and lupins: imperialism in the middle of the nineteenth century was dictated not by a nation's desire for power but by the unstoppable flows of migration. If roads had always been the tentacles of imperialism, in the 1850s they were also physical manifestations of the hopes and ambitions of colonists and settlers on the frontiers. Along with bridges and tunnels, few things symbolised so viscerally their triumph over nature or heralded their future prosperity. One example is the road from Wellington, New Zealand, to Wairarapa that climbed up and over the steep, rampart-like Rimutaka Ranges. Cutting the trail began in the 1840s and was delayed by Maori unrest, earthquakes and the forbidding terrain. When it was completed in 1856 the snaking Rimutaka Hill Road gave pioneers access by wagon to the interior and allowed them to export their primary produce.

Power over nature is often a euphemism for power over other people. Roads were forerunners of the expanding West, a development that might soon be followed by cities and railroads. More often than not they were deeply resented by people whose territory was invaded. A few years before American pioneers ventured west across their continent, the Boers of Cape Colony began their great treks into the interior of southern Africa; their wagon trains were soon followed by British settlers. By the 1850s ox-wagon trails were being converted into roads that crossed such formidable barriers as the Drakensberg Mountains, opening up landlocked regions to trade, elephant-hunting and missionary activity. These migrations destabilised the entire region, sparking conflict between freelance colonists and the Bantu people whom they displaced, and between the Bantu themselves as they competed for land and resources. Despite its reluctance, the British government was drawn into a series of frontier wars and/or treaty negotiations with the Xhosa, Basotho, Ndebele and the Zulu. Similarly, mass migration across the Great Plains dragged the US government into the region, burdening it with new responsibilities, and new powers.

The story of the surge across the continent is in many ways the story of the world in the 1850s. It is a tale of ecological devastation that led to the extinction of a way of life within a few short years.

The footprint left by the hundreds of thousands who travelled through the region over the course of the decade was ineradicable. The map would suggest that these narrow transit lines were minuscule pathways through a vast terrain, locally inconvenient but regionally negligible. But the map view is deceptive. The Plains were not untouched by history, as some imagined. They had been the scene of migration and competition since time immemorial; but in the years immediately prior to the great white migration the population there had doubled and the area was hotly contested. Over the last few generations the Cheyenne and the Sioux had been moving into the region. After a series of wars, the Sioux controlled much of the northern Plains; to the south, in the central Plains, the Cheyenne had become the dominant power. These fearsome nations overlapped on the Platte, North Platte and South Platte rivers. One of the lushest regions in the entire continent, the Platte valley was vital to the Lakota, the Sioux and other less powerful peoples; no one controlled it. In the summer months the Native American tribes hunted the buffalo on the highlands and returned every winter to the abundant shelter and sustenance that these valleys provided.

In good times there was just enough to provide 'a precarious supply of food'. Even when modest trickles of pioneers used the Platte valley in the 1840s the Indians complained of the damage done to the wildlife, timber and flora. But when those swelled to a raging torrent – 145,000 people and numerous military freight trains passed through in the first few years of the 1850s – the Native Americans returned every winter to a scene of permanent devastation. Thousands of iron-rimmed wagon wheels ground a pathway ten times wider than most roads, tearing up the vegetation. The overlanders' draught and stock animals had gnawed away grass four miles either side of the road; the timber was stripped bare; and the great herds of buffalo, elk, antelope and deer had been scared away. Added to that were the piles of rubbish, rotting carcasses and polluted rivers. By the middle of the decade, the once-verdant Platte valley had become, according to an army officer, 'a lifeless, treeless, grassless desert'. Its contamination forced Native Americans to seek new winter sanctuaries, and to compete for them with others.[18]

Given the upheaval to their way of life, the Plains Indians were remarkably restrained in their dealings with the overlanders. Most of the 400 migrants estimated to have been killed by Native Americans were attacked west of South Pass, beyond the Plains. Nonetheless, they blamed the whites for the destruction and sought compensation, some by raiding wagons, others by demanding tolls or tribute from overlanders who passed through their territory. During her trek to California Sarah Royce's caravan was stopped by Indians claiming tribute. Royce could not contain her outrage, holding that 'the demand was unreasonable! that the country we were travelling over belonged to the United States, and these red men had no right to stop us'. The Indians were answered with a show of firearms. It was not the money; it was the principle: many overland emigrants were fired with patriotic zeal and did not believe that the indigenous peoples had any rights to the land. They were the foot soldiers of Manifest Destiny.[19]

From the moment they left the banks of the Missouri, emigrants were on their guard against the treacherous and sadistic Native Americans bruited in the newspapers. 'I recollect that when I made my first overland trip my hand was constantly on the revolver in my belt,' wrote James Gamble, who travelled to California to establish its first telegraph lines. 'Twenty and more times a day I was ready to pull it out on the shortest possible notice, and lodge its contents in the first animate object that disputed our right of way.' People were advised to keep their revolver on the right-hand side of their belt, butt to the fore, so that it could be drawn quickly and shot from the hip: 'the difference of a second saves life'. After disease, the biggest cause of death on the road was from mishandling weapons – and there were plenty of inexperienced greenhorns showing off their new toys.[20]

Most never saw an Indian to threaten with a Colt. But there were other uses for the box-fresh firearms. 'A herd of buffalo was seen this afternoon at a comparatively short distance away,' Helen Carpenter recorded in her overland journal. 'This created general excitement and eight or ten of the company gave chase, some on foot, some on horseback, armed with muskets, revolvers and knives.' Even if trigger-happy overlanders could not seriously dent the 15 million buffalo, the roads that scythed across the Plains cut the herd in two and pushed them far away, making hunting much more difficult.[21]

With the Indians blaming the migrants for denuding their lands, the migrants responding with belligerence, and inter-tribal violence increasing in the competition for resources, the federal government was obliged to step in. The largest gathering of Plains Indians in history occurred in 1851 when upwards of 12,000 members of the Lakota, Arapahos, Cheyenne, Assiniboine, Shoshone, Arikara, Hidatsa, Mandan and Crow met with US agents near Fort Laramie. In exchange for fixing territorial boundaries and permitting the construction of forts, the tribes would receive $50,000 worth of merchandise and provisions each year. A similar treaty was made at Fort Atkinson two years later to protect the Santa Fe Trail along the Arkansas River.

The treaties expanded the military frontier in the Plains. That in itself exacerbated the problems. 'There is a decided aversion among all the wild tribes of Indians to the establishment of military settlements in their midst,' wrote the experienced trapper, trader, trailblazer and Indian agent Thomas Fitzpatrick; 'they consider that they destroy timber, drive off the game, interrupt their ranges, excite hostile feelings.' The treaties were also doomed to failure because nomadic warrior tribes dependent on hunting could not be confined to set areas. 'You have split the country and I don't like it,' the Lakota Sioux chief Black Hawk said of the Treaty of Fort Laramie. 'What we live upon we hunt for, and we hunt from the Platte to the Arkansas, and from here up to Red Butte and the Sweet Water [an area covering large portions of modern-day Colorado, Kansas, Nebraska and Wyoming] . . . These lands once belonged to the Kiowas and the Crows, but we whipped these nations out of them, and in this we do what the white men do when they want land of the Indians.'[22]

As far as Fitzpatrick was concerned the system of paying off the Indians with gifts and attempting to overawe them with inadequate military force was doomed to failure. 'Our relations with the wild tribes of the prairie and mountains resolve themselves into a simple alternative. The policy must be either an army or an annuity . . . Any compromise between the two systems will only be productive of mischief, and liable to all the miseries of failure.' By 1853 the damage done to the wintering grounds had become even worse. Fitzpatrick reported that the Cheyenne, Arapahos and Lakota 'are actually in a *starving* state. Their women are pinched with want and their children constantly cry out with hunger.'[23]

Every year the situation got more dire and dangerous. Things reached a head in 1854 when 4,000 Brulé Sioux were camped near Fort Laramie, awaiting the annual distribution of annuities stipulated by the 1851 Treaty. The federal agent was late arriving that year (an all too common occurrence), the Indians were starving and the braves were growing distrustful; the atmosphere was tense.

On 17 August a cow belonging to a westbound Mormon strayed into the camp and was promptly butchered. Chief Conquering Bear offered a horse or a cow in compensation, but Lieutenant Hugh Fleming, the commander of the fort, wanted the Indian responsible to be arrested. The brave in question, High Forehead, was a visitor from another tribe; Conquering Bear had no authority over him and did not want to violate the code of hospitality, so he refused to hand him over. The next day Second Lieutenant John Grattan, a young cavalry officer contemptuous of Indians, led out twenty-eight men to the Sioux camp, with no intention of negotiating.

Grattan formed a line and directed his two cannon at the Brulé camp. His drunken interpreter shouted that the army would murder all the Sioux. Conquering Bear reiterated his offer of horses; this too was garbled by the intoxicated translator. Shots were fired during the confusion, but still Conquering Bear urged the Sioux braves to remain calm. When Grattan's men fired a second volley the chief was felled by a bullet tearing into his back. All restraint was now lifted; Grattan and his contingent were killed and mutilated. The Brulé braves raided the trading posts and melted away.

The Grattan Massacre, as it was dubbed in the press, sparked outrage. The Superintendent of Indian Affairs, however, believed that the tragedy was not premeditated and praised the subsequent good behaviour of the Sioux. It was not enough. Public and political mood demanded revenge. Brigadier General William S. Harney led 600 soldiers out of Fort Leavenworth in 1855 to chastise the Sioux and present a show of force to the other Plains Indians. On 3 September Harney came across a camp of 250 Brulé Sioux at Ash Hollow on the Platte River, a pretty spot enclosed by cedars and bluffs. By the end of the ensuing battle, eighty-six Brulé, including women and children, lay dead among the wild clematis. Two years later Colonel Edwin Sumner led the First Cavalry on an extensive punitive expedition against the Cheyenne, who had been attacking emigrant trains. One half of the command scoured the North Platte, then headed along the Front Range to the South Platte; the other ventured to the same destination via the Arkansas River. Having circled the Plains, the re-united force set off into the central Plains, along the Republican and Solomon rivers, where they fought a band of Cheyenne. Sumner's men destroyed 171 lodges and quantities of buffalo meat in storage for the winter. These incidents marked the beginning of intermittent wars with the Plains Indians that were to last for much of the rest of the century.

Burton had sympathy for the Native Americans, writing as he bounced along the overland trail of 'those poor remnants of nations which once kept the power of North America at bay' but which are 'now barely able to struggle for existence'. But for all his admiration for the Plains Indians, he offered advice drawn from his experiences on imperial frontiers on how to subjugate them. He compared the outpost system of scattered garrisons adopted by the British in India and the Americans on the Plains unfavourably with the centralised military strategy of the French in Algeria. The French established large depots of men and ammunition, 'making them pivots for expeditionary columns, which by good military roads could be thrown in overwhelming numbers . . . wherever an attack or an insurrectionary movement required crushing'. Thomas Fitzpatrick would have agreed. Although he had spent much of his life among the Indians and sympathised with their fate, he wrote that skeleton companies of dragoons patrolling 400,000 square miles created 'a belief in the feebleness of the white man'. A large force was required to cow the Native Americans for ever.[24]

It wasn't that such men had lost patience with the inhabitants of the Plains or their compassion had withered. Rather, there was a sense of fatalism. The progressive forces of the 1850s were 'beyond human power to retard or control', they held, and it was the law of nature that the stronger people – with their weapons, methods of agriculture, machines and sheer numbers – would displace the weaker.[25]

It also came from fear: the Cheyenne, the Sioux and the Comanche were among the most feared mounted warriors in the world, capable of serious resistance, and they had been pushed to breaking point. 'The Indian race is becoming desperate, wild-beast like, hemmed in by its enemies that have flanked it on the east and west, and are gradually closing in upon it,' wrote Burton. '. . . The tribes can no longer shift ground without inroads into territories already occupied by neighbours, who are, of course, hostile; they are, therefore, being brought to final bay.' By the time he wrote, the refuges to which the Plains Indians resorted – such as the verdant Republican, Solomon and Smoky Hill river valleys – had been turned into busy roads by the new generation of Argonauts and the gigantic freight trains. Even worse, the fertile valleys on the Front Range of the Rockies that provided essential wintering grounds had suddenly become sites for cities and ranches. In the fall of 1859 William Bent met the leaders of the Kiowa and Comanche at the great bend of the Arkansas. They spoke peace, but Bent believed that 'a smothered passion for revenge agitates these Indians'. The reason was obvious: it was 'the failure of food, the encircling encroachment of the white population, and the exasperating sense of decay and impending extinction with which they are surrounded'.[26]

That sense of 'impending extinction' was being felt throughout the world at the same time. Every year the reports of the Aborigines Protection Society – a pressure group formed by British evangelicals who were veterans of the anti-slavery movement – detailed the precipitous population declines of indigenous people. For Australian Aborigines and New Zealand Maori, southern African Xhosa and Zulu, Canadian and American Indians, in Madagascar and Brazil, on the Pacific islands and the Pampas, the story was depressingly similar: local authorities and imperial governments lacked the will to restrain the tsunami of settlers. Australia offered the starkest example of calamitous population decline: there were 700,000 Aborigines in 1788 and 180,000 by the end of the 1850s. One of those who warned the world of the impending extinction of numerous indigenous peoples was the American showman and explorer George Catlin. In the Egyptian Hall in Piccadilly, London, he exhibited – alongside other exotic curiosities from around the world – Native American artefacts and members of the Canadian Ojibwas and American Iowas. At a well-attended lecture at the Royal Institution he called for a Museum of Mankind to record for posterity peoples who were facing oblivion. He reminded the great and the good that 'Great Britain has more than thirty colonies in different quarters of the globe, in which the numbers of civilised men are increasing, and the native tribes are wasting away – that the march of civilization is everywhere, as it is in America, a war of extermination, and that of our own species'.[27]

Despite the forgivable hyperbole of Catlin and others, indigenous people had less to fear from the genocidal behaviour of settlers (although that was horrendous enough) than from the indirect effects of rapid colonisation. In southern Africa the onslaught of settlers and the advancing frontier pushed the Kohekohe (then known as Hottentots) into new lands, where they came into frictious contact with other peoples; their population declined from over 200,000 to 32,000. On the American Plains, according to William Bent, the 'numerous and warlike Indians' were being pressed north, south, east and west by settlers and 'are already compressed into a small circle of territory, destitute of food, and itself bisected athwart by constantly marching lines of emigrants'. In the middle of the 1850s it was noted that among the Southern Cheyenne, Arapahos and Comanche women were outnumbering men three to two, a sure sign of intensified fighting as warriors competed for the remaining resources. And death stalked the land in another guise.[28]

Roads were vectors of germs. Cholera and dysentery thrived among the crowds gathered on the Mississippi tributaries preparing to travel; they carried them westwards, into the Plains. Epidemics felled up to 30,000 migrants, but these diseases, as well as smallpox, bronchitis and syphilis, devastated the Native Americans just as they had done Australian Aborigines. Some tribes lost between a quarter and a third of their populations. Disease inflicted death on a greater scale than wars and massacres in North America and elsewhere. Despite the loss of land and warriors in numerous frontier wars the Xhosa remained independent, untouched by Western diseases. However, in 1853 a European cow imported into the south-west Cape carried with it a virulent form of bovine lung disease. For two years the virus moved inexorably along the ox-wagon routes until it reached the Xhosa's lands in 1855. Cattle formed the bedrock of the Xhosa's wealth and power; the destruction of the herds threatened their existence as a people.

Just as the Xhosa were dependent on cattle, the Plains Indians relied on buffalo. The disruption of the great herds by the roads and then the expanding ranching frontier forced the hunting bands to travel further, exhausting their horses and reducing the amount of food for the villages to live on. The Xhosa and the Plains Indians were far from being the only peoples to have their lives turned upside down in the turmoil of the 1850s. To take another example, in New Zealand the Maori population was overtaken by European incomers in 1856. But it was not demography that was the main problem: whites were by and large confined to congeries of coastal urban settlements that left large tracts of inland territory untouched. The sheep population, however, exploded from thousands to millions; these multiplying merinos required extensive rolling pastures to graze. On South Island vast areas of tussock grassland, bracken and wetlands were burnt off and the native landscape transformed into a European pastoralist one, with imported grass varieties and enclosed runs. Incomers used and related to land in a very different way than indigenous peoples did. They wanted to fence off and *own* territory, to domesticate it; their methods were often extensive and extractive.

Rash it would have been to doubt the ability of peoples such as the Maori, Xhosa, Comanche, Sioux or Cheyenne to fight back and resist. The ecologies upon which they depended, however, were much more fragile. Nomadic pastoralists, hunters and warriors needed freedom to range vast tracts of land in search of dispersed and finite resources. The constriction of land, the encroachment of farming, the ravages of imported diseases and the environmental impact of roads entailed the end not of the people, but of the way of life that defined them.

Few would have predicted in 1850 quite how quickly these changes would play out. By the time the full implications were apparent it was too late to arrest the momentum. The 'exasperating sense of decay and impending extinction' provoked apocalyptic anxieties. A common response was for indigenous societies to rediscover old traditions and appeal to their deities. On the Solomon River in 1857, when Colonel Sumner's men prepared to attack the Cheyenne camps, two young warriors named Ice and Dark told their fellows that Maheo, the All Being, would reward their courage and faithfulness. Ice and Dark gave the warriors medicine that would stop the 1st Cavalry's rifles from firing and had them bathe in a sacred lake that would make them bullet-proof. When the cavalry charged with sabres drawn it became apparent that Maheo was punishing them. The Cheyenne warriors fled. In later years, as the advance of miners and settlers became unstoppable, groups of militant bands, desperate to cleave to the old ways of life, retreated further into the remaining empty spaces, away from contact or compromise with the whites, to make their last stand.[29]

The West's advance, collapsing ecologies and catastrophic diseases had a traumatic effect on peoples. After eighty years resisting white encroachment, the Xhosa suffered the onslaught of bovine disease in the aftermath of military reverses; extinction stared them in the face. It seemed to them that they were being punished for faithlessness; their land was polluted, they were bewitched by evil spirits. In 1856 a teenage prophetess named Nongqawuse commanded the Xhosa to slaughter their surviving 400,000 head of cattle and burn all their crops. For a people facing annihilation her prophecy chimed with ancient creation beliefs and newly introduced ideas of Christian resurrection. If they did as Nongqawuse said, they would be purified; their ancestors would return with herds of cattle that defied numeration; the whites would be swallowed up by the sea; paradise would be realised on earth. The outcome was more predictable: 40,000 Xhosa died of starvation (out of a population of 107,000) and another 40,000 fled as refugees to Cape Colony; their land was taken by white settlers.[30]

Richard Burton saw the American frontier of the 1850s as one part of a global Western frontier. If during the decade white people saw the world speeding up and distances closing, the feeling of imminent and complete change was experienced by millions of others. In America the white population had been pushing out Native populations and annihilating the buffalo east of the Mississippi for over two centuries. The vast extent of the Great Plains was supposed to be the last refuge. The wildfire pace of change in the 1850s left its inhabitants shell-shocked; they contemplated the end of their ways of life for good.

A Lakota Sioux told Thomas Twiss of the Upper Platte Indian Agency in 1859 that as a young man thirty years before he had headed east from the Platte to Minne Tonkah (the Mississippi). He continued into the lands of the Winnebago, in what would become Wisconsin, and then on to the Great Lake (Lake Michigan). He went home by way of the Irara Falls, known to the white people as the St Anthony Falls in Minnesota. In all that vast territory he traversed – a round trip of almost 2,000 miles – he glimpsed white men only once, a handful of isolated settlers on the shores of Lake Michigan in present-day Illinois or Wisconsin.

'Our "father" tells us the white man will never settle on our lands and kill our game,' he said; 'but see! the whites cover all these lands that I have just described.' They covered them to the extent of a cool million. Twiss wrote that the 'greatly increased velocity' of emigration in the 1850s meant that 'millions' would soon start settling the Plains as well, bringing agriculture and scattering the game. Already the great herds of buffalo, once so concentrated they made the prairie appear black as far as the eye could scan the horizon, were dispersed into dwindling bands fleeing 'the white man's rifle' and the road. As Twiss wrote, the great waves of migration were making the 'utter extinction of the wild tribes' inevitable. No wonder the mood was apocalyptic. 'Our country has become very small,' the Lakota leader said, 'and, before our children are grown up, we shall have no more game.'[31]

At sea the great clippers were shaving weeks off journeys. On land it was the same story. Within five months of leaving the banks of the Missouri, Richard Burton was in the Caribbean on board a ship bound for Britain. Not being in a hurry, he had stopped off for a month in Salt Lake City, mooched around Virginia City and San Francisco, gone sightseeing in Mexico. Even so, his meandering holiday took the same time that overlanders had taken to get from St Joseph to Sacramento only a few years before. The overland road, the cities and settlements, the steamer lines and the Isthmus railroad had transformed North and Central America. If he had hurried, Burton could have got to the Caribbean in a fraction of that time. He owed the swiftness of his journey not to the need to conquer distance, but to vanquish time; not to the value of conveying people, but the premium placed on information.

During the 1850s California remained isolated from the rest of the Union, far distant from the telegraph, railroad, canal and river networks that bound the other states together. The overwhelming majority of letters and newspapers sent to and from California went by steam via the Isthmus of Panama. The arrival of the mail was signalled to the people of San Francisco by a large black ball hoisted on Telegraph Hill. It was 'Steamer Day', when San Francisco descended into frenzy. This happened twice a month. News and vital business correspondence arrived in one go, like a deluge of rain after a prolonged drought, sparking flurries of frenetic activity in the worlds of commerce, trade, journalism and politics. Here was news from Congress, the latest stock market figures from Wall Street, up-to-date commodity prices and vital information on the state of the global economy. People queued – or paid boys to queue – outside the post office for days before. The departure of the mail was just as frenetic. Businessmen and bankers hurried to complete money orders and write instructions to their agents across the Union. Journalists finalised their copy to be dispatched to cities in the US and Europe.

It was hardly a satisfactory way of conducting affairs. For one thing, priceless information travelled through foreign countries, which might at any time become destabilised. For another, the service was infuriatingly infrequent and expensive. The goal was to establish uninterrupted communication across land controlled by the United States.

In 1857 Congress authorised the US Postmaster General, Aaron Brown, to contract out an overland mail service between Missouri and California. Brown, a Southerner, secured a route that described a gigantic arc from St Louis, Missouri, and Memphis, Tennessee, to San Francisco via Fort Smith (Arkansas), El Paso (Texas), Tucson (Arizona) and Los Angeles. The winner of the $600,000 government contract, the Butterfield Overland Mail Company (which was owned by four companies, including American Express and Wells Fargo), ran two stagecoaches each week between St Louis and San Francisco, and two in the opposite direction.

The mail coaches travelled twenty-four hours a day, changing horses at stations spaced at springs and water sources twenty miles apart. It was a bone-crunching race across Indian territory and the arid terrain of Texas and New Mexico during which the bouncing Concord coaches covered the 2,800 miles between St Louis and San Francisco in fewer days than the twenty-five stipulated in the contract. Californians now enjoyed postal deliveries twice weekly, an important development for a rapidly growing economy.

At over 100 miles a day the journey was fast, but it was also furious: according to Burton, 'passengers becoming crazy by whisky, mixed with want of sleep, are often obliged to be strapped to their seats; their meals, dispatched during the ten-minute halts, are simply abominable, the heats are excessive, the climate malarious'. The *New York Herald* journalist Waterman Ormsby, a passenger on the first coach that took the Butterfield route, wrote that 'I now know what Hell is like. I've just had 24 days of it.'[32]

When Burton made the journey stagecoaches raced with the mail across the Central Overland Route, where recently the fastest things going were ox teams lugging emigrant wagons a few weary miles a day along rutted tracks. He rested his behind on the leather seats of a luxury sprung coach made in Concord and, dosed with opium to combat the boredom, careened along well-made roads. This was one of the new scheduled weekly mail stagecoaches that raced from the Missouri River to Salt Lake City, where they connected with the California mail coaches, changing mules at the scores of stations that now punctuated the route. Mail sent from San Francisco reached the Missouri Frontier towns in an astonishing nineteen days, where it was then put on board a train or telegraphed east.

But even these swift coaches were old-hat when Burton took one. By then the Pony Express carried mail from the termini of the Californian telegraph system to that at St Joseph in an incredible ten days. The daredevil young riders, picked for their boldness and slightness of frame, rode in relays day and night across the familiar Central Overland Route. They changed horses every twenty-five miles or so, and raced between stations fifty to seventy-five miles apart. In this way information travelled 250 miles non-stop in twenty-four hours. The mail these men carried in their *mochila* – a light Mexican pouch – was charged at $5 per half-ounce. Not surprisingly, the time-sensitive business, political and newspaper correspondence that accounted for most of the Pony Express mail was written in microscopic letters on 'paper as airy and thin as gold leaf', like the airmail of a different age.

The story of the overland route magnifies many of the themes that characterised the 1850s in America but also across the globe: mass migration, communication, the dislocation of indigenous people and the need for speed. Not only did it show how great distances could be abbreviated, but it hinted at a major shift in global power.

Just as Cyrus Field dreamt of connecting Europe and the United States with cable, other entrepreneurs readily perceived that the westward road had the potential to become a gateway not just to the riches of California, but to the Pacific and on to Asia. In the words of Senator William H. Seward, 'the fiery course' of a railroad over the Sierra Nevada to San Francisco would 'send a new sensation through the world'. Britain's global commercial power would start to crumble. The reason was clear: 'Attracted to the great eastern station of the Continental [railroad], as steel to a magnet, a freight would roll down upon the States of the Union such as the [British East] India Company never saw, embracing the furs of the north, the drugs and spices of the south, the teas, silks and crapes of China, the cashmeres of Tibet, the diamonds of India and Borneo, the various products of the Japan Islands, Manchuria, Australasia and Polynesia . . . and unimaginable elements of commerce which would be brought into life from the depths of the sea.' In return, the US would export meat, breadstuffs and cotton as well as 'the Bible, the Printing Press, the Ballot Box, and the Steam Engine' to as yet 'vast and unregenerated fields'.[33]

As the history of the roads in America showed, new developments slashed journey times from months to days in a short passage of time. Yet for all that, impressive as fast stagecoaches and Pony Express riders were, they panted to keep pace with human ambitions. The vision was of iron tracks laid along the transcontinental trail, with a parallel telegraph line disappearing into the horizon; those twins possessed the power radically to transform the world. They would short-cut the transmission of goods and information across the planet and open vast virgin swathes of it to exploitation.

People could readily perceive that future. The land was surveyed; the demand was there; finance could be found. The desert, the canyons and the Sierra Nevada were barriers, to be sure; but the greatest, most formidable barrier was political. Time and again the scheme of a transcontinental railroad foundered on the rocks of sectionalism that bedevilled American politics. The North would not sanction a railroad from New Orleans to San Francisco; the South vetoed a route from Chicago.

The fate of the railroad was of intense interest to people in Europe and Asia. They were forced to wait for the Gordian knot of American politics to be cut. In the meantime, however, the remorseless advance of America into the heart of its own continent contained a profound message. In the 1850s it was becoming clear that the balance of power in the world was shifting away from its great seaports and shipping lanes into its great interiors.

5

Star of Empire

MINNESOTA

Continent and island, wilderness and jungle, forest and prairie, but a few years since the desolate haunts of the savage and of prowling and creeping things, are now alive with the busy hum of commerce . . . On that luxuriant plain, which but a few years since revealed no trace of humankind, save the wreathing smoke from the wild man's solitary wigwam, you see flocks and herds and golden crops surrounding the busy city, teeming with 'civilised life'.

Charles Hursthouse[1]

'Have the glories of Great Britain reached their climax, culminated, and begun to pale? Is England in her decadence?'

These were the questions with which Anthony Trollope began a tract in 1855, a work that would remain unpublished until the twentieth century. 'Is the time quickly coming when the New Zealander shall supplant the Englishman in the history of the civilization of the world?' he asked. The New Zealander Trollope had in mind would have been immediately recognised by his intended readership. In an essay published in 1840 the great historian Thomas Babington Macaulay had depicted some future age when a tourist from New Zealand would stand on the remnants of London Bridge and sketch the ruined domes and archaeological relics of the once-mighty metropolis, just as modern tourists recreated in their minds the glories of Rome's capital and mused on the remorseless cycles of history.[2]

Macaulay was not saying that New Zealand would surpass Great

Britain in the future. But, subconsciously or consciously, professional promoters of the New Anglo-Saxon world bent the historian's vision to suit their purposes. The tireless salesman for Australia and New Zealand George Butler Earp affirmed that 'Australia and New Zealand are rapidly becoming the culminating points of the Anglo-Saxon race . . . It is something to be amongst the founders of a new empire, such as is rising up in the southern hemisphere – to influence the destinies of the human race, when the old empires of the northern hemisphere shall live only in classical history.'[3]

And if the genius of the British people was being transferred to the southern hemisphere, in America the pendulum of civilisation was held to be swinging west. 'The centre of power, numerical, political, economical, and social, is . . . indubitably, on its steady march from the Atlantic border toward the interior of the continent,' wrote the prophet of westward expansion, Jesup Scott. 'That it will find a resting place somewhere in its broad interior plain seems inevitable as the continual movement of the earth on its axis.' The star of empire – of progress – of mass migration – tracked north-west.[4]

In the future the bustling cities of the Atlantic coast would decline; the cradle of the new republic would relocate to mighty river cities in the geographical heart of the United States, what we now call the Midwest but which was then the western frontier. Leavenworth, Minneapolis, Omaha, Kansas City, St Louis or embryo frontier cities as yet unheard of would soon eclipse New York, Boston, Baltimore, Philadelphia, Charleston and New Orleans as the metropolises of a new empire based on the Great Lakes and the Mississippi River System. Already Chicago, Cincinnati and Detroit had sprung up and were growing with gravity-defying velocity. People foresaw the Midwest becoming the most densely populated place on earth, interconnected with webs of railroads, rivers and canals, its fertile agricultural lands and mineral deposits sustaining industrial-commercial cities of 10 million inhabitants drawn from every part of the globe. Railroads that annihilated space, telegraphs that did the same to time, rapid urbanisation and mass migration would converge to shift decisively the hubs of civilisation from seaports to the interior.[5]

Such opinions represented, as it were, a rush of blood to the extremities. With economic growth and settlement being generated on such a colossal scale on the peripheries of the Anglo-Saxon world, in the American West and the British Antipodes, it seemed automatic that the future belonged to these places. The exhilaration of the 1850s boom ignited such prophecies and this frenzied excitement powered migration and economic development as much as, or more than, gold or trade.

If New Zealand came to be seen by some as the future repository of British values in the 1850s, its American equivalent was the recently opened territory of Minnesota, which was widely held to herald the future direction of the United States. Both were newly acquired outposts of empire: Europeans had set up whaling stations on the shoreline of New Zealand; the Minnesotan equivalent was remote fur-trading posts far beyond 'the borders of civilisation'. The former was annexed to Britain in 1840, the latter was organised as a territory of the United States nine years later. At the beginning of the 1850s New Zealand had a population of 22,000 white settlers who were outnumbered by 65,000 Maori; Minnesota contained about 6,000 whites and 30,000 Native Americans. By the Treaty of Waitangi the Maori transferred sovereignty to Queen Victoria; they were given rights to their property and, if they wished to sell, they could do so only to the Crown. The Crown snapped up 32 million acres by 1853, which included most of South Island, for the trifling sum of £61,847. In 1851, by the Treaties of Traverse les Sioux and Mendota, the US government paid sub-tribes of the Dakota Sioux $1,665,000 for 24 million acres of land, much of it in Minnesota.[6]

None of these treaties were satisfactory. It was almost impossible to translate concepts such as 'sovereignty' into Maori. Becoming members of the British Empire overnight – with all its implications – did not change things materially for the indigenous New Zealanders. They only had present reality to go on. Small, outnumbered settlements of Pakeha – as the settler community was known – offered opportunities for trade and land sales, which the Maori gladly seized upon. The land, even if it was sold, remained largely unchanged. Just 10 per cent of the Maori population lived on South Island, where the majority of land had been sold. In any case, Maori kept Pakeha in fear, by their numbers and warrior skills. The influx of settlers and troops in New Zealand and Minnesota would reveal, over time, the implications of the treaties.

Both New Zealand and Minnesota had been known as 'Ultima Thule' – a term used by medieval geographers to denote lands lying beyond the boundaries of the known world. Until recently New Zealand had been blackened as an island of cannibals, a remote and dangerous place for white settlers, its land too wild to cultivate. During the 1840s there had been wars with powerful Maori tribes and they had not resulted in British supremacy. Minnesota was labelled the 'American Siberia' because its climate and terrain were considered too harsh and desolate for European settlement. Yet by the 1850s both regions were being aggressively sold to publics in their parent countries and further afield as ripe for settlement. No longer were they harsh liminal zones far beyond the pale of civilisation: forget Victoria and California, they were the true El Dorados of the 1850s.

Selling new territories to potential migrants was not a new phenomenon. It was known as 'boosterism' and it reached new levels of hyperbole in the cases of New Zealand and Minnesota. Boosters had to mount sustained campaigns against the prejudices that existed about their regions. Potential migrants were consumers in a free market, with an array of worldwide destinations to choose from. The British were eager to populate their colonies, but they were equally wedded to laissez-faire. Emigrants were as free to go to the US as to Australia, Canada, Cape Colony, Natal or New Zealand. Indeed, America was closer to Britain, it was cheaper to get to, and land was inexpensive. Thus New Zealand was in direct competition with Minnesota and myriad other destinations in the States, just as it was with the goldfields of Australia, the Canadian prairies and promising Cape Colony. The extensive marketplace of migration made hard selling a necessity. After all, why go to New Zealand when Australia was rich in gold? What made Minnesota more alluring than California?

New Zealand, according to Charles Hursthouse, one of the colony's most enthusiastic champions, was 'bold and beautiful' with 'probably the finest agricultural wild land in the world'. Sarah Tucker depicted 'gardens rich with the fruits and flowers of central and southern Europe' and 'plains now waving with a golden harvest, or covered with grazing cattle'. Minnesotan boosters also emphasised the verdant beauty of their territory and described its agricultural potential as second to none. Its rivals in the burgeoning west – Iowa, Ohio, Michigan, Indiana, Illinois and Wisconsin – might have fine agriculture and dark, rich prairie soil, but Minnesota possessed that *and* waterpower *and* timber in abundance. Moreover, Minnesota 'stands at the head of the navigation of [a] wonderful natural channel of communication, unequalled in the world'. The Mississippi rose in Minnesota and the territory was bounded to the north by Lake Superior. 'She is thus . . . the centre from which radiates the greatest watercourses in the continent,' a conveyor belt that linked the Gulf of Mexico and the Atlantic via the Mississippi, the Great Lakes, the Erie Canal and the St Lawrence River. Minnesota was not some isolated Lapland, but a potential pivot of world trade. Might not this territory become Jesup Scott's final 'resting place' of American economic power in the interior, the site of the great megalopolis of the future?[7]

'Siberia' was thus transformed into the Garden of Eden in the countless books, pamphlets and newspaper articles that constituted the vigorous PR campaign waged in the US and Europe. Next to natural wonders and the unsurpassable productive qualities of the soil, climate was an important part of the booster's art. Minnesota was a tough sell in this regard, with its savage winters and broiling summer heat. New Zealand's boosters could make much of their country's mild climate and compare it with their rivals in the great business of migration – the extremes of temperature in America and Canada or the heat of Australia and South Africa. A clergyman imbued New Zealand's Goldilocks climate – not too hot, not too cold – with almost miraculous qualities: 'the sickly become healthy, the healthy robust'. Other writers said that crops grew quicker, fruit became bigger and animals got fatter – truly a Garden of Eden.[8]

One tactic for the Minnesotans was to downplay the coldness and reach for euphemisms such as 'bracing'. But boosters preferred to meet the challenge four-square and work it to their advantage.

Minnesota's climate held disease at bay, it was claimed, and cured the chronically ill with its 'strong and pure' atmosphere. One day doctors would send invalids to the territory for a rest cure instead of to Madeira. And better still, Anglo-Scandinavians flourished best in hardy climates. Balmy weather made people languid; contending with hostile elements made them industrious, ingenious and imbued them with republican virtues. One excited booster claimed that it was a provable fact that humans and animals worked harder in Minnesota than in any other western state. 'There seems to be a certain zone of climate within which humanity reaches the highest degree of physical and mental power,' declaimed the same writer. This imaginary 'zone' once lay in ancient Rome, when apparently the weather was 'bracing', the Tiber froze and the people were glorious, before moving north to animate the Saxon peoples. 'It is the good fortune . . . [of Minnesota] to lie not only within this zone, but within its very apex.' Now that really is making the best of the weather.[9]

'I speak in no boastful or vainglorious theme when I say there is largely more *character* in Minnesota than . . . in any of the older western members of our republican family,' said another booster. The blessings of climate, health and soil converged on the notion of character. Victoria and California might hold out the temptation of glittering gold. But rugged and hearty places such as New Zealand and Minnesota offered more substantial things. 'Sturdiness, self-reliance, and singleness of purpose' were the chief virtues of pioneers, and they were essential in the shaping of that great Victorian ideal – character.[10]

Almost all the boosters of New Zealand spoke of the creation of a country of independent yeomen who replicated, and perfected, British agrarian values and character in the wilderness. Victoria, by contrast, represented the worst of greedy, loutish, lawless British behaviour. If you were suddenly transplanted to a New Zealand city, the boosters said, the only reason you would know it was not an English market town was the prevalence of moustaches and beards.

But the promise of a Britain-lite was not enough for a successful booster. The hale, hearty immigrants they wanted must believe that they were relocating somewhere that would exceed the mother country. The young, idealistic aristocrat Thomas Cholmondeley wrote after a sojourn in New Zealand that it was 'like a pledge of new life' for Britain: 'There we may renew ourselves . . . New faculties may be bestowed upon us; fresh energy radiated from the inventive and constructive genius of a young society, as yet unbent by conventionalities, unbroken by excess.' In the 1850s new names were proposed for New Zealand, such as Britannia, Austral-Britain or South Britain. An Englishman emigrating there would, according to Hursthouse, find 'he has the same queen, the same laws and customs, the same language, the same schools, the same churches, the same press, the same social institutions and, save that he is in a country where trees are evergreen, and where there is no winter, no opera, no aristocracy, no income tax, no paupers, no beggars, no cotton mills, he is, *virtually*, in a young England'. In much the same way Minnesota was proclaimed the new New England, where a distinctively northern, liberty-loving, hard-working Anglo-Scandinavian culture would flourish and exceed its parent.[11]

'It is the strong and the bold who go forth to subdue the wilderness,' said Hursthouse. Emigration, he wrote, 'calls up pluck, bottom, energy, enterprise, all the masculine virtues. The feeble-minded, the emasculate, the fastidious, the timid, do *not* emigrate.' That was what would make New Zealand the new Britain – the endeavour and *character* of the pioneering generation that took the best of British values and rejected the worst. The social tyranny that writers such as Hursthouse and Earp depicted as strangling the true British character did not exist in the colonies. New Zealand was also more democratic, with a wider franchise and a relatively open political system. In an age of free trade and open markets, the British concept of empire was in a bout of transition. The idea was to ready its white settler colonies for independence as quickly as possible. The first step in the journey was responsible self-government, which New Zealand got in 1853. Free from history, places like New Zealand and Minnesota were clean slates, where nineteenth-century man could be born and Western civilisation renewed.[12]

What boosters had to do was make their target audience fall in love with the future. For that was where all the good things lay.

*

Among the myriad exhibits in the Crystal Palace in 1851 was Boyd's Patent Double-Action Self-Adjusting Scythe. Unlike the traditional scythe it had an adjustable blade and handle, so that it could be set up without the need for a blacksmith. It was marketed at emigrants, who would have to scythe through virgin grassland to create their homestead and to make a little sliver of Britain in the remote part of a distant land.

That picture accords with our notion of frontier life in the nineteenth century, whether it be the American West or Britain's settler colonies – a remote, primarily agrarian existence, a 'little house on the prairie'. But as commentators remarked, the pioneer of the 1850s was entirely different from those who had gone before. One American writer astutely noted that in the recent past new lands had been settled first by hunters and then farmers; they were followed over the succeeding years by traders, merchants, lawyers, bankers, capitalists and speculators, 'the ranks of civilization . . . [advancing] in succession'. But in the 1850s 'all poured in together'. And they wanted to settle in cities, not the remoteness.[13]

The result was breakneck urban development and a profusion of towns jockeying to be the great cities of the future. After the Great Exhibition and the Transatlantic Telegraph, there are few more potent symbols of the confidence and ambitions of the 1850s than this phenomenon, the 'urban frontier'.[14]

It was into the midst of this frenzy that the young British traveller Laurence Oliphant journeyed in 1854. He tracked west from Portland, Maine, into Canada before crossing the Great Lakes, taking railroads, steamboats and shooting the rapids on canoe. His goal was Minnesota, where the most startling manifestation of this urban frontier could be witnessed.

There can be few better guides than Laurence Oliphant. He is one of the most intriguing – and well-travelled – personalities of the Victorian heyday. Born in Cape Town to Scottish parents, he had, by the time he settled in London in 1851 at the age of twenty-two, lived and journeyed through Ceylon, Paris, Germany, Italy (where he witnessed the 1848 revolution in Naples), Greece and Nepal, and had, as a barrister in Ceylon, taken part in twenty-two murder trials. While studying for the English and Scottish Bars he published his first travel book, an account of his journey to Kathmandu.

In the same year, 1852, he went from Moscow to Nizhny Novgorod, telling everyone who would listen that he was headed for the Caspian Sea. That destination was a ruse to throw Russian spies off his scent. His real goal was Sevastopol, the new Russian naval base on the Black Sea that was sealed off to foreigners. Travelling in disguise, he made it there, where he secretly mapped its fortifications.

As Britain and Russia squared up for war, Oliphant's *The Russian Shores of the Black Sea* became required reading and his specialist knowledge in demand from statesmen and generals. In the next decade or so he would travel round the globe, from the Caucasus to Central America, from Peking to the Paris Commune. He was an eyewitness, as diplomat or war correspondent, to the Crimean War, the Second Opium War in China, the opening-up of Japan, the struggle for the unification of Italy and the Franco-Prussian War.

But all that was yet to come. In the early 1850s he turned down journalistic commissions and instead accepted the appointment of secretary to James Bruce, 8th Earl of Elgin, the Governor-in-chief of British North America. This was a tense time for the British Empire. Free trade exposed Canadian producers to the icy winds of global competition; there were calls for Britain's colonies to join the United States. Britain answered the crisis by sending Elgin to Washington to negotiate a reciprocal free-trade agreement between Canada and the US. British money also helped fund the creation of an enormous transportation network that would connect Toronto, Montreal and Quebec City with Detroit, Buffalo and Portland (Maine). With Canada's future bound up with that of the northern US, it was little wonder that Laurence Oliphant was drawn across the border to the miracle that was Minnesota.

On his way he stopped at Ontonagon on the Michigan shore of Lake Superior. A half-finished hotel – 'certainly more comfortable, and upon a grander scale' than the Adelphi, Liverpool's most luxurious hotel – towered above the log shanties that surrounded it. A deluxe, striking hotel was crucial for marketing your town. Its picture could adorn publicity material and attract investors and, above all, migrants who would buy your land and goods and develop your town in the quickest possible time. Next in importance was a newspaper (or two) in which to inform the world of the glories of your infant city. As one writer put it, in the 1850s '*capital* [goes] *in advance of population*': speculators arrive first, actual economic development comes later. Ontonagon was, for Oliphant, 'a perfect specimen of a backwood town in an embryo state'. Tree stumps littered the site and 'the old forest still seemed to dispute the soil with the settlers'.[15]

Forget the gold rush: the truly wise in the 1850s gambled on the cities of the future. The art was to buy land in the wilderness, start a town, and wait for the nineteenth century to catch up with you: its railroads, mills, telegraphs, steamboats, macadamised roads, schools and the inevitable tidal wave of settlers that would transform your infant city into a fully functioning giant. As anyone who saw the explosion of cities in the new New World could tell you, that future was never long in coming.[16]

The business model of the pioneers of Ontonagon 'consists in buying property upon the outskirts of the inhabited world . . . preparing themselves betimes for the inevitable influx of emigrants'. After a few years, during which the town would inevitably boom, they would sell up, move on to found fresh embryo cities beyond the frontier, and start all over again. They wore the ubiquitous clothes and accoutrements of hundreds of thousands of restless, rootless men the world over in the 1850s, from Ballarat to Kansas, Auckland to Sacramento: wide-brimmed hats, red flannel shirts and jeans or moleskin trousers with bowie knives and Colts stuck in their belts. They were, almost to a man, bearded.[17]

Writing from New Zealand, Charles Hursthouse enticed early migrants by saying that he had made his fortune merely by buying a hundred acres of land near a budding city that he had left uncultivated: 'I can do this by virtue of . . . a natural law, almost as certain as gravitation: – "*golden population" has flowed in around me, and doubled the value of every acre of cultivated land in the neighbourhood.*'[18]

It is little wonder that boosterism reached such a pitch in the 1850s. Land speculators could not just say that their new town was a fair prospect: they had to promote it as a future metropolis. You just had to look beyond the log huts, tree stumps and encroaching forest and *see* the future: the bustling streets, the shrill whistle of a railway engine, the smoking chimney stacks, the cross-hatchings of telegraph wires overhead and the mountains of grain and lumber piled for export.

Most often there were not even log huts upon which to hang your fantasies. There existed throughout the American West thousands of town sites bought by speculating companies. The street plans and splendid urban architecture, published and circulated to entice the flow of 'golden population', existed only on paper. In some cases the lots of the future cities were staked out on the soil, an eerie impression of streets, squares, parks, city halls, churches, schools, factories and houses patiently awaiting the future to breathe on them the life that would make them spring out of the wilderness.[19]

Laurence Oliphant visited one of these future metropolises, the city of Superior on the shores of the lake of the same name. Its inhabitants made large claims about Superior City: one day it would be a mega-city, the hub of a network of shipping lanes, canals and railroads that reached not just into the Midwest but across the continent to the Pacific. Oliphant expressed an interest in buying land in the great future city and he chose lots in the heart of the business district, fronting the main square, two doors down from the bank, near the Grand Hotel and opposite the wharves.

He was taken to see the lots by a slick, fast-talking real-estate agent fresh from New York: 'we commenced cutting our way with billhooks through the dense forest, which he called Third Avenue, or the fashionable quarter, until we got to the bed of a rivulet, down which we turned through tangled underwood (by name West Street), until it lost itself in a bog, which was the principal square . . . We did not think it worth our while cutting our way through to the business quarter.' Superior City, at that time, consisted of a single barn optimistically rebranded as a hotel and a huddle of tents.[20]

Oliphant, the unsentimental globetrotter, may have enjoyed unleashing his pen on such a comic situation. But the excitement that gripped the west was highly contagious; only a few built up immunity to it. Oliphant made the real-estate agent's day and purchased the town lot. If, he said, you could put up with the rigours of a new town in the west you would see your 'bright visions . . . realised in an incredibly short space of time'.[21]

That was not as far-fetched as it sounds – or at least such sentiments are understandable in the context of the time. Chicago numbered only 200 inhabitants in 1833; by 1850 the population had leapt to 29,963. But its take-off occurred in the 1850s; by the end of the decade it had 112,172 residents. Its growth was frenetic but it could not grow fast enough to keep pace with the tidal wave of settlers; on one day in September 1855 it was noted that although there were over 2,700 buildings under construction, there was an acute shortage of living space. Chicago's fortunes were built on the trans-shipment of grain, lumber and preserved meat from the north-west to the eastern seaboard of the US and on to Britain. In less than two decades it became the transportation hub of America, the centre of a web of canals, shipping routes on the Great Lakes and, from the 1850s, railroads.

That decade saw the great acceleration in American railroads. The mileage in Illinois, for example, increased from 110 to 2,867 in the 1850s; the total in the United States went from 8,571 to 28,820. Four major east–west routes linked the eastern seaboard to the Midwest; from 1851 a further eight lines pushed on to the banks of the Mississippi; feeder lines reached out across Missouri and the plains of Iowa towards the jumping-off points for the far west: St Joseph, Council Bluffs, Independence and Kansas City. In 1856 the first railroad bridge over the Mississippi was constructed at Rock Island, giving farmers in Iowa access to Chicago.

The greatest of all railroads in the 1850s was the Illinois Central, chartered in 1851. It was the longest line in the world, linking Chicago with the city of Cairo in the south of Illinois. Lines from Cairo connected to Memphis, Mobile and New Orleans. The costs of construction were estimated at $16.5 million (£3.3 million), and it was financed by the sale of bonds – over three-quarters of which were purchased by British banks and investors – secured by a mortgage on 2.6 million acres of land granted to the Illinois Central by the state and federal governments.[22]

Railroads were not mere *facets* of colonisation; they were the main *force* of colonisation. That is to say, in most places railroads came when new populations had established firm economic foundations. In the American West in the immediate antebellum years railroads often came in advance of settlement and set the conditions for colonisation and economic development.

The history of the Illinois Central exemplifies this perfectly. The company had 2.6 million acres of empty, remote land to dispose of – and it needed to sell it fast at the highest price to meet interest payments to its British bondholders. It also needed vast numbers of settlers to turn the prairie into productive farmland. These people would buy the company's land and, when the railroad was completed, become its customers, exporting the agricultural riches of the prairies and importing manufactured goods and farm machinery by rail. The company therefore spent considerable sums boosting Illinois in American, Canadian and northern European newspapers; publishing promotional pamphlets, placards, handbills and broadsides with the inevitable hyperbolic claims; and recruiting agents in the eastern states, in European countries and on the docks of New York to steer immigrants the right way. It also sold lumber to budding towns and went into the mortgage business, offering long-term credit to urban and farm settlers alike.[23]

Thanks to the colonising activities of the railroad there was a land rush to Illinois in the 1850s, the state's population more than doubling to 1,711,951. The Illinois Central had a global impact. Vast areas of some of the richest farmland on the planet, recently considered too remote for profitable cultivation, were opened up and settled like wildfire. The railroad made it possible for the pioneering farmers of the prairies to enter the global market. Their produce was transported to far-distant markets by an array of modern technology.

Nature put up barriers to successful exploitation of these lands. Prairie tall-grasses grow so dense and the soil is so tough that conventional wooden ploughs edged with cast iron simply broke. To meet the needs of pioneer prairie farmers a number of inventors and entrepreneurs from the east relocated to the west. John Deere's factory in Moline, Illinois, mass-produced 'The Plow that Broke the Plains', the first commercially successful steel plough tough enough for prairie farming. Reaping machines invented and built by the bitter rivals Cyrus McCormick based in Chicago, John Manny of Rockford, Illinois, and Obed Hussey of Cincinnati, Ohio, mechanised the harvest. Agriculture underwent a new revolution as these inventions became affordable and reliable in the mid-1850s. At exactly the same time market information pulsed along the new telegraph wires, reducing the risks and costs of exporting farm produce.

Grain is humankind's most important commodity, an almost sacred thing. But it is all very well harvesting it with the latest machines in vast quantities and connecting with potential customers thousands of miles away. Getting it to markets separated from you by a gulf of distance and innumerable natural obstacles is another matter.

When grain is put in sacks it becomes a great physical weight that has to be carried on the backs of thousands of labourers from farm to wagon, from wagon to wharfside and then onto a ship by block and tackle – an expensive, labour-intensive, time-consuming process.

The leap forward was to turn a solid mass into something that behaved like water. Grain elevators at farm stations along the railroads in Illinois, Indiana, Iowa and Wisconsin tipped prairie grain into the open railroad cars that transported it to Chicago's dockyards. There it was scooped out by the city's enormous, famous steam-powered elevators and warehoused high up in towers. When a trapdoor was opened gravity poured the grain along chutes directly into the holds of waiting ships.

At Buffalo, New York, the same process transferred the golden harvest from lake-going ship to canal boat and transported it down the Erie Canal and Hudson River to yet more elevator warehouses at Brooklyn and New York. Steam-powered elevators were invented in 1843 at Buffalo. But it was only in the 1850s that an interlinked network of elevators conveyed the grain from the west to the Atlantic ports. The grain never knew the inside of a sack; it moved over prodigious distances with the felicity of a liquid.[24]

When all twelve of Chicago's great grain elevators were in operation they could receive and ship half a million bushels in ten hours. That was the amount of grain the city had exported in a whole year in the late 1840s. Before the widespread adoption of the steam elevator in the 1850s a team of stevedores could take a week to load or unload a single shipment of grain – some 3,000 bushels. Now, in the 1850s, 26 million bushels of grain flowed like a mighty river through Chicago every year. Each day at harvest time saw 120 trains, some with forty fully loaded freight cars, arrive in and depart from the city. On one day in September 1857 the *Prairie Farmer* reported 272 vessels loading or unloading in the port.[25]

Those ugly, towering elevators and cavernous, dusty granaries changed the world. During the debates on the repeal of the Corn Laws back in 1846, Thomas Babington Macaulay conjured up a vision where Britain, fuelled by the foodstuffs of the world, concentrated on flooding the globe with its manufactures, gaining 'almost a monopoly of the trade of the world' while 'other nations were raising abundant provisions for us on the banks of the Mississippi and the Vistula'.[26]

This dream of offshoring food production would remain just that – a dream – in the immediate aftermath of repeal. The great costs of importing food from halfway round the world meant that British farmers retained their commercial advantage even though the tariff had been abolished. But by the end of the 1850s mechanised agriculture, railroads, telegraphs, screw-propelled ships and, of course, the elevators had slashed the costs of prairie grain; it flowed through Chicago and Buffalo and onwards to the rapacious markets of New York and Liverpool, where it was sold at competitive prices to power the industrial revolutions of North America and northern Europe. The Chicago experience announced a new, more efficient way in which commodities were traded around the world, annihilating distance.

It also meant that the gold of California flowed into the burgeoning west. On one day in 1855, for example, when news reached Wall Street that the French harvest had partially failed, a record amount of business was transacted: 450,000 bushels of wheat, 125,000 of rye, 100,000 of corn and 75,000 barrels of flour were traded on a single Saturday. Californian gold brought by railroad and steamer to New York via Panama was then sent to the west to pay for the vast quantities of grain. Lots of gold in circulation lowered the costs of borrowing, meaning more railroad construction, faster urbanisation and fevered property speculation.

The Times of London urged investors to take note of these developments. With $400 million worth of American railroad securities, not only did British bondholders virtually own railroads such as the Illinois Central, but they also lent vast sums to the western states. Demand from Europe meant more gold going west, more land coming under cultivation and more migration; in turn, the value of land owned by railroad companies went up and the volume of passengers and freight on the rails increased. A single day's trading in New York reveals an interconnected world: the effects of poor weather in France were felt on the prairies, in Chicago, on the wharves of the Atlantic ports, in the docks of Liverpool and at the London Stock Exchange.[27]

The globalisation of food production heralded the beginning of a revolution. When that stream became a torrent, when refrigerated freight ships transported meat across oceans, it would fundamentally alter the economies and social structures of Europe. By the 1870s income from agriculture was falling in Britain, undermining the wealth of the aristocracy. Many noble families turned to another American export: wealthy brides.

Why should not the magic that had transformed Chicago and Illinois in the 1850s spread elsewhere and at a quicker pace? Superior City and Chicago were roughly the same distance from Buffalo by steamer across the Great Lakes. When railroads and grain elevators came to Minnesota and Wisconsin – which was surely imminent – a great metropolis (it might be Superior City) would spring up in the region as the commercial hub of a newly opened and rich agricultural region and follow or surpass the trajectory of Chicago. Laurence Oliphant reckoned his investment would make him rich beyond his wildest dreams.

His dreams were shared by many, many others. Minnesota's population of 6,000 in 1850 exploded – there really is no other word for it – to 53,000 by 1855. Two years later an extra 100,000 had been added to the total. It was a rate of growth comparable to California. No other western territory had grown so fast so quickly. That was what attracted Oliphant there in 1854, to witness the incredible transformation.

From Superior City he travelled by Indian bark canoe through the little-known rivers and lakes of northern Minnesota, guided by natives. The headwaters of the Mississippi, not too distant from its source, swept them south. They shot the rapids in their canoes and paddled downriver, bound for the new urban centres of Minnesota, the frontier of civilisation. Oliphant looked at the lands, still populated by Indian tribes, and pictured in his mind the cities, sawmills, canals, steamboats, railroads and fields of wheat stretching to the horizon that would turn his investment in Superior City into a fortune.

And then he arrived in St Anthony Falls, the first of a chain of new cities lying along the headwaters of the Mississippi. St Anthony and St Paul, a few miles downriver, had, in well under a decade, erupted out of the ground. In the year of Oliphant's tour another city, Minneapolis, began its life in the vicinity. Waterpower and tourism were the mainstays of these settlements: the Upper Mississippi provided 'the most beautiful river scenery in the world' as it cut between wooded bluffs which seemed 'like the ruined walls of some gigantic fortress' and its force powered a profusion of sawmills. Oliphant described St Anthony as a pleasant town set in scenery of 'extreme beauty'. Picturesque riverside villas sat alongside sawmills and foundries, from which arose 'an incessant hubbub' – for Oliphant the 'delightful music' of progress. This was a decade when 40 million acres of forest in the US – the size of England – were stripped bare.[28]

Timber, it was said, 'is the fundamental element of colonial growth'. It was ravenously consumed in thousands of boom towns throughout the Anglo world, in Australia, Canada, New Zealand, California and across the vast extent of the west. It was particularly needed in the treeless prairies and plains for building houses and fencing land. It was required for railroad sleepers, telegraph poles, bridges and as fuel for locomotive and steamboat engines. Coopers were in need of timber to make the barrels that were to contain the foodstuffs that fuelled settlement. Minnesota's forests of towering white pine – the most desirable north-country timber – were described as 'perhaps the most extensive in the world'. The territory's many rivers made the lumber highly accessible.[29]

Throughout the winter crews of lumberjacks from Maine, the traditional home of the timber industry, and also from Britain, Ireland,

Lumberjacks freeing a logjam in Minnesota

Quebec, Scandinavia and Germany, braved freezing weather and primitive living conditions to fell the mighty white pine. This was winter work, when the bogs were frozen and the ground hard enough for horse or ox teams to drag the trunks to slipways on the banks of the rivers. The roving crews of lumberjacks inhabited a rough, lawless world far beyond the frontier; they epitomised the wildness of the west as surely as cowboys would in the succeeding generation.

When the frozen rivers flowed again, when the winter snows melted and swelled those waterways, they became clogged with hundreds of thousands of logs – a dangerous, tumultuous rush of timbers crashing and bashing against each other as they coursed downriver. Sometimes an ominous roaring sound could be heard for miles around the river. It emanated from logs grinding against each other, a stuck mass of wood, sometimes with log piles thrust thirty feet into the air above the water, and backed up for miles. These logjams could only be liberated by daredevil men who ventured onto the angry, rumbling, shifting mountain to remove logs or place dynamite among them. Then they had to clear fast before the logs burst free and cascaded in deafening confusion on their way again.

At last the tide of pine was collected at immense booms and the logs – their ownership indicated by brand marks – were sorted and sent to the appropriate sawmill. In 1857 140,000,000 board feet of logs massed at the St Croix River boom alone. Water brought the white pine to the booms, and the same force made it into something useable. The sawmills of St Anthony and other Minnesotan river towns transformed the lumber into planks, shingles and laths, which were then rafted downriver to markets in the Mississippi valley.[30]

St Paul, the capital of Minnesota, a little further downriver from St Anthony, stood on a high bluff overhanging the Mississippi, 'its handsome houses and churches crowning the heights, and a fleet of steamboats moored at their base'. St Paul was just five years old and for Oliphant it was 'the best specimen to be found in the States of a town still in its infancy with a great destiny before it'. The city boasted four daily and four weekly newspapers, more than Manchester and Liverpool put together. Wooden dwellings were giving way to brick buildings, some with three or four storeys; there were innumerable warehouses along the wharves and as good a selection of shops as anywhere in the US. St Paul was booming because it commanded the head of the navigable Mississippi. It was from there that the produce of Minnesota was exported by way of the river to the South or the booming frontier towns, or to Chicago via the railroad at Dubuque.[31]

Property prices in St Paul were skyrocketing. One booster told of a friend who had invested $3,000 in real estate and within a few years had sold half his portfolio for $60,000. 'My ears at every turn are saturated with the everlasting din of land! land! money! speculation! saw mills! land warrants! town lots! &c &c' recorded a journalist from Pittsburgh. 'Land at breakfast, land at dinner, land at supper, and until 11 o'clock, land; then land in bed, until their vocal organs are exhausted – then they dream and groan out land, land!'[32]

He was witnessing one of the most frenetic real-estate booms in

Minnesota at the height of the boom: a real-estate dealer in Minneapolis open for business and waiting for customers, 1856. Such a scene of smooth-tongued, sharply dressed estate agents on the frontier was typical across the American West and Australasia.

history. In 1854 speculators put pressure on Congress to sell land on a reserve on the west side of the river. St Paul gained 4,500 acres and Minneapolis 20,000, speculators paying just $1.25 an acre. As a result the latter city mushroomed, industrialism took off and speculators made fortunes. Seemingly everyone was in on the game. Harriet Bishop, for example, came to Minnesota to teach, but she was infected with the mania. She made numerous trips back east, with a clutch of deeds for sale and glowing descriptions of Minnesota. During this incredible building boom a sawmill at St Anthony could sell 20,000 shingles, 25,000 laths and 50,000 board feet of timber in a single day.[33]

The value of land was rising so quickly not just because of the influx of migrants. St Paul was expected to be the railroad hub of the west very shortly, with lines extending north to the Great Lakes at Superior City, east to Chicago and New York and south to New Orleans, transporting the bounty of the Midwest to distant markets. The first survey of a future railroad to the Pacific, undertaken in 1853, followed a route from St Paul to Puget Sound. The city bid fair to become the crossroads of American trade and the great gateway metropolis of the west. With a future like that, well might Oliphant describe the diverse population of St Paul as 'happy in the anticipation of fortune making'.[34]

There was already evidence of 'opulence and luxury' in St Paul mingling with the roughness of a typical juvenile frontier settlement. Oliphant sampled chilled cocktails in the bars of St Paul – mint juleps, gin slings, cobblers, red lions and white lions – which he sipped through straws as he conversed with Minnesota's public officials, politicians and well-to-do businessmen. He also visited a bowling saloon frequented by the 'roughest characters from all parts of the West'. There, from morning to night, shouts of hoarse laughter, imaginative swear words and 'the booming of the heavy bowls' were intermingled: 'you come out stunned with noise, and half blinded with tobacco smoke'. Another journalist called St Paul 'the liveliest town on the Mississippi . . . continually full of tourists, speculators, sporting men, and even worse characters, all spending gold as though it were dross'. In one year alone the city's four hotels registered 28,000 guests. In the streets Oliphant heard a babel of languages and patois – Yankee,

Scots, Irish, English, French, Chippewa, Sioux, German, Dutch and Norwegian.[35]

The contrasts that these cities afforded clarified the experience of progress. Oliphant noted the 'utter wildness' of the country surrounding St Paul. There could be no stronger testimony to the breakneck growth of the city, he said, 'than the fact that the country in the immediate vicinity is still in a savage state of nature'. Food and groceries, household items and clothes, luxuries and necessities arrived on the hundreds of Mississippi steamboats that came upriver every year. It seemed that such cities represented islands of civilisation parachuted into the emptiness of the wilderness, a truly urban frontier, one that was fed by umbilical cords that extended from settled areas.[36]

But it was the juxtaposition of modernity and indigenous people that really impressed visitors. One Minnesotan booster called St Paul 'the dividing line of civilized and savage life'. You could look across the river, he said, 'and see Indians on their own soil. Their canoes are seen gliding across the Mississippi, to and fro between savage and civilized territory.' Henry Wadsworth Longfellow's *The Song of Hiawatha* was published in 1855, and it was seized upon by boosters in Minnesota to publicise their territory and imbue it with a magical allure.[37]

In the early years of St Paul's existence a lively – and indispensable – trade was carried out between Indian tribes and the urban pioneers. Blankets and guns were traded for food, lumber and furs. Just before Oliphant arrived in St Paul a party of Chippewa braves had ridden into town to trade and then departed immediately on the warpath. Visitors to such towns liked nothing more than to see the vivid juxtaposition of Native Americans and fur traders in Indian garb with the ultra-modern. It provided the subject of countless paintings and prints. Nothing else seemed to emphasise so starkly the remorseless nature of progress.

On the other side of the world a similar urban phenomenon was at work. Neither Australia nor New Zealand saw the same kind of explosion of towns. The British preferred colonialism at a more measured, controlled pace. In other words, the American experience was one of unrestrained laissez-faire, while Britain – supposedly the home of free-market values – kept the free play of market forces tightly reined in. Crown lands were sold at high prices, to the intense annoyance of people who wanted cheap land (as in the US), so that these colonies could attract a class of yeomen farmers. Land in Australia and New Zealand was sold at a high rate, in part to raise revenue to subsidise the travel costs of poor migrants, but also to keep out American-style 'land sharks'. The policy encouraged the concentration of population in a handful of gigantic cities such as Adelaide, Melbourne and Sydney. These metropolises contained at least a third, perhaps as much as 50 per cent, of the total populations of their colonies. Small towns and small farms were rarities.

Over in New Zealand the experience of most pioneering colonists was urban, not rural. In the 1850s the situation was one of 'congeries of communities dotted about along the coasts, separated from each other by hundreds of miles of sea and long, interminable land tracks'. These cities were founded and planned by emigration associations. The New Zealand Company created Wellington, New Plymouth and Nelson. Dunedin on South Island was founded by the Otago Association, a company representing the Free Church of Scotland. Dunedin is the Gaelic name for Edinburgh, and the city was intended to recreate an idealised, godly Scottish metropolis in the Pacific. The Church of England, working through the Canterbury Association, founded Christchurch. The site of New Zealand's first capital, Auckland, was chosen by the governor. This was systematic top-down colonisation at work: settlements, in time, would radiate out from predetermined urban hubs and colonise the wilderness.[38]

Economic interaction with the Maori was essential for survival during these formative years. This was not the case in Australia, where the Aborigines were crushed under the juggernaut of Western expansion. Feeding nascent Pakeha cities was lucrative business, however, for Maori. They did all they could to foster the development of urban centres. Thousands of Maori trading canoes supplied Auckland with potatoes, maize, wheat, flour, green vegetables, fruits, fish, oysters and meat, along with fuel, building materials and straw in the early 1850s. They brought items for export as well, such as kauri gum (used for varnish), flax and whale products; both the Californian and Victorian gold rushes were in part sustained by Maori-grown or produced foodstuffs. Some Maori began settled agriculture or built flour mills; others purchased ocean-going ships to trade between colonial port towns. A testament to their success is the fact that Maori paid 60 per cent of North Island's import duties. In 1852 the governor wrote that Europeans and Maori 'form one harmonious community, connected together by commercial and agricultural pursuits'.[39]

That was to look at things through rose-tinted spectacles. But it is true to say that the nature of colonisation in New Zealand and Minnesota in the 1850s created a unique situation. An urban frontier did not encroach on ancestral lands to the devastating extent that a farming frontier or a logging frontier or a mining frontier did later. Instead, concentrated Pakeha populations lacking subsistence provided numerous opportunities for profitable trade. As in Minnesota, the spectacle of indigenous peoples on city streets was a much-discussed exotic feature of the decade. Sarah Tucker wrote with pride of the bustling New Zealand cities, their lines of English houses, shops filled with English merchandise, and grand neo-gothic public buildings. What added piquancy to the scene were, as she put it, 'those noble-looking men of a darker hue' who freely mingled with the European settlers and were, in her eyes, amalgamated. What impressed her most was that only a few years ago these very English settlements had been the scenes of bloodshed. Here, it seemed, was the flourishing of multi-racial co-operation and mutual profit under the aegis of the Empire and the Church.[40]

The tens of thousands of sightseers who, like Oliphant, passed through St Paul every year in the 1850s were swept along with the excitement of the urban frontier: here were cities entirely of the nineteenth century, unencumbered by history; here the future was being written in front of your eyes. St Paul, Minneapolis and St Anthony embodied the restlessness, the exuberance of the era as much as, or more than, Melbourne or San Francisco. St Paul, to my mind, is the spirit of the 1850s in microcosm. Here people from all parts of the globe came and settled, not for something tangible like gold, but full of confidence that they stood in the vanguard of turbo-charged progress. With its swagger and overweening confidence, its combination of frontier roughness and metropolitan elegance, St Paul epitomised the heyday.

6

The Hashish of the West

KANSAS

Only those who lived through the 'flush times' will ever know what they were. Everybody seemed inoculated with the mania, from the moneyed capitalist to the humble laborer.

J. Fletcher Williams[1]

The productions of slave labor have advanced the United States a century ahead of what they would have been without it. It has built up cities and towns in America and Europe, where there were none, and imparted energy to commerce, trade and civilization over the world.

New York Herald[2]

'Time's money, time's money!' muttered a tall, thin man with an enormous dagger concealed under his ill-fitting suit as he agitatedly paced the steamboat. He had barely enough time to conduct business in Alabama before racing home to east Texas to oversee his slaves for the cotton-growing season. His fatigue and impatience, the 'pale and harassed' demeanour of his wife, were testaments to a pace of life dictated by the ferocious cotton boom of the 1850s. Irritated by the delay in loading his bales of cotton and the vessel's frequent stops, he growled, 'Time's worth more'n money to me now; a hundred per cent more.'[3]

As this careworn couple, and thousands of other white enslavers, could have testified, the Southern states were every bit as energetic and dynamic as those of the north. The vibrant frontiers of the world – such as Australia, California or the Midwest – found their dark reflection in the expanding cotton frontier of the American South. A quarter of a million free white people trekked across the continent to seek new opportunities in the glorious golden west; an identical number of African-Americans migrated across the States. They were marched or shipped off as slaves, uprooted from their homes and torn from their families, in order to turn the south-west corner of the US into one of the most flourishing regions in the global economy.[4]

Like the northern and middle west, huge amounts of land in the American south-west were being opened up for exploitation by continually expanding networks of railroads. In the previous decade the Northern states built 7,000 miles of rail, the South just 1,479. But the 1850s saw 10,000 extra miles laid in the slave states. 'Upon the face of the globe,' declared one triumphant Southern journalist, 'there is not so stupendous a railway network as that in embryo, which is to embrace in its circle and ultimately develop every foot of southern soil.'[5]

Rather than declining in the face of modernity, as many believed (and still believe), slavery was reaching its grim apogee in the 1850s. During the decade demand for cotton exploded. By its end, over a billion pounds of raw cotton left New Orleans, Mobile and Charleston bound for the 2,500 textile factories clustered in Lancashire. And still the worldwide demand remained unfulfilled. Only by coercing and whipping millions of black men and women could this growth be sustained. Slaves in the 1850s picked cotton three times faster than their grandparents; free labourers simply would not put up with such a ferocious pace of production, whatsoever the wage. The *Economist* pointed out that the southern American planter was one of the most entrepreneurial and fiercely competitive capitalists on the planet.[6]

But that journal was rare in its bluntness. According to Josiah Nott, the Alabaman physician and racial theorist, few people in the west liked to admit that 'the immense superstructure of wealth and power' that sustained the heyday of the 1850s, and the 'vastness of [the] industrial fabric' that was being created in northern America and Europe was 'reared upon the foundation' of American slavery.[7]

Two million enslaved African-Americans, Nott argued, set in motion the cogs of capitalism, stimulating heavy industry, employing 7 million free workers in northern America and Lancashire, and making fortunes for the brokers and ship owners of Liverpool, the Manchester millocracy and the financiers of the City of London and New York. The cheapness and abundance of American cotton sparked rapid industrialisation first in Britain, then America, France, the Low Countries and Germany. 'What raised Liverpool and Manchester from provincial towns to gigantic cities?' asked the British civil servant and Oxford historian Herman Merivale. 'What maintains now their ever active industry and their rapid accumulation of wealth?' His answer was simple: slavery. Liverpool's and Manchester's 'present opulence is as really owing to the toil and suffering of the negro, as if his hands had excavated their docks and fabricated their steam-engines'.[8]

Increasing exports of cotton helped enrich the West in the 1850s; the lusty growth of the decade had as direct an effect on millions of African-Americans. The bodies of these enslaved people provided the collateral for credit raised on Wall Street and in the City of London. The creation of complex financial instruments by the world's money markets during the boom transformed the South's economy from relative decline in the 1840s to boisterous growth in the 1850s: awash with credit, the cotton frontier pushed onwards, land was cleared, track laid down and production doubled to 4 million bales. And the world wanted more with every passing year. 'Slavery on the North American continent has extended, is extending and will extend,' adjudged the London *Times*.[9]

Investing in slaves, directly and indirectly, was a shrewd bet in this decade. Most investors in New York or London never burdened their souls by lending explicitly on slaves; their investments were repackaged and passed down the chain till the credit reached Louisiana or east Texas. With so much money bound up in cotton futures, and interest payments to keep up with, the potential of every scrap of land had to be ruthlessly extracted; time really was money for those who found themselves sucked into the vortex of modern capitalism in the 1850s. The result, for millions of African-Americans, was horrendous: forced migration tore apart families, diseases on the cotton frontier killed thousands, and the muscle-power of slaves had to be exploited harder and harder in order to realise the heavy investments made on it. 'When the price [of cotton] rises in the English market,' wrote the escaped slave John Brown in 1854, 'the poor slaves immediately feel the effects, for they are harder driven, and the whip is kept constantly going.'[10]

Growth on the gargantuan scale of the 1850s depended only partly on the enterprise of white Anglo-Saxon men and amazing new technologies. There was, said Josiah Nott with chilling candour, 'an indissoluble cord, binding the black [slave] . . . to human progress'. Abolish slavery, he argued, and cotton yields would crash, dragging down the world economy with it.[11]

Nott wanted, with one breath, to elevate American slavery's global importance and warn against the dire effects of tampering with it. His words reveal something very important about the heyday. Rapid progress came with new technologies, long-distance migration, gold discoveries and territorial expansion, to be sure. But the upward trajectory of growth was vulnerable to sudden upsets: a crisis in one part of an increasingly interdependent world could reverberate across the planet. Investors, brokers and manufacturers in northern England always looked with an uneasy eye on the volatile internal politics of the US. In the same way, the health of the American economy depended on the vitality of its key export markets. The year 1854 marked (in hindsight) a transition from the freewheeling confidence of the early decade to a darker, more uncertain time. Events in the United States and in Europe threatened to bring the good times to an abrupt end.

When Laurence Oliphant sat and drank cocktails in St Paul or entered the smoke-filled bowling alleys in 1854, he listened to, and joined, heated and sharply polarised discussions about the future of America. The thunderous arguments focused themselves on one particular place: Kansas.

Prosperity in the American West depended upon continual expansion, railroad construction and city-building. The same was true of the South: the cotton frontier had to expand at the same sort of galloping rate or suffer decline. In common with Josiah Nott, many in the South saw themselves as the motor of worldwide economic growth in the heady 1850s. As the poet John Greenleaf Whittier put it in 1854, cotton had become the 'hashish of the West', a plant capable of creating hallucinatory visions of the South's glorious destiny. With cotton booming with unparalleled intensity, 'there never was a time full of hope for the South and for the maintenance and the extension of slavery', as the *Montgomery Mail* put it.[12]

But at the moment of its explosive growth, the restless cotton empire found itself trapped. It was alike restrained from expanding its frontier and facing political eclipse by the North. The problem went back a generation, to the Missouri Compromise of 1820. According to the Compromise, slavery was excluded from any new territories or states created north of a line drawn west of Missouri along parallel 36° 30', and permitted south of it.

The Compromise worked well enough for thirty years or so. Every time a new free state was created, a slave state joined the Union so that North and South were exactly balanced in the Senate and the presidential Electoral College. But the velocity of change unleashed in the 1850s was of a different order. The first test came with the admission of California as a free state in 1850, which tipped the political balance away from the South. In order to get Southern support for the new state, it was enacted that slavery would be legal in the newly acquired territories of New Mexico and Utah. The South was further mollified with a draconian Fugitive Slave Act. As the tides of emigrants surged westwards in the 1850s it was clear that there would be a profusion of states created north of the Missouri Compromise line. More and more senators and congressmen would come to Washington from these new states. And, in the imaginations of the Southerners, they would bring with them abolitionist sentiments. Outnumbered politically, the South would be powerless to defend its interests. The South had become, the *Montgomery Mail* boasted, the motor of American economic growth for the first time in the 1850s. 'For the first time, too,' the paper continued, '[the South] sees her fanatical enemies clothed with full power to do their will, in the House of Representatives of the United States.'[13]

The test came in 1854 when the vast chunk of territory west of

Missouri was settled as two territories to be known as Nebraska and Kansas. Given the pace of migration, they would soon possess the requisite population to become states of the United States; as this entire area lay north of the Missouri Compromise line, they would automatically become free states. The South might have been economically strong, but demographic growth was most intense in the north-west. No territory south of the Missouri Compromise line possessed the population that would ensure that new southern states were created in parallel to northern and western ones. The South faced the possibility of losing its carefully fortified – and immensely strong – national political power, becoming a rump of the United States. Senator David Atchison of Missouri (a slave state bordering Kansas) said he would rather see the whole area 'sink in hell' than for it to become an outpost of the expanding Northern empire. Southern senators vetoed any attempt to organise these territories.

Kansas and Nebraska could not be left in limbo, however. The great transcontinental railroad to the Pacific would pass through this land, and it had to be settled before that could happen. Senator Stephen Douglas of Illinois, heavily invested financially and politically in the Pacific railroad, realised that Southern senators would only unlock the trans-Missouri West if the old Compromise was broken. Under the Kansas-Nebraska Act of 1854 the issue of slavery in all new territories would be left to popular sovereignty – settlers would chose for themselves whether their states would be free or slave at the ballot box. The Act effectively repealed the Missouri Compromise that had ensured the stability of American politics for thirty years until the convulsions of the 1850s. The South had what it wanted: the principle that slavery could expand once more across the frontier.

The Act sparked outrage in the North. Those opposed to slavery who looked at the map showing slave states and potential slave states beheld a daunting sight: Slave Power – the political dominance of the South – had, since 1850, burst the bounds set in a previous generation and seeped out west and north and occupied a gigantic portion of territory, dividing two islands of freedom in the north-east and the Pacific coast. Kansas and Nebraska were unsuited to cotton-growing, however. But cotton politics was heavily involved in the fight for the

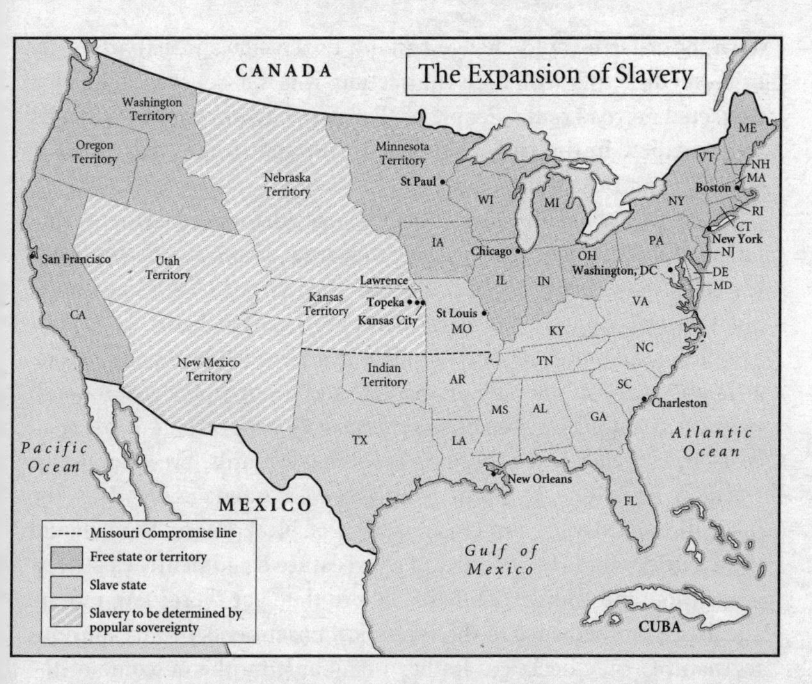

future of this enormous territory. Whether slaves would ever go there or not was not the main point: this was a bid to shore up the political power of the slave-owning South in Congress. With the future of slavery left to democratic decision, thousands of Missourians led by Senator Atchison crossed the border into Kansas. The pro-slavery 'Border Ruffians' elected a legislature that sat at Pawnee.

But then migrants began arriving in Kansas. By May 1856 over 100,000 had taken the rails from Chicago; this was population growth on the scale of Minnesota at the same time. Migrants kept on coming – as many as 1,000 a day. They were attracted by the agricultural richness of Kansas, to be sure, but the territory was ripe for speculation. Railroads connecting the territory to the main US networks, to the Pacific and the Gulf of Mexico, were being incorporated. Like Illinois the railroad map of Kansas would resemble the craquelure on the surface of an aged oil painting, and these lines would open up the territory to the global economy. Better still, these rails would soon head off towards the Pacific. Land speculation became a feral beast: a town lot or farm could change hands a dozen times in sixty days, its price skyrocketing with each transaction, if it was believed to lie on a projected railroad route. People of all classes borrowed immense sums to participate in the rush to riches. Investment from other parts of America and Britain flowed in.

Along with the multitudes of frontier speculators and farmers came fanatical abolitionists, their passage to Kansas sponsored by anti-slavery societies in New England. Of the great crowd of migrants, nine out of ten were opposed to slavery. But most settlers hated slavery not primarily as a moral abomination but as a retarded economic model, a derangement of the labour market, and a corrupt, conspiratorial political system. The stain of slavery must be kept off new lands at all costs, or else they would become economic laggards. Freedom meant freedom for whites to profit from the west unencumbered by the institution of slavery, not emancipation of black slaves. Why should these settlers abide by the rules of a legislature fraudulently elected by a few thousand 'Border Ruffians' before they got there? 'We owe no allegiance or obedience to the tyrannical enactments of this spurious legislature,' they declared. Rather than abide by the outcome of the 'stolen election' they drew up a constitution that outlawed slavery and elected their own legislature that sat at Topeka.[14]

From the end of 1855 the people of Kansas were drawn into two sharply opposed and militant factions. Abolitionists in New England began sending rifles and cannon to defend the free-soilers; groups in the South raised fanatically pro-slavery volunteers to go and enforce the authority of the territorial government. Kansas seethed with violence.

On 22 May 1856 Thomas Gladstone, a correspondent of the London *Times*, was spending the evening in the bar of a hotel in Missouri, on the western frontier. 'I had just arrived in Kansas City,' he wrote, 'and shall never forget the appearance of the lawless mob that poured into the place, inflamed with drink . . . Men, for the most part of large frame, with red flannel shirts and immense boots worn outside their trousers, their faces unwashed and unshaven . . . wearing the most savage looks, and giving vent to the most horrible imprecations and blasphemies; armed, moreover, to the teeth with rifles and revolvers, cutlasses and bowie-knives – such were the men I saw around me.'[15]

The 'drunken, bellowing, blood-thirsty demons' who crowded round Gladstone at the bar, shouting for whisky, were Missourian Border Ruffians just returned from the city of Lawrence across the border in Kansas. The Border Ruffians went to Lawrence with the purported aim of reasserting the authority of the pro-slavery legislature. Their smoke-blackened faces and clothes attested to the means by which they went about their business. The Missourian posse destroyed the town's printing presses, pummelled the Free State Hotel with cannon-balls, and burnt buildings.

A day after the Sack of Lawrence Congressman Preston Brooks entered the floor of the Senate and sought out Senator Charles Sumner of Massachusetts. Sumner had made a speech called 'The Crime Against Kansas' in which he had impugned the honour of the elderly senator from South Carolina, Andrew Butler, comparing his pro-slavery attitude to Kansas to the rape of a virgin. Brooks was Butler's cousin, and he avenged his kinsman by beating Sumner on the head with his gutta-percha cane.

A few days later the zealous abolitionist John Brown – who saw himself as the agent of Jehovah – attacked pro-slavery settlers at Pottawatomie Creek. Brown and his followers hacked five settlers to death with their broadswords. In August hundreds of Border Ruffians set out to attack and burn free-soil towns, including Topeka and Lawrence. They began their assault on Osawatomie, where they met stiff resistance from John Brown and his forty or so followers. Nonetheless, Osawatomie was burnt and looted.

During the summer of blood in Kansas the sun over the Minnesotan prairie was obscured and the sky darkened by a great glistening cloud that moved as fast as wind and from which emanated a menacing rasping noise. Then those unlucky enough to be outdoors under the cloud felt its first rain thudding onto their heads: not water or snow or hail, but a horrible, large, brown, furiously fluttering thing.

The cloud was a moving mass of millions upon millions of Rocky Mountain locusts (*Melanoplus spretus*). The air seemed alive with these vile insects as they beat en masse against buildings and bodies and oozed through cracks between windows and walls. Wheat fields were devoured whole, grassland reaped bare, trees stripped of foliage. Their excrement turned water brown and rendered it undrinkable by man or beast. The harvest was consumed and livestock perished from want of forage. Even worse, they laid billions of eggs that would hatch the next year.

Plagues of locusts always presage dark times. But the Minnesota of 1856 was as cocksure as ever. If Oliphant thought he was seeing the height of the boom in the St Paul of 1854, he would have been amazed by the scene in 1856. 'Only those who lived through the "flush times" will ever know what they were,' wrote Fletcher Williams, an early chronicler of St Paul. '. . . Farmers, mechanics, labourers, even, forsook their occupations to become operators in real estate, and grow suddenly rich, as they supposed.' They borrowed at exorbitant rates. Fast-talking real-estate dealers with 'no office but the sidewalk, and no capital but a roll of town-site maps and a package of blank deeds' made quick fortunes selling lots in paper cities, which they spent as rapidly on 'fast horses, fast women, wine and cards'. It was a 'mad, crazy, reckless' time.[16]

The idea of infinite progress itself had become the main focus of economic activity. The sawmills of places like St Anthony cut lumber that would construct thousands of boom towns across the west, including Kansas. In 1856, 1,026 steamboats arrived at the levees of St Paul in the 198 days that the Mississippi was open for river transport between the cold seasons. Many of the new arrivals to the city did not want to begin the back-breaking work of clearing land for a farm. They became waiters in the tumultuous hotels humming with tourists and speculators; they took high wages as labourers at the levees or warehouses carrying the vast amounts of goods brought upriver on the steamboats; they drove wagons; they added their muscle-power to the building boom. The literate accepted jobs as clerks and cashiers and in other professional roles in the grocery, banking, legal, insurance, railroad or real-estate sectors. In other words, the thousands of immigrants typically found employment in the industries of urban growth, preferring the effervescent boom town to the remote and labour-intensive farm. The local economy was buoyant on imports and commercial activity. The demand for money was such that it was as safe a bet for investment; the *New York Daily Tribune* recommended Minnesota 'as a good place for poor folks to lend money'. 'Perhaps in no city in America,' wrote Williams, 'was the real estate mania, and reckless trading and speculating, so wild and extravagant, as in St Paul.'[17]

Only a few voices were raised to express concern that speculation did not necessarily add up to economic progress. Cotton may have been the South's hashish; elsewhere property prices created similar intoxicating delusions. A St Anthony newspaper said that 'We must have an outlet [for exports], and an outlet by rail, and this as speedily as possible, or we are nowhere.' In 1857 the Minnesota legislature issued twenty-seven railroad charters. One of them was to the St Paul and Pacific Railroad, which received a stupendous tax-free land grant of 2.5 million acres (one of the largest in the United States) and a $5 million bond issue. The territory was in delirium. Maps were being sold in great numbers 'marked with lines in every conceivable direction, each line representing a Railroad, with towns all along each route'. Soon, Minnesota would be criss-crossed with railroad tracks, prosperity would come, and the gamble would pay off.[18]

Places such as Minnesota and Kansas depended utterly on confidence. They needed repeated waves of settlers to keep on coming with cash in their pockets to keep the price of land high. 'Bleeding Kansas' and the feast of the locusts hinted at the precarious nature of the boom: the natural and the political worlds alike had the power to throw up unpredictable limits to rampant expansion. What would happen when confidence began to ebb?

When Laurence Oliphant was in Minnesota, a Texan told him that in the wake of the Kansas-Nebraska Act the 'glorious institution' of slavery 'will shortly pervade our beauteous territory; slave-grown corn will wave upon our hill-sides'. Unlikely as it might have been, the fear was that slavery would indeed pervade much of the west as it had done elsewhere throughout the growth of America. Why go to the trouble and expense of migrating from Britain or the eastern states – always a risky business – when slave labour could undercut free labour? Why settle in lands scarred by civil war between free-soilers and enslavers? After 1856 the numbers of people emigrating to Kansas dwindled to almost nothing as terrorism took hold.

This was not a matter just for America, but for much of the rest of the world. In Kansas the two great forces of the 1850s in the States – mass migration and the cotton empire – collided with an almighty bang. British financial houses at this time had £80 million ($400 million) tied up in American railroads. They also loaned money, directly and indirectly, to faraway places such as Minnesota. Between 1850 and 1854 the value of British exports almost tripled to £97.3 million; the American and Australian markets were the fastest-growing and the most important. As America boomed, so too did Britain.

The increasing prosperity and development of the US was one of the primary engines of growth in the 1850s. But, as was so obviously the case in Minnesota and Kansas, growth was based on faith in future prosperity; it rested on an inverted pyramid of debt. It does not take much to turn heady confidence into headlong panic in the twinkling of an eye. There were plenty of portents of doom in 1856. With fewer people migrating west because of uncertainty over the future spread of slavery and outbreaks of violence, the railroad companies' income was being hit hard.

Any prospect that confidence was beginning to waver, or that America was about to be engulfed by violence, would send shockwaves through the global economy. As North and South became increasingly hostile, and when speculation became ever more risky, the possibility was all too real. Was the heyday of the 1850s about to encounter its limits?

But while the world boomed, such intimations of disaster were muted; paper fortunes accumulated and optimism ruled. Of more obvious concern to the wider world was the political instability ushered in by Kansas-Nebraska. Its fallout destroyed the second-party system and gave rise to an explicitly anti-slavery party in the North, the Republicans. As a result of their support for the Kansas-Nebraska Act, the Democrats were virtually wiped out in free-soil states and became the party of the South.

Events in Kansas and violence in Congress were watched with a sharp eye by the British press. 'Every breeze that blows to us from America comes laden with the shouts of contending parties,' commented a contributor to the *Edinburgh Review*. '. . . Mingled with the storm of words sounds yet more portentous fall upon the ear – the noise of bludgeons and the cry of Senators stricken down . . . and from the distant frontiers of the Republic the clash of arms and the fury of civil war. These are not the ordinary accompaniments of political conflict in a free state.'[19]

The potential for American politics to become shipwrecked, and for the great cotton empire to go down with it, would have global ramifications. The mills of Manchester imported 77 per cent of their raw cotton from America – a whopping 2 million pounds *a day* in 1854. As one German newspaper put it, 'the material prosperity of Europe hangs on a thread of cotton'; if slavery was abolished or America degenerated into civil war cotton production would fall by 75 per cent 'and all cotton industries would be ruined'. The industrialised world's dependence on American cotton was well known; events in Kansas, and the fallout of the conflict, revealed how dangerous this dependence was. Slavery may have made cotton cheap and abundant; but by the mid-1850s the system of coerced labour was endangering the supply of cotton because of its political toxicity and reckless bids for expansion. For many observers of international trade, after Kansas-Nebraska it was no longer a question of *whether* slavery was to be abolished in the States, but *when* emancipation would occur. And when that happened, it was no great leap of logic to imagine the hurricane that would blow through the global economy.[20]

The London money market was a sensitive barometer of impending storms; Southern states and railroad companies found that they were paying substantially higher interest rates on their loans than their Northern counterparts. 'This mistrust arises,' said the *Westminster Review*, 'from a shrewd calculation of the dangers, in both a moral and physical sense, which hang over a state of society whose foundations are laid in injustice and violence.' In the aftermath of Bleeding Kansas, the mill owners of Manchester began the long process of weaning themselves off complete addiction to Southern cotton, their hashish. The dominance of America in the supply of cotton was comparatively recent, they said, and they wanted to return to a situation where the staple came from diverse sources. And so they founded the Cotton Supply Association 'with a view to having a more abundant and universal supply'.[21]

The Manchester Cotton Supply Association sent fact-finding missions across the world, carrying cotton seeds, cotton-gins and advice on cultivation to potential producers. They sought samples of cotton from Algeria, Morocco, Malta, Sicily, the Greek islands, Turkey, Egypt, Palestine, West Africa, the West Indies, the hinterlands of Buenos Aires and Montevideo, from Peru and Tahiti. But it was India that offered the best possibility for a supply of raw cotton that would displace the American South. Before that could happen, however, there had to be an economical way of getting India's white gold to the industrialised world. Already railways, roads, bridges, tunnels and ports were being constructed on the subcontinent by the British as a way of connecting Indian producers to the world market. The hope, according to the London *Morning Advertiser*, was that 'if we could materially relieve ourselves from dependence on [American cotton], we should leave the slave-drivers with so inadequate a market for their produce as to damage their craft, and render slave-mongering a profitless pursuit'. The threat was felt in the American South, most notably by James H. Adams, Governor of South Carolina: 'Whenever England and the continent can procure their supply of raw material elsewhere than from us, and the cotton States are limited to the home market, then will our doom be sealed. Destroy the value of slave labor, and emancipation follows inevitably. This, England, our commercial rival, clearly sees, and hence her systematic efforts to stimulate the production of cotton in the East [Indies].'[22]

The Kansas-Nebraska Act, and the violence it ignited, prompted the industrialised world to seek a way out of the Faustian deal it had made with American enslavers. But the task of securing an alternative supply of raw cotton – the lifeblood of mid-century industry – on a massive scale would take years. And the Southerners would retain an overwhelming advantage even if cultivation was successfully introduced elsewhere and infrastructure constructed to bring it to the globe's industrial belt. American cotton had captured the world market in recent decades because it was cheap; and the reason for that was the immense political power possessed by its producers, their absolute control over labour and their access to bottomless reservoirs of credit. No other region in the world's cotton zone enjoyed these three mighty advantages. As one Southerner put it: 'It will be found that no large, continuous supply is possible without such a command of compulsory labour as is only to be obtained in the United States.' A British statistician concurred: the 'probability of our ultimately obtaining even the greater part of our cotton without the use of slave labour, remains . . . small'.[23]

Far from being alarmed at the world's evident distaste for slave-produced cotton, the American South became convinced of its centrality to civilisation and its invincibility. The developed world, many Southerners were coming to say, should awake to the fact that many of its staples – not just cotton, but coffee, sugar, tobacco and rice – were the produce of slave labour and could not be obtained by any other means. 'It should be England's policy to encourage the growth of cotton in this country,' declared the *New York Herald*, 'instead of wasting her means and energies in vain attempts to develop its impracticable culture elsewhere.' The industrialised world panted for more cotton with every passing year; but in order to get its fix it would have to pay a price. Demand for cotton was beginning to run ahead of supply because the South's enslaved labour force was not growing fast enough by natural means. If it was not replenished, the price of cotton and sugar would increase and industrial expansion in Europe would tail off; already the price of an enslaved African-American was spiralling upwards. In other words, the Atlantic slave trade – made illegal earlier in the century – would have to be reopened to resupply the cotton plantations in order to prevent the boom from fading.[24]

The South believed that it could make such demands because it had the rest of the world over a barrel. 'The English will soon require an annual supply of two millions of bales of cotton, and must have it,' declared the *Herald*. 'Her operatives must have bread and clothing. Her ships must find employment, to supply which she must have our cotton.' James Hammond, a senator from South Carolina, told the Senate that the South had the wealth of resources 'to make an empire that shall rule the world' through the power of cotton. The North might try and stymie the advance of slavery in Kansas, he continued, but 'can you hem in such a territory as [the South]? . . . How absurd . . . Without firing a gun, without drawing a sword . . . we could bring the whole world to our feet.'[25]

Blocked in Kansas in 1854–6, the South sought fresh fields for the expansion of the mighty empire of cotton. The boom of the heyday of the 1850s had many unpredictable results; not least was the way it emboldened the South to broaden its horizons. Its bid for a great tropical empire encompassing the Caribbean and Central America was a direct reaction to slavery's territorial containment in the United States. In doing so, it blundered unexpectedly into a war being fought far away, on another continent.

The Crimean War (1853–6) impacted on the entire planet, in direct and many surprising indirect ways. The Australian boom was halted when the clippers and steamships that fed it with migrants, news and goods were requisitioned for the war. In southern Africa the millenarian excitement gripping the Xhosa was stoked by the rumour that the Russians would come and liberate them. Wheat growers (and prospective farmers) on the prairies might have been only dimly aware of it, if at all, but the increased value and volume of their exports were being inflated by Britain and France's declaration of war on Russia early in 1854. With the Baltic and Black Sea ports under blockade, the two Western European powers turned across the Atlantic for the grain that usually came from Poland, Ukraine and the Danube delta. Exports of wheat from the US to Britain surged from 36 million bushels in 1853 to 53 million the following year, when the latter went to war with Russia. In 1854 the US exported 31,848 barrels of flour to Britain and 2,096 to Continental Europe; the following year it was 391,734 and 423,021 respectively. France had an even greater need for prairie breadstuffs than Britain, with hundreds of thousands of peasants uprooted from the land to fight on the Crimean Peninsula. Gripped by excitement and the surging demand for produce, not many people in the American West stopped to ask whether this situation was permanent as they rushed to establish farms and build railroads to carry off the harvest to distant markets. Conflict on the Black Sea fuelled the boom as much as gold.

The global significance of the Crimean War, and the way it impacted on the progressive impulse of the 1850s, is at the heart of Part II of *Heyday*. The narrative begins in the epicentre of the war, but it spirals out to less obvious places that were touched by it, including Japan, Manchuria, Hong Kong, Persia, India, Cuba and Nicaragua. The violent upheavals in Kansas coincided with war on the Crimean Peninsula. One of the least-known side effects of the war was the clash between the aggressive cotton slave empire and the British Empire in the Caribbean (described in Chapter 8). Such ripple effects provide vivid evidence of the growing interconnectedness of the world in the 1850s, in which events in Europe affected wheat growers on the prairies, businessmen in Australia and tribes in southern Africa.

But the effects of the Crimean War would be felt more profoundly and permanently by millions of people from the western tip of Europe to the islands of the Pacific. In the decade and a half following the war the map of Europe had to be redrawn several times. Just as significant, Asia was engulfed by the tidal wave. Aftershocks of the Crimean War were felt across the Middle East, the great sweep of Central Asia, India, China and Japan. The tectonic plates of world politics shifted decisively in the middle of the decade.

PART II

Fault Lines: The Age of Silver

7

The Ramparts of Freedom

THE CAUCASUS

Seven inland seas and seven great rivers
From the Nile to the Neva, from the Elbe to China,
From the Volga to the Euphrates, from the Ganges to the Danube
This is the Russian Empire, and it will never pass away.
Fyodor Tyutchev[1]

After weeks on a British warship observing from the sea with frustration 'the most exquisite coastal scenery to be found anywhere' on the planet, Laurence Oliphant was at last able to disembark. 'We landed under the ivy-covered battlements of the old castle of Zikinzir . . . The country was everywhere clothed in the richest verdure. Here and there the hill-slopes, waving with long rich grass, terminated abruptly in precipitous walls of rock.' Behind the castle rose wooded mountains and beyond them a range of forbidding snowy peaks. 'It was a fairy-like scene.'

Oliphant, fresh from Minnesota, was in one of the world's most romantic and – for Westerners – untrodden regions: the Caucasus Mountains. The dramatic scenery Oliphant glimpsed from the paddle frigate HMS *Cyclops* 'gave rise to a longing desire to penetrate into the mysteries of their gloomy recesses'. And pretty soon he got the chance to scale those peaks. He gave a bird's-eye view:

Upon our left rose in majestic grandeur the snowy peaks of the towering Caucasus, and a flood of golden light bordered their irregular outline.

> Lower down, the glaciers met the dark green of the pine forest . . . From these gushed boiling torrents, and forced their way through narrow gorges, which expanded at our feet into winding valleys . . . hamlets were embowered amid fruit trees and orchards; and the streams, like threads of silver, no longer swept seethingly beneath overhanging rocks, but rippled calmly under the drooping foliage which kissed the water.[2]

The breathtaking scenery of the Caucasus and its impenetrable mysteries had acted on the human imagination since the dawn of recorded history. The mountain range marked the end of the known world for the ancient Greeks, nature's very own frontier. Chained to a wall of rock in the Caucasus, his liver pecked out by an eagle each day and for all eternity, Prometheus received his punishment for giving mortals the gift of fire. It was to this mountain fastness that Jason ventured to wrest the Golden Fleece from King Aeëtes. The elusive female tribe of Amazonian warriors resided somewhere here. Beyond the Caucasus, according to the Qur'an, lay the evil hordes of Gog and Magog; when the great metallic wall in the mountains crumbled these people would breach the barrier and flood the world, heralding the End Times. The mountains checked Alexander the Great's conquest of the world.

The Caucasus had lost none of its romantic allure or exotic strangeness in the nineteenth century. It was one of the great frontiers of the world: the barrier between Europe and Asia, between Christianity and Islam. On one side was the Russian Empire; on the other Persia and – further still – vast and little-known tracts of land known as 'Tartary' that stretched to China and the Pacific. The Caucasus was the 'gate of the Orient'.[3]

No less enticing were the peoples who lived here, the Islamic mountain tribes. The mountain range contained several distinct groups. In the east, near the shores of the Caspian Sea, were the Chechens and the tribes of Dagestan. On the western part of the range were Mingrelia, Abkhazia and, most famously, Circassia. For Europeans, all the many tribes of the Caucasus often went under the general heading of Circassians, whether they came from that region or not.

Western writers romanticised the peoples of the Caucasus as rugged, noble highlanders who led lives of frugality and whose men were chivalrous mountain warriors. Their irenic fortified stone villages – known as *auls* – clung magnificently to the southern slopes of the mountains.

It was with considerable excitement that Laurence Oliphant got his first glimpse of the fabled Circassian mountain warriors in the western margin of the Caucasus. Their dress was instantly recognisable from countless written descriptions and engravings. Most distinctive were their fur caps, as tall as a guardsman's busby, and the *chokha* – a tight-fitting, high-necked, knee-length woollen tunic with rows of ivory cartridge holders sewn onto the chest. Pistols and daggers (known as *kindjals*), embellished with silver decorations, and a sabre known as a *shashka* hung from their belts. They wore red or yellow trousers and red leather slippers, like an American moccasin. No male Caucasian was seen without a rifle slung over his shoulder. They were famed as sharpshooters and reckoned among the best fighting horsemen in the world. A Circassian boy was taught to ride like a centaur, it was said, to gallop up and down steep precipices and 'to leap the chasm and swim the torrent'. He was expected to unsling his rifle while riding at full tilt, hit a distant target and return the gun to his back, all without slowing. A mounted Circassian avoided enemy fire by hanging low from the saddle; he could swoop down to grab an object from the ground without reducing speed.[4]

Oliphant journeyed to this remote region in the autumn of 1855, not as a sightseer but as a secret agent at the height of the Crimean War. When war was declared between the planet's great superpowers on 28 March 1854 it was widely regarded, in Britain, as a war of the 1850s: at stake was free trade versus autocracy; modernity versus backwardness. Britain with its open trading policies, industrial strength, steam power, communications technologies, railways, rifles and revolvers believed it was taking on a relic of the past and one of the most stubborn barriers to global progress. Both empires contested a fault line that ran eastwards from Western Europe to North America, from the Baltic to the Pacific, scrapping for control over a succession of empires – the Ottoman, Persian, Chinese and Japanese – and numerous countries between. Each believed that it had a special role to play as the regenerator of these Eurasian and Asian societies; each saw the other as incompatible with its world vision. The war started for specific reasons – which we will come to – but for both sides it was fought with these global, even epochal, considerations in mind.

Oliphant travelled to the Caucasus because, for many strategists in London and St Petersburg alike, it was the key to the global struggle between Britain and Russia. The fate of the people who occupied the mountains – which one writer called the 'rocky ramparts of freedom' – was bound up for a brief, glorious moment with the geopolitics of the 1850s. And to us, looking back on the struggle between Russia and Britain, the Caucasus Mountains – commanding the intersection of Europe and Asia – provide the ideal location to assess the war in its worldwide context. For me the Caucasus, like Minnesota in a different context, is an unlikely eyrie from which the developments of the 1850s become more sharply defined.[5]

Laurence Oliphant went to the Caucasus to meet a guerrilla fighter who had become one of the most talked-about and venerated personalities of the 1850s: Imam Schamyl. Short of stature and with a full red beard, Schamyl was known to the world as the 'Lion of Dagestan' and the 'warrior prophet' of the nineteenth century, a modern Mohammad.

In the breakaway changes that engulfed the world in the 1850s it is natural that attention fixes on those places that burned the brightest – the American West, Western Europe, Britain, Australia, New Zealand. But Russia too was enjoying its own very different heyday and, like the other great powers, rediscovering itself and developing a civilising mission on its frontiers.

The Russian Empire was expanding at a phenomenal rate in the nineteenth century. Over the previous decades its frontier had pushed west by 800 miles, advanced 450 miles towards Constantinople, 300 miles against Sweden and within 1,000 miles of India. It occupied much of Poland. Thousands of miles away and on a different continent, Russia secured control over what would become Alaska. As China fell into disorder after the Opium War the Russian army in Eastern Siberia began to weigh up what territorial gains could be made in East Asia from the chaos besetting the region. Russian claims

on the Kuril Islands established a frontier with Japan, one that provokes tensions to this day. The archipelago gave Russia a springboard into the world's greatest ocean. In 1853 General Nicolai Muravev, Governor of Eastern Siberia, declared that 'It is highly natural . . . for Russia if not to rule all of Eastern Asia, so to rule over the whole Asian littoral of the Pacific Ocean.'[6]

But key to understanding Russia in the nineteenth century are its fault lines with the Muslim world. Russian identity was bound up with its wars of conquest on its frontier – the frontier of Orthodox Christianity and Islam. Throughout the eighteenth and nineteenth centuries the Russian Empire pressed against the weakening Ottoman Empire, a vast power that encompassed in the nineteenth century a semicircle of territory from the Balkans through Romania, Bulgaria,

Turkey, the Middle East, Egypt and Libya. Under Catherine the Great the Crimean Khanate was wrested from Ottoman suzerainty and Novorossiya – New Russia – was created; Russia was now a power on the Black Sea. She populated this large and strategically vital conquest with Russian, German, Ukrainian, Romanian, Bulgarian, Serbian, Greek, Polish and Italian colonists. To crown her victory – that of Christian civilisation, as she saw it, over Islam – Catherine founded modern cities built in rococo and neo-classical styles. Odessa was the queen of these new cities; Sevastopol became Russia's naval base on the Black Sea. Bessarabia (in Moldova and Ukraine) was taken from the Ottoman Empire in 1812. In 1829 Russia's dominance in the Black Sea was sealed when it gained control over the Danube delta.

It was Russia's role in the history of the world – as it saw things – to rescue beleaguered Christian communities and bring civilisation to the benighted areas of Eurasia. The Caucasus – standing as it did at the intersection of Europe and Asia with its embattled Christians and fierce, fanatic Muslim tribes – was central to this sacred mission in the world.

The Caucasus was a patchwork of numerous different tribes, khanates, principalities, religions, languages, ethnicities and social structures. For centuries appendages of the Persian and Ottoman Empires, these lands were seized by Russia in the early part of the nineteenth century. In the lowlands on either side of the mountains the established kings, khans and princes were assimilated into the Russian Empire without serious resistance. Georgia and Armenia were Christian and, in the main, welcomed Russian Orthodoxy in place of Islamic rule. It was in the high altitudes and dense forests of the Caucasus that Russia faced serious opposition.

Born to the Avar people in 1797, Schamyl was in his mid-thirties when proclaimed Imam and 'the second prophet of Allah'. He united the tribes of his native Dagestan and those of Chechnya into a state based on Sharia law, and proclaimed *jihad* against the invading Russians. Cossack settlers and Russian armies numbering in the hundreds of thousands burnt villages and destroyed crops in a series of brutal campaigns to defeat Schamyl and secure control over the region. But Caucasian guerrilla fighters, with their prowess as horsemen and sharpshooters, kept the empire at bay in the forests and mountains. Soldiers in punitive expeditions might have heard the sharp report of the musket and the terrifying hiss of the bullet as it flew past their heads; they sometimes saw the puff of smoke; but they rarely glimpsed the Chechen or Circassian guerrilla hidden in the trees or behind mountain rocks who had picked out and felled their comrade.

In 1845 Russia embarked on another incursion into the mountains of the northern Caucasus, led by the outstanding Field Marshal Mikhail Semyonovich Vorontsov, Viceroy of the Caucasus. It was a splendidly equipped and magnificently uniformed force numbering 10,000 troops; the array of nobles on the staff brought with them all the luxuries civilisation could require and all the champagne thirsty aristocrats could drink. This was no outdoor party, however: the punitive force numbered some of the Tsar's crack troops.

The Russian progress was largely unopposed. Schamyl's mountain guerrillas were nowhere to be seen and their *auls* had been abandoned and burnt in the face of the triumphant invaders. But in the exuberance of victory the Russian army had advanced beyond its supply lines. Just as Vorontsov was about to attack his main target – the *aul* of Dargo, Schamyl's stronghold – a column had to be sent back a few miles to escort provisions to the main army. It reached the wagons and pack mules without trouble.

What happened next, when the detachment returned to Dargo, passed into legend. The hitherto unseen enemy now attacked from the darkness of the ravines and forests, picking off Russian soldiers and then charging into the dazed and disorganised column, slashing with their *shashkas* and *kindjals*. The supply wagons were tipped from the sides of cliffs or abandoned as the Russians scrambled to escape the massacre and rejoin the army. Their way to safety was barred by trees felled by Schamyl's men and the bodies of their slaughtered comrades. Over the course of a day and a night 556 men (including two generals) were killed in the forests; 800 were wounded and three field guns taken.

With his food lost to the mountains and his force depleted, Vorontsov had no choice but to retreat from the gates of Dargo. The march (if it can be called that) took six harrowing days and nights. The army was menaced continually by Chechen ambushes and roadblocks; it was weighed down with wounded, numbering at first 1,100 but growing all the time. The rate of retreat was painfully slow – about four miles a day – and slaughter at the hands of marauding Chechens remorseless. The army fragmented into confused bands of soldiers reduced to near-starvation. The horrors were made all the more sinister because they were carried out to the sound of Russian military music mockingly played on drums and instruments captured by the Chechens. Just before their ammunition ran out completely, the routed, surrounded army was saved by a relief column. The Russians lost 195 officers (including three generals) and 3,433 soldiers dead and wounded.[7]

Schamyl had lured the Russians into the mountains by giving them easy victories. He waited until his enemy was high on illusionary triumphs, overextended and cut off from its supplies. Then he pounced with relentless fury. After the Battle of Dargo, Schamyl led thousands of his Dagestani and Chechen warriors deep into Russian territory. A Russian general wrote that Schamyl's state had developed a 'religious-military character, the same by which at the beginning of Islam Muhammad shook three-quarters of the globe'.[8]

A place of numinous beauty and unimaginable cruelty, of noble adventure and savage struggle, the mountains haunted the writings of Alexander Pushkin, Mikhail Lermontov and Leo Tolstoy. War against the Muslim guerrillas had lasted a quarter of a century. The hundreds of thousands of soldiers who endured long years of privation, boredom and moments of purblind terror guarding the vast military lines and remote outposts so vividly described by Tolstoy often followed in the footsteps of fathers, uncles and grandfathers. The war sucked in generations of noble families; aristocrats went to the mountains seeking adventure and glory and returned scarred by the brute realities of guerrilla warfare. Like the Vietnam War did later for America, the never-ending campaign in the Caucasus seared itself into the Russian imagination. People were alike repelled and seduced by the mountain grandeur, the heroic resistance fighters, and by the empire's arch-enemy, Imam Schamyl.

The war was not just against Islam; it was the means to a very important end. Schamyl's stout resistance and sharp lunges tied down huge armies in stalemate. The mountain range – the 'greatest fortress in the world' – remained the barrier to Russia's global imperial ambitions. Once the Imam was defeated, the empire could advance towards Persia, the khanates of Central Asia and on to Afghanistan and China. That future seemed a long way off.[9]

When Tolstoy arrived in the Caucasus in 1851 as a young officer, the Russians were more desperate than ever to bring an end to the blood-soaked saga. Under the direction of Field Marshal Vorontsov, Russia began to reorganise those parts of the region under its control. Vorontsov had succeeded in rapidly modernising Novorossiya when he was governor there. He brought the same zeal to the Caucasus: schools were set up and integrated into the imperial system; libraries and museums were instituted; and a new administrative structure was brought into being.

The centrepiece of the civilising frontier was Tiflis (now called Tbilisi), the capital of Georgia. It was transformed from an outpost garrison town into a flourishing urban centre, remodelled with boulevards and grand European-style buildings on Rustaveli Avenue. Most symbolically of all were cultural centres and places of Orthodox worship, demonstrating Russia's commitment to enlightenment and religion. The interior of the Sioni Cathedral was renovated in the 1850s, with new murals and a stone iconostasis. The famous Tiflis opera and ballet house opened in 1851 with a performance of Donizetti's *Lucia di Lammermoor*. The national library building opened in the same year. Tolstoy described Tiflis at this time as a 'very civilized town which to a great extent apes St Petersburg and greatly succeeds in the imitation'.[10]

Vorontsov had, at Dargo, been bloodied by Schamyl. The tactic of holding the line and launching annual punitive raids on the untameable mountaineers was clearly not working. The time had come for a new strategy. And Vorontsov had a man for the job. Prince Alexander Ivanovich Baryatinsky, known as the 'Muscovy Devil' to his Muslim enemies, was appointed commander of the Left Flank in the Caucasus in 1851.

Baryatinsky was a towering figure whose pronounced limp attested to the wounds he had received almost two decades previously as a mere boy when he led his men against a Caucasian tribe and returned as the sole survivor of a vicious fight. He was just thirty-six years old in 1851 and his scars, closely cropped hair and deliberately shabby uniform emblazoned with medals and decorations bespoke a dedicated, austere warrior. But this was a recent conversion. After his youthful heroism he had won renewed fame, this time as a dissipated noble and seducer *par excellence*, the terror of the husbands and fathers of St Petersburg. The prince was at the centre of Tsar Nicholas I's court; he was particularly close to the Tsarevich Alexander and to the Tsarina. When he reached his thirties he longed to be distinguished for martial rather than marital conquests. Subduing the Caucasus, vanquishing the Islamic revolt, was not just a matter of honour or realpolitik for the Tsar; it was a religious duty. If Schamyl was waging *jihad*, Nicholas was on a crusade. Baryatinsky came to understand his master's will; he devoured every book on the Caucasian War and worked tirelessly with the military staff. His dedication to the cause, and his newly acquired soldierly mien, reflected the determination to win at all costs.[11]

Baryatinsky realised that victory would come one battle at a time. Winning the Caucasus was like besieging a gigantic fortress, and it had to be approached in that way, as a game of patience and will. Above all, he should avoid head-on confrontations with Schamyl's Mujahidin, battles that were always bloody and rarely successful.

The new commander won over tribal leaders antipathetic to the prophet warrior. Those who came over to the Russians were treated with respect; their *auls* were rebuilt and their farmland secured. Most dramatically, Baryatinsky negotiated the defection of Schamyl's lieutenant, the legendary Hadji Murad, in 1851, a story told so vividly by Tolstoy.

That was the carrot, the battle for hearts and minds. The second prong of Baryatinsky's approach was to remake the geography of the battlefield – literally. Across the Caucasus the ravines echoed to the rhythmic thuds of axes wielded by Russian sappers and 100,000 soldiers drafted in to deforest the region. When the thick mountain beeches resisted the axe they were blown out of the mountainside by explosives. The smoke of burning branches hung over vast swathes of mountain where tangled forest had been turned into a desolate wilderness of tree stumps and bare earth. In his short story 'The Wood-Felling' Tolstoy describes a landscape that has become unrecognisable: 'Instead of the thick outskirts of the forest you saw before you a large plain covered with smoking fires and cavalry and infantry marching back to camp.'

The legions of troops engaged in this war against nature were defended by a ring of artillery and cavalry. Every tree that crashed to the ground deprived the Chechen sniper of cover. Schamyl's wooden ramparts began to recede in front of his eyes. If he wanted to fight the invaders he would have to do so in the deforested plains and be mown down by regular army units.

Nature subdued, Baryatinsky could deploy his full strategy. New military roads connecting beefed-up forts and outposts cut deep into Schamyl's territory. Bridges spanned the ravines. In some places there were tunnels. The army was now a mobile force and deforestation spared it incessant, morale-sapping sniping from the guerrillas. Communities allied to the Russians settled where the forest had been hacked away, to defend the area and provide food for the army as it shuttled along these new roads.

The tactic paid off. Schamyl was prevented from securing control over the entire highland spine of the Caucasus range from the Caspian to the Black Sea; Vorontsov and Baryatinsky drove a wedge between him and his would-be allies in Circassia. Those who came over to the Russians were treated well; those who continued to resist were massacred or simply cleared off the land. It was a means of war we know today as ethnic cleansing. Tolstoy wrote in his diary that this kind of war was so 'ugly and unjust that anybody who wages it has to stifle the voice of his conscience'.[12]

Schamyl survived in his redoubt. But his days of resistance were numbered as the forests shrank and his enemies closed in. In 1853 Vorontsov's success in the Caucasus was apparent. One after another, Chechen villages deserted Schamyl's independence movement so as to be spared ethnic cleansing. Nicholas I was able to draw down his troops in the Caucasus and transfer them to the Danubian frontier for a new crusade against the forces of Islam.

In June 80,000 Russian troops crossed the Pruth River into the Ottoman principalities of Moldavia and Wallachia (modern-day Romania). Nicholas believed he was acting on behalf of the 12 million Christians who suffered, as he saw it, under Islamic tyranny in the Ottoman Empire. More immediately, he was responding to events. The Ottoman Sultan had, in the face of French naval threats, transferred the keys of the Church of the Nativity in Jerusalem from the Orthodox to the Catholic Church. This was a major humiliation for Orthodoxy. 'I cannot recede from the discharge of a sacred duty,' an enraged Nicholas told the British Ambassador. His troops occupied the Danubian principalities to force the Turks to accept Russian protection of the Sultan's Christian subjects and restore the rights of the Orthodox Church in the Holy Land.[13]

Nicholas was prepared to fight all of Europe. He acted with the single-mindedness and self-delusion of a true believer. He did not wage war, he told the King of Prussia, 'for worldly advantages' or territorial gain. While Britain and France were happy to cut cards with Islam he alone was ready to 'fight under the banner of the Holy Cross'. And in doing so he would win or 'perish with honour, as a martyr of our holy faith'.[14]

The Sultan met the Russian crusade with a holy war of his own. On one flank, the outstanding general of the war, Omer Pasha – an Orthodox Christian Serb born in Croatia who converted to Islam and worked his way up the ranks – confronted the Russian army on the Danubian front. On the other, the Sultan reached out to Imam Schamyl and urged him to join him in *jihad* against Russia and rid the Caucasus of the infidel invaders.

After years of reverses, the Lion of Dagestan was resurgent, a warrior unchained. According to the British consul in Erzurum, Schamyl was at the head of 20,000 Mujahidin who were sweeping down from the mountains to take Tiflis. He was going to link up with the Ottoman army attacking Russia in its trans-Caucasian lands. The Ottoman Black Sea Fleet was mobilised to support the holy war and throw Russia back beyond the mountains for ever. In November the Russian navy attacked and destroyed the Turkish ships at the port of Sinop.

The Ottomans, for their part, seemed unconcerned and unsurprised by their crushing defeat. They knew that a Russian naval victory in the Black Sea, which presaged its dominance in the eastern Mediterranean, would act as a red rag to the British.

They were right: public opinion in Britain ignited for war over the winter of 1853–4. It was what Napoleon III had intended when he bullied the Sultan into provoking Russia. The new French emperor craved a famous victory. He wanted to restore French prestige, lost in 1812 during the retreat from Moscow. He wanted to destroy the Concert of Europe, the balance of power on the Continent established at the Congress of Vienna in 1815, whereby Russia, Prussia and the Austrian Empire suppressed revolution and prevented the redrawing of national boundaries. War against Russia would fulfil all the wishes on that list: it was weak enough to provide an easy triumph; the defeat of 1812 would be expunged for ever; and it was the power that loured over Europe, suppressing all movements for change. But Napoleon III needed Britain as an ally, and Britain would only join him if its position in Asia was under threat and Turkey was in mortal danger.

By the end of 1854 he had helped create the perfect situation. Fear and suspicion of Russia had a long history in Britain. Not just in Europe and Turkey, but across much of the world the two empires regarded each other with hostility. While the Great Exhibition was in full swing the Hungarian freedom fighter and hero of liberty Lajos Kossuth visited England. Kossuth's bid for Hungarian autonomy within the Austro-Hungarian Empire in 1848 had been crushed by 200,000 Russian troops. He told a packed, enthusiastic meeting at Southampton that 'Russia is the rock that breaks every sigh for freedom, and this Russian power is the same which England encounters on her way, on every point – in Peking and in Herat, at the Bosporus and on the Sound [of Denmark], on the Nile and on the Danube, and all over the continent of Europe.'[15]

The British did not need to be told of the Russian threat to their world system. From the Baltic to the Danube, from the Balkans to the Black Sea, from the Caucasus to Afghanistan, in Turkey, Persia, China, Korea, Japan and the north-west Pacific, 'everywhere we find [Russia] a successful and persevering aggressor', wrote Sir John McNeill, formerly British minister in Tehran. Look at the map, he said, and one would see 'that the plains of Tartary have excited her cupidity, while the civilized states of Europe and Asia have been dismembered to augment her dominions. Not content with this, she has crossed into America.' All these gains 'have been injurious to British interests'.[16]

British free trade and Russian autocracy were seen as incompatible – and in a death grip as they struggled for world domination. The *New York Herald* certainly saw things in this light: Britain feared the burgeoning commercial and military dominance of Russia, particularly in Asia, and wanted to destroy it. For the British, this represented commercial freedom versus militarised aggression and economic protectionism with the entire globe as the battlefield. 'England with her Great Exhibition, and Russia with her great armies,' wrote the *Illustrated London News* in 1851, 'are, to use a common expression, "fulfilling their mission". We cannot doubt which of the two principles will ultimately prevail.'[17]

By the middle of the decade it was clear who was winning. Karl Marx and Friedrich Engels told readers of the *New York Tribune* of the 'commercial battlefield' that Britain and Russia were contesting from the Danube to the River Indus. The bazaars of distant and little-known places such as Khiva, Samarkand, Bukhara or Tashkent (in modern-day Uzbekistan) had until recent years been monopolised by Russian traders. The markets of Central Asia, and the resources it held, were seen as the future of Russian industry. But by the 1850s, according to Marx and Engels, 'Russian trade, formerly venturing out as far as the limits of England's Eastern Empire, is now reduced to the defensive on the very verge of its own line of custom-houses.'

The reason was clear enough. After the British East India Company's victory over Ranjit Singh's formidable military empire in the Punjab in 1849, its control extended all the way up to Peshawar and the North-West Frontier. Goods bound for Central Asia found an inlet through the Indus valley. They also came via Turkey. Back in 1838 Britain had imposed upon the Ottoman Empire a Tariff Convention that opened the vast extent of Turkish land in Europe, Asia Minor and the Levant to free trade. In 1853 Colonel Hugh Rose, the Chargé d'Affaires at the British embassy in Constantinople, could report that in that year alone 1,741 British ships had passed through the Dardanelles into the Black Sea, and a London merchant house was investing a cool £3 million in a Turkish bank. Marx and Engels wrote that the familiar hubbub of English and German voices heard every day at the Manchester Exchange was being joined by Greek, Slav and Armenian dialects. These merchants were importing huge quantities of British manufactures into the port of Trebizond in north-eastern Turkey, where they were loaded onto caravans and carried to the markets of Central Asia. In a mere decade and a half Turkey had become one of Britain's most important trading partners and its gateway to the interior of Asia.[18]

And with a liberalised economy came liberal political and social reform. Or so the advocates of free trade liked to believe. Under British tutelage, Turkey was gradually modernising. During the Tanzimat reforming period, officials began to dress in Western styles; censuses were held; the financial and legal systems reorganised; academies and universities instituted; and greater rights granted to the empire's millions of oppressed Christian subjects. In the 1850s the Ottoman Empire stood as a poster boy for the regenerating effects of free trade, a model for other ancient empires across Asia, and the centrepiece of Britain's new world order.

But for all these successes in the commercial sphere, Britain had much to fear from Russia's military might and its predatory moves on the Ottoman Empire. A strong, independent, modern Turkey was at the heart of Britain's geopolitical strategy. It provided the bulwark against Russian expansion. So too did Schamyl. The Caucasus was the 'inexpugnable fortress set up by nature against Russian aggression'. As the London *Morning Post* put it: 'Russia may be a great territorial boa-constrictor; but she has not been able to digest her Caucasian "conquests".' As long as Schamyl held back the Tsar, Russia's advance towards the Ottoman Empire, Persia, Afghanistan, India and the great trading road from Trebizond to Tashkent was stymied.[19]

During the 1840s Schamyl's name started to creep into the newspapers as Russia-watchers began to appreciate his global importance. Freelance agents smuggled guns to him and the Circassian resistance fighters. For an American writer Schamyl evinced 'a career of heroism nowise inferior to that of the most famous champions of classical antiquity', and his 'war of independence' against Russia was among 'the most glorious struggles recorded in the annals of liberty'. Books and newspaper articles in the 1850s detailed his bravery and brilliant tactics against vastly superior Russian armies, with the majestic scenery of the Caucasus providing the backdrop; they told of his miraculous escapes from the enemy, his clever wiles and his great victories. Even before the Crimean War started, the British government sent the Imam modern revolving rifles for his snipers.[20]

As far as the Russians were concerned, Britain was propping up a corrupt, immoral and decrepit Muslim empire in Turkey and supporting a fanatical terrorist in the Caucasus. Russian soldiers and officials came to exaggerate the role of Britain in the region, believing that their enemy intended to seize Circassia for itself and even that Schamyl was in fact a British secret agent. If Britain feared Russia, Russia saw Britain as aggressive and predatory, a barrier to its global ambitions. Every move Britain made in China as well was seen as inherently anti-Russian. According to General Muravev, the British were 'to the detriment and reproach of all Europe, disturbing the peace and well-being of other nations'. Their claim to be a disinterested power, dedicated to spreading the neutral benefits of free trade and international law, was just so much cant and humbug. In the view of Muravev, the British were prescribing 'from their little island their own laws in all parts of the world . . . laws not the least aimed at the benefit of mankind, but only at the satisfaction of the commercial interests of Great Britain'.[21]

The world's two greatest empires were locked in mutual suspicion along a frontier that ran across most of its land mass. There were flashpoints all along this great arc; but the Ottoman Empire was the most likely place for the two to come to blows. In his dispatch of 1853 Rose informed the Foreign Secretary that 'Turkey's fall would be the signal for general war and confusion, the triumph of socialism and anarchy, and the ruin of British trade and interests.' If Russia defeated the Ottoman Empire and subdued the Caucasus it would have an open door to Persia and the khanates of Central Asia. What then? A resurgent Russia could grab land from China and occupy naval bases in Japan. British power in Asia would wane as Russia's waxed. It would lose many important markets – not least the valuable trade being built up in Trebizond – and it would be forced to pour troops into India at vast expense as the Russians pushed towards Afghanistan. Little wonder that Schamyl, the man who tied up vast armies, was hailed as the brake on Russia's inexorable advance into Asia. As Oliphant observed, 'the balance of power in Europe depends upon the balance of power in Asia'. When Russia marched on Constantinople, the British weighed up the global ramifications.[22]

The British public, then, was primed for war after years of propaganda about Russia's plans for world domination. But before French and British troops could get involved, Russia was defeated by Omer Pasha's army in July 1854 and evacuated Wallachia and Moldavia. The war might have ended then, had not the publics of Britain and France been so enthusiastic for complete victory. Offensive operations transferred to the Crimean Peninsula. It would be a short war, the public believed, over by Christmas at least with the capture of the Russian naval base of Sevastopol.

The Crimean War was bound up inextricably with the ideas of progress and regeneration that defined the 1850s. The coming war represented for many the confrontation of two incompatible world systems. It was the armed version of the Great Exhibition – the victory of free-trade liberalism and modern technology over autocracy and conservatism in an epochal struggle. The Tsar might command an army a million and a half strong. But Britain matched massive armies with white-hot technology: telegraphs, railways, screw-propeller steamships, gutta-percha bivouacs, shell-firing guns, and many more modern inventions. The British army, equipped with the latest Enfield rifles and Colt revolvers, would be meeting an old-fashioned army that still fought with the musket. British forces went into battle with portable field telegraphs to connect them to HQ via gutta-percha-insulated copper wire. These field telegraphs had another use: they sent pulses of electricity to detonate explosives remotely. Victory over the forces of repression and backwardness would be quick and decisive. *The Times* said that Britain would defeat the limitless resources of Russia 'by making the century our ally'.[23]

After such heightened – almost millennial – expectations, it was little wonder that failure was very hard to accept. Not for the first or last time a superpower taking on what it regarded as a backward opponent found that war was more complex, messy and unpredictable than armchair strategists allowed.

Landing the army at Kalamita Bay turned into a fiasco. What was supposed to be a quick attack on Sevastopol got bogged down in a protracted land campaign. Victories at Alma and Balaclava were not followed up by the allies, so by the winter of 1854 the British and French were encamped on the heights above Sevastopol.

Here the full miseries of war became apparent. Although Laurence Oliphant had warned how cold the Crimea was in winter in his *The Russian Shores of the Black Sea* – the most up-to-date book on the region – the British army neglected to pack warm clothes or arrange suitable accommodation. Disease nibbled away at a force that had left its ambulance corps and medical supplies in Bulgaria. To compound matters, the British failed to construct a road between the harbour at Balaclava and their camps on the heights. At first horses and mules laboured up the slopes carrying food and clothing; by winter supplies were being carried by hand. Sevastopol was supposed to have been taken by surprise in a matter of weeks; instead the poorly equipped and starving British army had to endure a siege that lasted 346 miserable days.

The shock of unfulfilled expectations and the unprecedented way in which the war was reported proved to be a toxic brew. Details of suffering and military debacles were splashed across the British papers without censorship. No other military campaign in history had been picked over and exposed in this way. 'The excitement, the painful excitement for information,' opined Lord Clarendon, the Foreign Secretary, 'beggars all description.' It was not just the mass of details that was so shocking; it was the speed at which it reached the presses from the front. The government often had to turn to *The Times* for the latest news, and the Russians found the British papers the best source of military intelligence. Clarendon moaned that 'if *we* could get such information, or a tithe of it, it would be worth thousands to us'.[24]

By early 1855, with the army ground down in the Crimea and hopes

of an epoch-defining victory dashed, Britain was facing political crisis. The generals were lambasted as old and dithery. The fate of the Light Brigade, which had suicidally charged into the teeth of the Russian artillery at the Battle of Balaclava on the basis of ambiguously worded orders, exemplified, for the public, the noble sacrifice of the men and the ineptitude of their commanders.

The machinery of war and government was criticised as hidebound and sclerotic, unbecoming of a hyper-modern country. *The Times* complained about 'the cold shade of aristocracy' that was holding back the energies of the people. The paper was utterly dominant in 1854–5. Its correspondents reported on the day-to-day debacles of the war, and its sulphurous editorials tore into the government, aristocracy and military without mercy. War, it declared, 'has become an affair of science and machinery, of accumulated capital and skilful combination'. At another time it thundered that 'This great commercial and mechanical country is governed by an official body comparatively ignorant of the mechanical arts.'[25]

The government fell at the end of January 1855 as a result of public fury. After exhausting every other alternative, the Queen eventually asked Palmerston to form a government. It was the popular choice. Palmerston spoke for the frustrated, disillusioned and very angry public. Famously bombastic and bullish, he promised new energy in prosecuting the war and fulfilling the dreams of the year before.

By the spring of 1855 Britain at last seemed to have strengthened its sinews of war after a calamitous start to the campaign. It appeared to have cracked modern warfare with the help of business, industry and private initiative. Sir Joseph Paxton – the architect of the Crystal Palace – organised 2,000 navvies into the Army Works Corps to build an all-weather road from the British military port at Balaklava Bay to Sevastopol. Isambard Kingdom Brunel sent out a prefabricated hospital, with air conditioning and drainage, for 1,000 patients. The former *chef de cuisine* at the Reform Club, Alexis Soyer, went out to the Crimea at his own expense to teach the army how to cook and supply the troops with his pioneering field stove. The firm of Peto, Brassey & Betts built the Grand Crimean Central Railway, the first military railway. It brought supplies up to the British forces besieging Sevastopol and took the wounded back down.

Modern communication came to the battlefield as well. R. S. Newall & Co., the company responsible for the Channel Cable of 1851, laid a 310-mile-long submarine telegraph on the bed of the Black Sea from the British military headquarters at Balaclava to Varna in Bulgaria. At Varna the British submarine cable met the French landline to Bucharest, the outpost of the European telegraph network. The marvels of modern invention meant that Whitehall was separated from the front line by a mere day.

The pioneering railway and the submarine telegraph heralded a new way of waging war. This was the military manifestation of the Great Exhibition: the arts of peace, *The Times* said, had been translated into the science of war. The get-up-and-go attitude of private enterprise – including nurses such as Florence Nightingale – at last vindicated Britain's claims to greatness after a halting start. American military experts who examined the front line brought home insights into how war could be fought in the age of railways and telegraphs.

'Doubtless freer governments are *slower* machines than absolute ones,' commented the *Edinburgh Review*, 'but their ultimate means are incalculably vaster . . . England, clumsy, constitutional, almost democratic, but opulent even to plethora – is only just warming to her work, whereas our autocratic ally [France] and our despotic adversary [Russia] are alike breathless and bleeding.'[26]

This brings us back to Laurence Oliphant and his mission to Schamyl. As things started to go their way at last, the British began to think big. Even before Sevastopol fell in September 1855 Palmerston was planning to make the war global: 'The main and real object of the war is to curb the aggressive ambition of Russia.' That meant destroying Russia's power for good, not just in the Black Sea but in the Baltic and the Pacific. It should be thrown out of Poland; its territorial ambitions in Central Asia, China, Korea and Japan should be terminated by force. Palmerston's world war would end with the establishment of 'a long line of circumvallation to confine the future extension of Russia'.[27]

Key to all of this, however, was re-establishing the Caucasus Mountains as an insurmountable barrier to Russian expansion in Asia. Laurence Oliphant told the public that the 'opportunity of attacking Russia at a remote and most vulnerable point may never again occur'. Throughout the war the stock of Schamyl had been rising, so that he was a household name in Europe and America. During the dark days of the siege of Sevastopol the public read satisfying accounts of the Russians being bloodied in the Caucasus. Adulatory newspaper profiles and books celebrated his victories over the Russians, past and present. Patriotic citizens displayed engravings of the Imam on their walls. Crowds roared their support for the Lion of Dagestan and the Circassians at public meetings. A racehorse named after the Caucasian warlord competed on the turf, and a new dance was invented and honoured with his name. Theatregoers were treated to a play called *Schamyl: The Warrior Prophet* in which mountain scenery was recreated on the stage, a Circassian ballet performed and Russian soldiers slaughtered. At the same time another West End theatre was showing *Schamyl: The Circassian Chief*. When news of Schamyl's advance on

Tiflis came through, the price of government bonds on the London Stock Exchange spiked. 'In Caucasia, we might at any time, and for a trifling expense,' wrote one very excited commentator, '. . . equip 20,000, or more, of the bravest troops in the universe, capable of carrying fire and sword . . . to the very gates of Moscow!'[28]

The exploits of Schamyl offered the public psychological release at a frustrating time. He also offered the possibility of delivering the Tsar a knockout blow and reshaping the geopolitics of Asia. The British government had been supplying Schamyl and the Circassians with the latest military technology since 1853; Napoleon III sent 12,000 muskets. The Royal Navy had blasted the Russians out of their forts on the Black Sea coastline of the Caucasus, opening up the Circassians to the world again. But in 1854, Schamyl's advance on Tiflis came to nothing because the Ottoman army in Asia Minor failed to make any headway in the borderland between Turkey and Georgia. In the spring of 1855 the allies began to explore new ways of bringing Schamyl into play. The London *Morning Post*, an organ that spoke for the British government, declared that 'we may be sure that [Schamyl] will play no common part in the present great struggle of civilisation against a dark despotism'.[29]

That was why Oliphant was in the Caucasus in the summer of 1855. He had penned a pamphlet in which he argued that if the British army invaded Georgia it would score a more decisive blow against Russia than ever it could in the Crimea. The British could raise the Christian populations of the Transcaucasian provinces in revolt against the Tsar and, with the help of Schamyl and the Circassians, drive Russia back over the mountains and bolt the gate behind them.[30]

Oliphant's mission was to travel through the mountains to Chechnya, evading Cossack guerrillas, and make contact with the enigmatic Schamyl. Once in the Imam's camp he would co-ordinate the Mujahidin and the Ottoman army. The aim of the campaign was to catch Russia's Transcaucasian provinces in a pincer movement. Omer Pasha would invade Georgia from the port of Sukhumi and march on Tiflis while his son, Selim Pasha, led another army overland from Erzurum in north-eastern Turkey. The Russian Caucasian army under the command of General Nikolai Muravev – numbering just 21,000 men and eighty-eight guns – was at that time besieging the Turkish border fortress at Kars. Kars was defended by 18,000 troops – including a legion of émigré revolutionaries from Poland, Italy and Hungary – under the command of a British general. With Omer Pasha advancing south-east with 40,000 men, Schamyl streaming down from the mountains with 20,000 vengeful warriors, and Selim's 20,000 advancing north from Erzurum, Muravev would surely be destroyed.

The Times believed that the moment of Russia's expulsion from the Caucasus was nigh. 'The fall of Sevastopol, carried by breathless messengers to every cottage and cavern of the Caucasus, has shaken the trust of those who made submission to the Czar, while it has encouraged the more independent to obstinate resistance.' Britain would create a protectorate over the region, allowing tribal leaders such as Schamyl to carve out their own states. The paper commented with glee that 'the cloud is gathering from which the tempest is to burst on the Transcaucasian provinces of Russia'.[31]

There were only two things standing in Britain's way: its determination that other people do the fighting, and the attitude of the French.

After the fall of Sevastopol France was feeling the strains of war and was not capable of fighting the kind of global conflict that Palmerston wanted, a world war that would have lasted years. In any case, Napoleon III had had his victory; now he looked forward to a time when he could forge an alliance with Russia as a means of redesigning Europe and unifying Italy. Helping the Circassian independence movement and aiding Schamyl would benefit Britain without any gain for France. Napoleon III did not want to antagonise his future friend by hitting him in his most sensitive region.

The French commander-in-chief, Marshal Aimable Pélissier, delayed releasing the army of Omer Pasha to invade Georgia. Oliphant waited impatiently, realising that the time to act was slipping away. At last Omer Pasha arrived at Sukhumi, his army conveyed by British ships and made up of a diverse assortment of the peoples of the Ottoman Empire – Turks, Egyptians, black Africans, bashi-bazouks from Albania, and Bulgarians. The march to Tiflis began.

Oliphant longed to head off through the mysterious mountains and enter the stronghold of the legendary Schamyl. But by the time he got there, Schamyl's stock was falling fast in the West. He had kidnapped two Georgian princesses during one of his raids. These princesses were the wives of officers in the Russian army, they had been ladies-in-waiting to the Tsarina, and they were cherished by St Petersburg society. But here they were imprisoned in the seraglio of a mountain brigand. There were reports that the Imam was going to give the princesses and their female attendants to his followers as the spoils of war. Nothing did more than this rumour to suggest that Schamyl might not be the gallant warrior of the Western imagination but a dangerous zealot.

The incident lost Schamyl many of his admirers. One British agent on the ground feared that Schamyl was desirous of 'founding a fanatic Moslem Empire in the Caucasus' and advised the Foreign Office to have no part in stirring up the forces of Islam. Lord Stratford, the Ambassador to Constantinople who had authority over British policy, now called Schamyl 'a fanatic and a barbarian' and an unsound ally. *The Times*, a former champion of the Imam, labelled him a 'gloomy fanatic' at the head of a 'blood thirsty' horde. The British government reprimanded Schamyl and ordered him to refrain from attacking women and children. The Imam was deeply insulted; he had done everything asked of him, only to be let down; now he was being chastised like a toddler.[32]

It was clear that the politics of the Caucasus were too labyrinthine for British policymakers. Even if Oliphant had succeeded in meeting Schamyl he would have had nothing to offer. Defeated in his plans, Oliphant transferred, overnight, from British agent to *Times* special correspondent covering Omer Pasha's advance through the southern Caucasus towards Tiflis. The plan was to force Muravev to abandon the siege of Kars and return to defend the Georgian capital.

But it was too late in the year. The march to Tiflis stalled in November as the rivers became torrential. After a desultory scrap with the Russians, Omer had to retreat to the Black Sea with winter nipping at his heels. Selim's advance through northern Turkey was just as desultory; his force dwindled and he lamely broke off the relief of Kars.

Rather than becoming trapped in a noose, Muravev came off with a stunning victory. The garrison at Kars was facing starvation; when Omer and Selim limped for home the British general had no choice but to surrender the town to the Russians. Muravev was a great general; he was also a wily operator. In return for freeing the Georgian princesses, Schamyl was offered the repatriation of his son Jamal al-Din, captured sixteen years before. Schamyl had pined for his beloved heir for all that time; he eagerly awaited his reunion with Jamal, his successor in the never-ending war against Russia.

The man who returned to the mountains was not the eight-year-old boy who had been carried off all those years ago. He was nervous and uncomfortable at the reunion. His baggage contained books, atlases, writing materials, paintbrushes and just about everything a St Petersburg aristocrat would need in exile.

Jamal had been raised in splendour by the Tsar, who took a paternal interest in the young warrior. He was thoroughly Russified – virtually a member of the royal family, the darling of St Petersburg society and a dashing officer in a lancer regiment. Being sent back to a primitive *aul* high in the mountains to live with illiterate Muslim brigands was his idea of hell. Jamal would rather have gone back with his regiment to fight his own people in the name of Mother Russia. But he submitted to the Tsar, to whom he was fanatically loyal. He had met and danced with the princesses; he was prepared to sacrifice himself for their deliverance. Now the Russians had their man sitting at the right hand of the enemy; at their last tearful meeting the Tsar told Jamal, 'Try always to do good among your father's people and never forget your true country – Russia, which made you a civilized man'. Muravev used Jamal as a conduit to open negotiations with Schamyl.[33]

The Lion of Dagestan was thus neutralised at the crucial moment in 1855. His guerrillas did not swoop on Tiflis. Earlier in the war Russia had thought of evacuating all its possessions in Eurasia because it had too few troops to contain the Muslim tribes and hold back the Ottoman army. But after Muravev's triumph against the odds the Russians were not only stronger than they had ever been in the Caucasus, but they had captured more enemy territory than the allies had managed to take from them in the whole of the Crimean War. Britain's nightmare was realised: in the contest for the crossroads of Europe and Asia, Russia had clearly won. If Britain had been less hesitant and more honest in its treatment of Schamyl, the second half of the nineteenth century would have been very different. A victory in the Caucasus would have inflicted a far greater blow on the enemy than ever was achieved in the Crimea. As it was, in the winter of 1855 the British abruptly abandoned Schamyl to the vengeance of Russia.

From his mountain fastness Schamyl no doubt understood the wider implications of the Crimean War and the opportunities it offered him. But he was not prepared to become the cat's-paw of a distant country, a bit player in an imperial game. As he wrote much later of his decades-long resistance: 'We waged war on oppression and enslavement . . . we did not fight against the Russians or the Christians: we fought for our freedom.'[34]

As Oliphant limped home from the desultory Caucasus campaign in the winter of 1855 the Crimean War was far from over. It was about to spill out into a much wider area, into East Asia and North America.

8

El Presidente

NICARAGUA

The path of our destiny . . . lies in . . . tropical America [where] we may see an empire as powerful and gorgeous as ever was pictured in our dreams of history . . . an empire . . . representing the noble peculiarities of Southern civilization . . . having control of the two dominant staples of the world's commerce – cotton and sugar.

Edward A. Pollard[1]

The English piously believe themselves to be a peaceful people; nobody else is of the same opinion.

William Ewart Gladstone[2]

War with the United States would be 'neither difficult nor dangerous', growled *The Times*. Britain would destroy the US navy and sweep her commerce from the sea. Then the Royal Navy would blockade the eastern seaboard and burn to the ground several cities, including New York. Next, the States would be engulfed in a slave rebellion. Lord Palmerston held that 'a British force [made up of Jamaican soldiers] landed in the Southern part of the Union, proclaiming freedom to the blacks would shake many stars from their banner'.[3]

Russia was, for the British, an expansionist power that had to be contained where its frontiers came into abrasive contact with their empire, in its formal or informal manifestations. The same kind of fault line existed with another expanding empire. Palmerston was planning war with the United States even as conflict with Russia rumbled on, at exactly the same time as Britain contemplated expelling the Tsar from the Caucasus with the help of Imam Schamyl. Indeed, the contest with America would be virtually another front of the Crimean War.

Britain was concerned that the US was going to take advantage of the war to seize territory in the Caribbean and Central America. There was also the ever-present fear that the US would ally with Russia and turn the Crimean War into a global conflict. Don't be fooled by failures at Sevastopol, the British press loudly proclaimed. Britain was more than capable of fighting in two hemispheres if so roused by America's 'unjustifiable arrogance and unprovoked hostility'.[4]

Events in the Caribbean and North America during the Crimean War often merit a footnote at best to the great conflict. But this exceptional passage of history, during which Britain and the US came to a hair's breadth of war, reveals much about the two countries in the middle of the century. More pertinently to the theme of *Heyday*, it brings together many of the dominant aspects of the 1850s – migration, transportation and gold, slave power and the mighty cotton empire, trade and colonisation, public opinion and the extraordinary global reach of the Crimean War – with an almighty bang.

Even before British boots hit Crimean soil, it was suspected that America was going to exploit the situation for its advantage with the connivance of the Tsar. The arrival of one of the most flamboyant nineteenth-century diplomats at Madrid in 1854 signalled America's mischievous ambitions. Pierre Soulé, the new American Ambassador to Spain, was a Frenchman by birth, a radical republican and senator for Louisiana. Within two months of taking up his post both he and one of his sons had fought duels. First Nelvil Soulé overheard the Duke of Alva at a ball making saucy comments about his mother's décolletage. The young man and the duke fought by sword. Not to be outdone, Pierre Soulé challenged the French ambassador, the Marquis de Turgot, to a duel because the marquis had hosted the ball. The American ambassador shot his French counterpart in the knee and crippled him for life. The Soulé family appeared to be waging a personal battle against European aristocracy.[5]

Soulé had other mischief in mind. The French and Spanish believed, with good reason, that he was dabbling with revolutionaries and republicans in their countries. It was a habit he shared with other American diplomatists. Over in London the American consul George Sanders held a dinner party on 21 February 1854 to mark George Washington's birthday. The company round his table was a roll call of the radical heroes of the 1850s. Lajos Kossuth was there, as were the three revolutionary titans of Italy, Mazzini, Garibaldi and Felice Orsini. Also enjoying Sanders' hospitality were the icons of Continental republicanism and socialism, Alexander Herzen of Russia, Arnold Rudge from Germany and the Frenchman Alexandre Ledru-Rollin. James Buchanan, US Ambassador to London, graced the occasion with his presence. He sat next to Mrs Sanders and jokily asked her 'if she was not a little afraid the combustible material about her would explode and blow us all up'.

Sanders was owner and editor of the *Democratic Review*, the eloquent mouthpiece of the Young America movement. The Young Americans advocated free trade and modernisation through railroads, telegraphs and canals. They also believed that America had a divinely ordained mission to regenerate the world through republicanism. The first step of this crusade would be the expansion of the United States southwards, into the Caribbean and Latin America. Sanders and Soulé appeared to be signalling a new move in American foreign policy – towards the support of European socialism and anti-monarchical revolution. The United States would be a new force in the world – a republican, revolutionary force that would challenge the absolutism of Europe and the imperialism of Great Britain.[6]

Around the Sanders' table that night the conversation no doubt frothed and bubbled with the inevitable union of Young America and revolutionary Europe; with the prospect of the fall of British liberalism and the triumph of republicanism and nationalism; with the renovation of the world and the epoch of the People.

In October 1854, as the siege of Sevastopol was getting under way, Soulé travelled to Ostend, where he met with the American

Ambassadors to London and Paris, James Buchanan and John Mason. After that he went to London and announced he wanted to return overland to Madrid via France. The French were having none of it: the erratic Soulé had crippled their ambassador and he was clearly about to stir up revolutionary trouble. He was turned back at Calais.

Washington was incandescent. The French were not prepared to back down, even if it meant war. At exactly the same time, the British learnt that the United States was about to establish a military base on Santo Domingo, the first step in a conquest of the Caribbean. The fear was confirmed in a spectacular way by a scoop in the *New York Herald*: Soulé, Buchanan and Mason had met in Belgium on the orders of President Franklin Pierce (a member of the Young America movement) and Secretary of State William Learned Marcy to discuss one particular subject. The result of their deliberations became known as the Ostend Manifesto.

'It must be clear . . . that . . . Cuba,' the ambassadorial troika asserted with extraordinary bluntness, 'is as necessary to the North American republic as any of its present members, and that it belongs naturally to that great family of states of which the Union is the Providential Nursery.'[7]

Earlier in that year the South had scored a victory in the passage of the Kansas-Nebraska Act. But Kansas was unsuited to growing cotton. If the great cotton empire wanted to expand, therefore, the South had to look elsewhere. The object of desire was Cuba, with Central America running a close second. The island was close to Florida; it was large and underdeveloped. 'I want Cuba,' declared a senator from Mississippi. '. . . I want Tamaulipas, Potosi, and one or two other Mexican states; and I want them all for the same reason – for the planting and spreading of slavery.' A Southern newspaper said that with Cuba and Santo Domingo the produce of the tropics – cotton, sugar, tobacco and coffee – would be theirs and with it 'we [would control] the commerce of the world, and with that, the power of the world'.[8]

Most important of all, a constellation of slave states in the Gulf of Mexico would shore up Southern political power in perpetuity, providing a counterweight in Congress to the newly created free-soil states in the North and the far west. Imagine a United States, said the hotheads, with 100 states rather than thirty, the majority of them concentrated in the South and the Gulf of Mexico. The expansion of the United States into the Caribbean was the next step in the fulfilment of Manifest Destiny, America's moral duty to spread republican democracy throughout the entire extent of the continent.[9]

'The policy of my administration will not be controlled by any timid forebodings of evil from expansion,' declared Franklin Pierce at his inauguration in March 1853, announcing a new direction in American foreign affairs that caused champagne corks to pop throughout the South. 'Your future is boundless,' he told the American people.

Soulé was dedicated to the seizure of Cuba by any means possible. So was the Pierce administration. It empowered Soulé to offer $120 million to the Spanish for the purchase of the island. There was a new urgency in the matter.

Britain was threatening Spain with the withdrawal of naval protection and the compulsory payment of interest due to bondholders in the City of London if it did not emancipate all the slaves in Cuba forthwith. The Marqués de Pezuela, Captain-General of Cuba, was already encouraging racial intermarriage and raising militia from emancipated slaves while disarming the white, pro-American plantation owners. This was seen as a direct threat to the stability of the slave-owning American South, which was after all just ninety miles from the island. Southerners had always been sensitive to regimes close to their border that emancipated slave populations, believing that such moves would engender slave rebellions on their own territory.

Britain's and Spain's actions were, for Southerners and Secretary of State Marcy, an attempt to 'Africanize' Cuba. It must be stopped. Britain's humanitarian concern with the slave population of Cuba and its pressuring of Spain over the issue were seen as deliberately anti-American: the British were attempting to wreck the US from the outside. The Ostend Manifesto accused Spain of trying to 'seriously endanger our internal peace'. If it carried on, then 'by every law, human and Divine, we should be justified in wresting [Cuba] from Spain'.[10]

That was an explicit threat of annexation. Throughout the 1850s

Southerners had supported expeditions to 'liberate' Cuba from Spain. The term 'filibuster' became popular in this decade. It came from the Spanish word *filibustero*, a corruption of the Dutch word *vrijbuiter* meaning pirate or privateer, and it was applied in the sixteenth and seventeenth centuries to buccaneers who raided Spanish territories in the New World. The English-language equivalent was 'freebooter'. In the mid-nineteenth century the title 'filibuster' was given to a growing band of freelance adventurers who set out, without the support of a recognised national government, to foment revolution in various West Indian and Central American countries as a prelude to taking them over.

The Venezuelan filibuster Narciso López was a hero in the South – a kind of Garibaldi of the Americas. It was from there that he had raised men, money and weapons to invade Cuba in 1850. He had failed, but he was cheered on by the likes of Pierre Soulé. At the time that the ambassadors met at Ostend, a new filibustering expedition was being fitted out by John Quitman, quondam Governor of Mississippi, to capture Cuba. Previous filibustering ventures had been regarded as illegal by Washington. Quitman, by contrast, enjoyed the blessing of President Pierce. 'Now is the time to act,' urged one champion of the cause, 'while England and France have their hands full' with the Crimean War.[11]

But that was to underestimate the determination of the British to police the frontiers of their empire even as the siege of Sevastopol engrossed world attention. In an editorial *The Times* thundered that the American attitude to Cuba was identical to that of Tsar Nicholas towards Poland and Turkey. The British Cabinet was notified by the Foreign Secretary, Lord Clarendon, that any attempts by the US to annex Cuba would mean war. John Crampton, the British Ambassador to Washington, informed the president that Britain was 'an American power' and it would not tolerate the United States empire-building in the Caribbean. At the same time, British warships paid a visit to Havana. There was little doubt why they were there.[12]

And there was very little doubt that British high-handedness in America was deeply resented. 'Sympathy for Russia,' admitted *The Times*, '. . . [is] almost universal, expressed from one end of the Union to another.' That was borne out by Laurence Oliphant when he was in Minnesota just before the Crimean War began. Over cocktails in a hotel bar in St Paul a colonel in the Texas Rangers lectured him: 'Wal, you Britishers air 'cute – you go on the high moral ticket. You call annexation "robbery" and "territorial aggression"; but there ain't no power in creation that's swallowed more of other people's country without choking than you have when nobody was looking perticler. And now you're a-going to fight for "civilisation", by protecting the most barbarous power in Europe, and for "liberty" by allying yourself with a French despot and a Mahommetan tyrant.'[13]

The colonel went on to connect the cause of Cuba and the fate of Russia. According to a rumour circulating about the States, if Britain captured the Crimea then it would intervene militarily to spur on Spain's liberal reforms and emancipation programme in Cuba. It would mean war with the US. 'But there's no fear in that,' the colonel concluded with relish. 'The Roosians will whip you into ribbons . . . Why, they've got the sympathies of our country with them.'[14]

Such sentiments were received with malicious glee in St Petersburg and Moscow. The plan was to convert that resentment at Britain and those pro-Russian sympathies into a desire for war. And there were plenty more points of friction between the British Empire and the United States to exploit.

Eduard de Stoeckl knew his way around the corridors of power in Washington. Tall, handsome and with impressive side-whiskers, he had the unmistakable bearing of a dashing Russian aristocrat. It also helped that de Stoeckl – the Russian Chargé d'Affaires – had an American wife who was 'stately as a queen and beautiful as a Hebe'. The de Stoeckls charmed their way into the social whirl of Washington, favourites alike with powerbrokers and political hostesses. Eduard de Stoeckl was on intimate terms with congressmen, senators and members of the Cabinet; the door of the Secretary of State's office and the drawing rooms of the White House were open to him.

'Russia,' he wrote home, 'must be ever watchful, must never lose an opportunity to fan the flames of hatred' that existed between London and Washington. That he did in the American capital (where he had many willing listeners), in the press, and through secret channels.

Cuba was a cause of friction between Britain and the US. So too were the Sandwich Islands (Hawaii), which America was desirous of acquiring. But the most combustible issue of all involved shipping. Britain's aggressive policing of the seaways, particularly operations to suppress the slave trade, had for years incensed Americans. Russia, de Stoeckl believed, should find some way of provoking a confrontation between Britain and America on the high seas which would hurl the two countries to war.[15]

De Stoeckl controlled a secret fund to hire American ships to run arms to Russia in violation of the allies' blockade. His legation was besieged with eager applicants for letters of marque so that they could plunder British merchant shipping on behalf of the Tsar. He was also offered a force of 300 riflemen from Kentucky who wanted to fight the British in the Crimea.

The plan was coming together. Sooner or later a ship flying the Stars and Stripes would be captured by the Royal Navy as it attempted to trade with Russia. American public opinion would be ignited by the outrage. President Pierce and Secretary Marcy were pro-Russian, as were swathes of the public. All it would take to push them into military support for Russia would be an outrage on the high seas. And indeed Lord Palmerston was blunt in his reaction to the threat. 'The United States have no navy of which we need be afraid,' he said in his usual belligerent manner, 'and they might be told that if they were to resort to privateering we should, however reluctantly, be obliged to retaliate by burning all their sea coast towns.'[16]

A clash on the high seas seemed inevitable, given America's refusal to be cowed and Britain's notoriously effective bully-boy tactics. 'We desire more sincerely to remain neutral,' said President Pierce with doubtful sincerity, 'but God alone knows whether it is possible.'[17]

It was possible. Britain stuck to its faith in free trade throughout the war, unwilling to antagonise the US or undermine the basis of its wealth and global power. American ships brought weapons to Russia and secured much of the carrying trade that had previously belonged to Britain without incurring the wrath of the Royal Navy. The refusal of the United States to respect the blockade and the country's enthusiastic support for Russia caused resentment in Britain. There were over 500 American citizens serving with the Russians as soldiers and engineers; some forty US doctors were at work in the medical department of the Tsar's army. Over 10,000 Colt revolvers had made their way to the Crimea. The resentment continued to build throughout the war.

Having acted carefully to avoid provoking the United States during the early part of the war, Britain at last made a mistake. Its war effort was hampered by the fact that its strength was primarily naval. It could project its power into every ocean, sea and river system on the planet, but it simply did not have a big enough army to field in the Crimea, making it heavily dependent on the French and the Ottomans. One solution would be Continental-style conscription; but that would be incompatible with British liberties. The answer was the Foreign Enlistment Act. By the summer of 1855 the army was augmented by 13,000 mercenaries, drawn from such places as Germany, Switzerland and Poland. The quality of these recruits was doubtful. What the British desired was a Foreign Legion made up of men of an 'indolent but restless disposition . . . [who were] scattered about the towns or wandering the prairies' of the United States.[18]

To recruit directly from the 'warlike population' of the US would be to infringe its Neutrality Act of 1818 that prescribed three years in prison for enlisting citizens for a foreign war. Ambassador Crampton got round it by opening up barracks beyond the US's northern borders – at places like Niagara, near Buffalo, and Windsor, Ontario, within sight of Detroit – where potential recruits could be formally enlisted on British soil. A network of agents across the States was to direct eager mercenaries to these depots. The scheme might have worked had it been done secretly, or at least discreetly. Instead, Crampton's agents opened recruitment offices in major towns, distributing handbills and offering a bounty of $30.[19]

In the early months of 1855 the US authorities began to round up and arrest the British recruiting agents. Caleb Cushing, US Attorney General and member of the Young America movement, was leading the rolling-up operation. The whole story came out in the subsequent trials: there was a large-scale, covert operation being conducted on

American soil by agents of a foreign state. Crampton denied to Marcy's face that he or any British consul had anything to do with the scheme. Back in London Clarendon made the same avowals to Buchanan.

This was not a trivial matter. For America, the British had not just flagrantly broken the law; they had violated its national sovereignty and then its highest officials had been caught telling barefaced lies to their counterparts. It was a grievous insult to American honour. The storm built throughout the summer and autumn of 1855; from November, people in both countries were talking of war.

The press in Britain was convinced that the American government was hammering on the war drums because of the upcoming presidential election. Franklin Pierce had been hobbled politically by the Kansas-Nebraska Act and the subsequent violence. Dragging America to the brink of war and flexing the country's muscles in the face of British arrogance was a way of rescuing his discredited presidency. There were many who responded to Pierce's and Marcy's warlike pretensions. The *Pennsylvanian* newspaper said that the 'haggard voluptuary Great Britain' was 'drunk with the blood of other nations'. She had been 'bullying and bribing' the world for centuries: 'Her insolence is astounding' and it was about to receive a check. Caleb Cushing wrote in a newspaper article that if the two countries came to war, 'we could raise three hundred thousand men for the invasion of England with less trouble than she raises thirty thousand for the invasion of Russia'.[20]

'Is it to the predominance of Celtic blood that America owes her love of petty little quarrels?' sneered *The Times*. But the paper feared that this little quarrel could swirl into a hurricane. 'The smallest questions have caused implacable wars,' noted a journalist elsewhere. America's belligerent response to the recruitment controversy – regarded as impossibly trivial in Britain – and its brinkmanship were lumped together with the country's continued support for Russia. There was incomprehension that while Britain was engaged in a millennial struggle with evil, the Americans were not merely failing to help but were the active 'accomplices of tyranny'.[21]

'The seat of war will yet, we fear, be transferred from the Crimea to America,' prophesied the Melbourne *Age*. The British had word that

Marcy was discussing the whole matter with his close friend, Eduard de Stoeckl. So much for what Palmerston called America's 'pseudo-neutrality'. Perhaps it wanted to divide the world with the Tsar. Already the two countries were discussing how America might exploit Britain's heavy commitment in the Black Sea to seize Hawaii and, with Russia's help, supplant British merchants in the Persian Empire. The British had it on good authority that the Russians were about to commission a super-fast American privateer to prey on British ships carrying gold from Melbourne to England. Clarendon believed also that the US government was about to covertly hasten the passage of several hundred Irish-American filibusters, or terrorists, to Ireland in order to incite rebellion. The headline in the *New York Herald* on 17 November 1855 announced the convergence of events that signalled the slide towards armed conflict: 'GREAT WAR EXCITEMENT/ The United States the Battle Ground of European Parties/The Great Celto-American Invasion Project/The Russo-American Privateers/. . . The Increase of the British West India Fleet.'[22]

Clarendon wrote to Crampton: 'We don't want a quarrel with the U.S. . . . but if one is thrust upon us we shall be willing and *are ready* to meet it.' The Royal Navy had been considerably augmented with lots of new state-of-the-art steam gunboats designed to bombard Russian naval bases and cities in the Baltic. The Royal Navy, Clarendon told Crampton to feed to American policymakers, 'is now at the service of the U.S. if they choose to give it employment'.[23]

This was an unexpected blowback from the Crimean War. And there was something else to add to the volatile mix.

The culprit was 'as unprepossessing-looking a person as one could meet in a day's walk': a thirty-one-year-old lawyer from Tennessee turned Californian journalist, short, very skinny and with mousy hair, a freckled face and almost pupil-less grey eyes. A brooding, taciturn figure who rarely, if ever, laughed, William Walker had megalomania on a colossal scale, charisma that mesmerised those he came into contact with and unflinching personal courage that belied his nondescript appearance. Gold lured him to California; but the riches that state offered were not enough. He wanted to be the Kossuth or Mazzini of the Americas. In 1853 he persuaded forty-five men to help him conquer Baja California from Mexico. They seized the city of La Paz and Walker became President of the Republic of Lower California.

For one glorious moment it looked as if he might become the head of a golden new republic. More footloose Forty-niners joined him and Walker led his raggedy filibuster army over the mountains and across the Colorado River to conquer gold and silver-rich Sonora. Poorly supplied, inexperienced and attacked by superior Mexican forces, Walker's army fell apart. The president fled back over the US-Mexican border, hobbling on his only remaining boot. He was arrested and put on trial in San Francisco for waging an illegal war. Despite his manifest guilt the jury acquitted him after deliberating for eight minutes. Californians loved his buccaneering style and admired his quixotic adventure. The American South had cause to hero-worship him too: his short-lived republic adopted the civil code of Louisiana, which permitted slavery.

Walker's strength – and weakness – was his ability to take decisive action while most missed the moment in reflection. Just over a year later he led another army – this time numbering fifty-eight men known as 'the Immortals' – to Nicaragua. That Central American republic was engulfed by civil war and the liberal faction invited Walker to intervene on their behalf against the Legitimist Party. This time General Walker had the backing not only of President Pierce but, crucially, two businessmen named G. K. Garrison and Charles Morgan who financed the expedition.

Walker's second filibustering venture was stunningly successful. He was aided by local Indians whose legends spoke of a man with grey eyes who would save them from oppression. From then on Walker referred to himself as the 'grey-eyed man of destiny'. By October 1855 he had captured Granada, the capital city, and was de facto ruler of Nicaragua with a puppet president in nominal control.

Walker's fame spread across the world. Thousands of Americans set out to join his movement to help him establish an Anglo-Saxon republic in Central America and transform an underdeveloped region into a flourishing, wealthy country. Walker dreamt of repeating his success in neighbouring countries such as Costa Rica, El Salvador, Honduras,

Guatemala and, ultimately, Mexico. He would form alliances with liberal factions, foment revolution and invade. The goal was a Central American empire, with Walker as dictator. But he was trampling on the toes of the British, fanning the flames of the escalating crisis besetting that country and the United States.

Nicaragua was one of the strangest areas in Great Britain's informal empire. Since the end of the seventeenth century Britain had given military protection to a group of people known as the Mosquito Indians, who occupied the entire Caribbean coastline of Nicaragua and parts of Honduras and Costa Rica. The Mosquitos – or Miskitos – were the mixed-race descendants of indigenous peoples and escaped or shipwrecked African slaves. In the 1840s the Nicaraguan government kicked out the Mosquito Indians from San Juan del Norte. British forces reclaimed the town and returned it to the Mosquitos; it was renamed Greytown.

This miserable collection of huts might have seemed far away and remote. The discovery of gold in California changed all that. Greytown was strategically placed on the mouth of the San Juan River. Every month, the 2,000 or so migrants and travellers who paid Cornelius Vanderbilt's Accessory Transit Company to convey them from New York to San Francisco transferred from ocean-going ship to river steamboat here, the first part of their journey across the Isthmus. In the near future the commodities and luxuries of world trade would pass through Greytown as well, for the San Juan River was to form part of the ship canal that would cut through the Isthmus. When that happened Greytown would command the gateway between the Atlantic and the Pacific and become one of the rivets in the chain of great port cities that girdled the planet. 'Central America,' wrote one British diplomat in a private memo, '. . . is . . . becoming the most important spot . . . in the whole world . . . We cannot, we must not see it American.'[24]

The United States government would not tolerate Britain exercising its power over so important a part of its world. Washington asserted the sovereign rights of Nicaragua over San Juan/Greytown, and the two powers nearly came to blows over the village. The compromise was the Clayton–Bulwer Treaty of 1850. Britain and the US pledged themselves to facilitate the construction of the canal and to guarantee its neutrality when it was opened to oceanic traffic. Neither country was allowed to fortify or colonise anywhere in Central America.

Fast-forward five years and Greytown was centre-stage in the escalating crisis between Britain and America. In his State of the Union address in 1855, President Pierce accused Britain of violating the Treaty by occupying the Mosquito coast from Honduras through Nicaragua to Costa Rica. That was part and parcel of increasing British aggression in the Americas, inseparable from the recruitment controversy.

Although the British did not, strictly speaking, occupy the Caribbean coast of Nicaragua, they did interpret the concept of 'protection' rather generously. British warships stood off Greytown, enforcing the payment of port and custom duties to the Mosquitos. When a British warship fired upon a steamer owned by the Accessory Transit Company for non-payment of port tariffs to the Mosquitos, among the passengers was the company's owner, one of the richest men in the States, Cornelius Vanderbilt himself. The municipal authorities of Greytown had repeatedly tried to exercise jurisdiction over Vanderbilt's business operations. When things turned nasty, as they often did, the Transit Company's property was attacked by local mobs. The Americans believed that the British were behind this interference and that their consul was de facto dictator of Greytown.

The possibility of open violence between Britain and America was always present. American warships were often to be seen off Greytown, there to protect US citizens and business interests. On one notorious occasion in May 1854 the captain of one of the Transit Company's steamers murdered a resident of Greytown. It just so happened that one of his passengers was Solon Borland, the US representative in Nicaragua and an advocate for US annexation of the country. Rifle in hand, he prevented the arrest of the captain, on the grounds that the authorities of Greytown had no jurisdiction over US citizens. Borland was cut on the face by a bottle thrown by the seething mob. He returned to the United States and brought the matter to Secretary Marcy. A few weeks after the fracas USS *Cyane* bombarded Greytown in reprisal and a party of marines burnt the town to the ground. It did

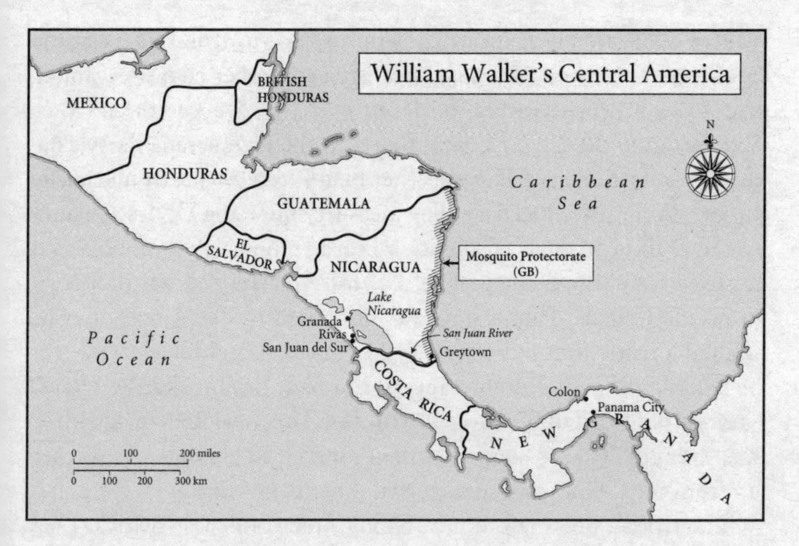

so in the presence of a Royal Navy schooner, HMS *Bermuda*, which did not open fire on the American ship because it had only three guns to the *Cyane*'s eighteen.

The British government was incensed. Clarendon called it an outrage 'without parallel in modern times'; Palmerston talked about seizing Alaska. But nothing happened, except to sour relations between the two powers: Britain was embarking on the Crimean venture and would not be diverted.[25]

Between them, the British and Americans, in their rivalry, had made this part of Nicaragua the scene of violence and anarchy. They had created a fault line in their relations. Behind the scenes members of Franklin Pierce's administration boasted about inflaming the recruitment controversy to beat Britain back in Central America. Now, in the winter of 1855, Walker's takeover of Nicaragua made the situation fraught with danger. He asserted Nicaragua's complete and unlimited sovereignty over the Mosquito Coast and Greytown.

For the British, William Walker was no better than a pirate. He was, it was suspected by the state and public opinion, the stooge of the US government, sent to Central America as a freelance conqueror to overthrow British interests while the Crimean War diverted attention.

Walker's adventure was regarded as the prelude to a massive American land-grab in Central America; Nicaragua and other countries would, like Texas a decade earlier, be taken over by adventurers and then absorbed into the Union as new states. At the Democratic Party's national convention in 1856 none other than Pierre Soulé, troublemaker *par excellence*, introduced a policy measure endorsing US 'ascendancy in the Gulf of Mexico'. Clarendon heard rumours that America was pushing for war over the issue of Central America and was preparing to invade Canada. Palmerston was infuriated by the Americans and wanted a battle fleet to be sent to the Americas and for Greytown to be blockaded. That did not happen; if it had, Britain and the United States would certainly have gone to war. The West Indian squadron was strengthened as a show of Britain's intent. Regiments were picked up from the Crimea and transported directly to Canada.[26]

The British press talked of ravaging American coastal cities. They did not want war with the US, to be sure. But if Britain was pushed to it, it would destroy its transatlantic cousin. Lord Clarendon said that he was afraid that peace was impossible 'as the President is evidently now careless of consequences and means to go the whole hog'.[27]

America was depicted as out of control in the pages of the newspapers – Bleeding Kansas and the caning of Sumner provided evidence of a country that had become dangerously unstable and wildly unpredictable. The coming conflict would not be glorious or pretty – much like an adult beating a child; but it would be quick and decisive. The War of 1812, when the British burnt the US Capitol and the White House, was within living memory: Lord Palmerston had held the post of Secretary at War back then and it was rumoured that he secretly longed to see Washington in flames again before he left office.[28]

In the late spring of 1856 things got even more heated. Walker became President of Nicaragua after a fraudulent election, and the regime was recognised by President Pierce. At about the same time, Ambassador Crampton was thrown out of Washington. Countries have gone to war over lesser matters. The time had come, declared Clarendon, 'for making a stand'. The British government ordered warships to be deployed off Greytown to prevent Walker from seizing it. The Americans were preparing to respond to the naval build-up in the Caribbean. The points of conflict were not confined to that sea, however, but everywhere that America was beginning to exert its power. Thousands of miles away the British naval squadron in Chinese waters was strengthened and reinforced in preparation for the war. The Pacific Fleet was put on notice. These were areas where the Royal Navy possessed an overmastering advantage and where US interests could be snuffed out with ease. 'The horizon,' wrote George Train, 'is as black as an inkbottle.' Britain and America teetered on the brink.[29]

For the reckless politicians in Washington and Westminster, the Crimean War had provided an ideal smokescreen for ramping up their ambitions in Central America. The White House exploited the recruitment controversy in order to rouse American patriotism against the British. Downing Street and the Foreign Office likewise baited the press with the recruitment debacle and America's pro-Russian activities. What they were trying to conceal was that they were fighting over Central America in the name of trade. Both governments realised that when the canal was constructed Nicaragua would become the gateway to Asia; both were convinced the other harboured designs to control it, the Americans via Walker and the British because of their hypocritical concern for the Mosquitos.

Added to that, Britain had been obliged to accept Napoleon III's settlement of the Crimean War, one that fell far short of its expectations and left it seeming powerless in the eyes of the world. Britain's prestige was badly dented in the spring of 1856. A humiliated superpower is an unpredictable creature. Having come out of the Crimean War a diminished military power, Britain was not prepared to be seen to be pushed around by the upstart United States. Clarendon and Palmerston were hot for a confrontation with Pierce's government; if it did not result in actual hostilities they wanted a show of devastating force in the western Atlantic, the Pacific and in East Asian waters that would chastise America and warn the republic off their turf – or turfs. The looming crisis was being driven by questions of national honour. Officials on both sides at the highest level were prepared to draw the sword and throw away the scabbard.[30]

When the moment for action came, however, Lord Clarendon fumed at the British public for its 'cowardly feeling with respect to war with the United States'. The cotton princes of Manchester dreaded the ruin of their trade; the *Daily Telegraph* depicted 'a cloud of clipper hornets' let loose from the States 'from which our commerce could not escape' after they ravaged British merchant ships. New York might be successfully burnt, but the Americans would raze Toronto, Montreal and Quebec City in reprisal. A fifth of British exports crossed the Atlantic and a half of American foreign trade entered Liverpool. There was an immense amount of British capital invested in the States. As one Boston newspaper put it, if Britain burnt New York it might lay waste to $100 million worth of property, but it would be $100 million of its own property. It was apparent that the dispatch of a powerful fleet across the Atlantic would result in the fall of Palmerston's ministry. War was averted because the public cared more about money than the politics of Central America.[31]

The government did a screeching handbrake turn, consigning traces of the bruited war to forgotten newsprint and official papers locked in the archives. It pretended that war had never been intended at all, taking instead a supercilious moral tone. 'We thought it right to be strong,' Palmerston told the House of Commons on 16 June; 'but being strong, we shall not abuse our strength, we shall not be the aggressors.' It was up to the over-mighty protagonist, he continued, to 'act with calmness, with moderation, and with due deliberation upon a matter of such great importance', especially when the two countries were bound together by 'so many mutual interests'. No fuss was made over the expulsion of Crampton. Pierce's government reciprocated with moderate language. London and Washington agreed to negotiate over Central America and, if necessary, settle their dispute by arbitration.[32]

Wars that do not happen are frequently entirely forgotten; in the light of history they seem impossibly trivial and the chance that they could have broken out at all outlandish. But in this case the determination of both sides not to back down was all too real. Public opinion, in the end, was too powerful a force to deny. It compelled hot-headed politicians to calm down and see sense. Peace was a close-run thing.

*

The world had it wrong about Walker. Or so said Laurence Oliphant. In 1856, fresh from his adventure on the trail of Schamyl in the Caucasus, he was following his reporter's nose to another distant part of the globe and another tantalising, enigmatic incendiary of the 1850s.

The riddle that was William Walker brought Oliphant to a port 'unlike . . . any other port in the world' – New Orleans. 'We picked our way across these extensive wharves, between barrels of sugar and molasses, through lanes formed by bales of cotton, past tobacco from Kentucky and Missouri, amid bags of corn and barrels of pork from Illinois and Iowa – in fact, through all that varied produce which is grown for two thousand miles upon the banks of [the Mississippi] . . . and which finds its port and export at New Orleans.'[33]

But the export awaiting transport on the levees that most intrigued him was a human one. There were Hungarian, Italian, Polish and Prussian heroes of the revolutions of 1848–9; French veterans of the Algerian campaign; British artillerymen who had recently bombarded Sevastopol. Among the Americans were filibusters bloodied in Cuba and 'Border Ruffians' who had taken arms and marched off to the civil war in Kansas. Most of the company wore the uniform of world adventurers of the 1850s – red flannel shirts and jackboots; others dressed in 'seedy black . . . as though they belonged to a sort of church militant'; some looked highly respectable, others 'detestably shabby and ragged'.[34]

Oliphant joined 300 of these men on the steamer *Texas*. 'The spirit of adventure was the moving cause of nearly all.' Most of them had tasted it at least once before and they wanted a fresh dose. They were the footloose soldiers of fortune of the 1850s with thousands of stories to tell. And they were bound for Nicaragua, rallying to the glorious cause of President Walker.

These battle-scarred men, professional revolutionaries and experienced freebooters, were allowed to leave American soil only because they claimed to be California-bound gold seekers. They kept their obligatory rifles, revolvers and bowie knives hidden away in coffin-shaped boxes.

As the *Texas* steamed out of the Mississippi delta in thick fog, Oliphant reflected that while the magnificent ship represented 'the product of a high state of civilization', the men on board were the heirs of the pirate captains Kidd and Morgan, who had plundered these same waters. But while the buccaneering freebooters of the seventeenth century sacked cities and robbed churches, those of the nineteenth 'profess to have nobler and higher aspirations'. They went not in search of gold and silver baubles, but of fertile and underdeveloped lands where they would replace the 'inefficient government', 'regenerate' the state, bring prosperity to the people and give 'the world at large' a new 'profitable market'.[35]

These musings came as a result of a series of meetings between Laurence Oliphant and Pierre Soulé in New Orleans. Oliphant might have been as ill-disposed to Walker as the rest of his countrymen when he arrived in the southern port, but Soulé gave him a thorough working-over.* The young Briton was converted to Walker's cause, fully taken with the dream of creating a 'new Anglo Saxon republic' in a benighted corner of the world. The artful and persuasive Soulé recruited Oliphant. In return for free passage to Nicaragua, a meeting with the enigmatic potentate himself and a choice Nicaraguan hacienda, Oliphant was to return to London and use his influence in Westminster and the drawing rooms of the great country houses to soften the British policy towards President Walker.

Oliphant admitted later that the offer of a great estate in Central America 'may possibly have acted as a gentle stimulant'. But it was unnecessary: like so many others he had fallen under Walker's spell (although he had not met him), intoxicated by the romance of the whole thing, the lure of adventure and, not least, the *idealism* of Walker's mission in Central America.

Or was he? What makes Oliphant so fascinating to us now, reading his travel books, is the knowledge that he was playing elaborate games in them. With the restless exuberance of a Labrador puppy and boyish naïvety, combined with irresistible charm and the gift of the gab, he was the perfect travel writer, the kind of man who encourages guarded people to drop their reserve and talk a bit too much. He was equally

* Oliphant was seemingly connected to everyone: he had assisted at the wedding of Nelvil Soulé.

personable and witty in the company of statesmen such as Palmerston, Count Cavour or Otto von Bismarck as he was with grizzled settlers in the American West, guerrillas in the Caucasus or filibusters in the Caribbean. A bed in a great English country estate or a tent in the wilderness held equal attractions for him.

These qualities fitted him for other kinds of work. The young man was a frequent guest at, among other great houses, Broadlands, the country home of Lord Palmerston, who spotted in Laurence Oliphant characteristics that were of use to him.

Throughout his career Oliphant used his reputation as an eager travel writer and *Times* reporter to open doors. His bounding energy and happy-go-lucky attitude made him seemingly sweetly innocent of politics. But in most cases these aspects of his personality covered up his real role as Lord Palmerston's secret agent. He had spied on the Russians at Sevastopol before the war; it is possible that he was gauging American intentions towards Canada when he visited Minnesota to witness the boom; and he almost became Britain's man in Schamyl's camp. Later he would act as a secret agent in war-torn Europe. But now, in the Gulf of Mexico, he was on an undercover mission to infiltrate William Walker's regime and get as close as possible to the elusive president. Perhaps it was not Soulé who had worked Laurence over; perhaps it was the other way round.

Oliphant's idealism might have been feigned. But for many others such feelings were genuine. Tens of thousands of migrants of all nations had turned Nicaragua into a highway in the early part of the decade; they cast their eyes on a hitherto unknown part of the world as they scuttled to California. And what they saw fully accorded with the racial notions of the day. A self-proclaimed 'enterprising and progressive race' marvelled at the fertility and ample resources of Nicaragua and looked askance at 'the apathy and incapacity' of the inhabitants who left their rich lands underdeveloped.[36]

Walker had taken matters into his own hands while others had merely pondered what the country would be like 'if it were only under the Stars and Stripes'. Once he seized power he set about tapping the dormant wealth of his new domain and advertised the country to white colonists as an El Dorado. Settlers were enticed with the promise of

250 acres of land on which to cultivate sugar, rice, tobacco, coffee and cotton, and with the rumour that Nicaragua abounded in gold. The watchword was 'regeneration' – the regeneration of Central America by white Anglo-Saxon men acting in the name of civilisation.[37]

The fight was on for the great Central American republic of the future. The existing Central American republics saw from the beginning that Walker's gargantuan appetite did not stop at the borders of Nicaragua. He was a freelance empire builder and if he was successful in Nicaragua thousands of heavily armed Americans and Europeans would stream to his banner and begin fresh wars of conquest on the Isthmus and then in the Caribbean. Walker's Nicaragua was invaded in 1856 by troops from Costa Rica; they were followed by those from the neighbouring republics.

In December his capital, Granada, was besieged by Honduran, Salvadoran and Guatemalan troops. The British-born General Charles Henningsen (a veteran freelance guerrilla fighter who had cut his teeth battling against Russia in the Caucasus) attracted the admiration of the world in holding off a much larger besieging force for some time. The president ordered the ancient and once-glorious city to be burnt and abandoned. Left impaled in the smouldering ruins of the 'antique, sun-embrowned, Moresco architecture' by General Henningsen was a lance inscribed with the words *Aqui fue Granada* – Here was Granada.[38]

But this was a temporary setback. Never had Walker's prospects looked as good as they did at the end of 1856. He had unshakeable faith in his own military genius and in the ability of elite Anglo-Saxon forces to overcome much larger armies of degenerate Central Americans. After all, his small, ragtag band of gringo expeditionaries had swept him to power in an incredibly short time the year before. His popularity across the United States was high. According to the *New York Herald*, Walker's Nicaragua movement was 'an expression of American enterprise, precisely at the right point': it gave security to the transit route; it bound California more firmly to the Union; 'it stations a power to the southward of Mexico, and opens to that dissolving republic a means of renovation'; and it offered a future supply of tropical produce. 'In short,' the paper summed up, 'the Nicaragua movement is wholly Anglo-American in character, orderly in administration, just in principle, and beneficial in results.' At Purdy's National Theatre in New York City a play named *Nicaragua* recounted Walker's deeds. Among its cast of characters were 'a young and ardent democrat', 'a roving Yankee' and William Walker himself, 'the hope of freedom'. The play included a medley of patriotic songs and advertised itself with the slogan: 'Now for Fun and Insurrection'.[39]

Now Walker was to be augmented by Border Ruffians from Kansas, Texas Rangers, frontiersmen who had fought Indian tribes, European veterans of revolutions and wars and men who had learnt the art of guerrilla warfare in the Caucasus. Moreover, the enemy coalition encamped in the ruins of Granada was in disarray and the Costa Rican invasion force was lacklustre and poorly led. Walker's army made a strategic withdrawal to the lakeside city of Rivas and awaited the reinforcements, rifles and cannon that were being sent from the United States in profusion.

Walker had achieved this kind of support by one of his characteristically bold, bridge-burning moves. He issued a presidential decree legalising slavery in Nicaragua. The aim was to bind his future to the interests of the American South, realising the dream of a tropical slave empire in the Caribbean. It worked. Walker's Nicaragua – and his future Central American republic – would be the new frontier of King Cotton, a prelude to seizing Cuba and Mexico. Walker was lionised in the South; Pierre Soulé began to raise men, materiel and money for the president. According to the *New York Daily Tribune* volunteers had, with the tacit backing of presidential hopeful James Buchanan, 'gone out to sustain the cause of slavery in Nicaragua, and to prepare that country for admission into the Union as a slave-holding State'. By the time Oliphant boarded the *Texas* at New Orleans in December 1856, mercenaries and Southern zealots fresh from the fighting in Kansas were rallying to the rescue of Walker's regime and the defeat (and possible conquest) of Costa Rica, El Salvador and Honduras.[40]

The approach of the *Texas*, packed with weapons and experienced soldiers, was essential for Walker's campaign. But when the steamer arrived at Greytown the situation was ominous. On shore there were Costa Rican troops and on the river one of the steamboats owned by

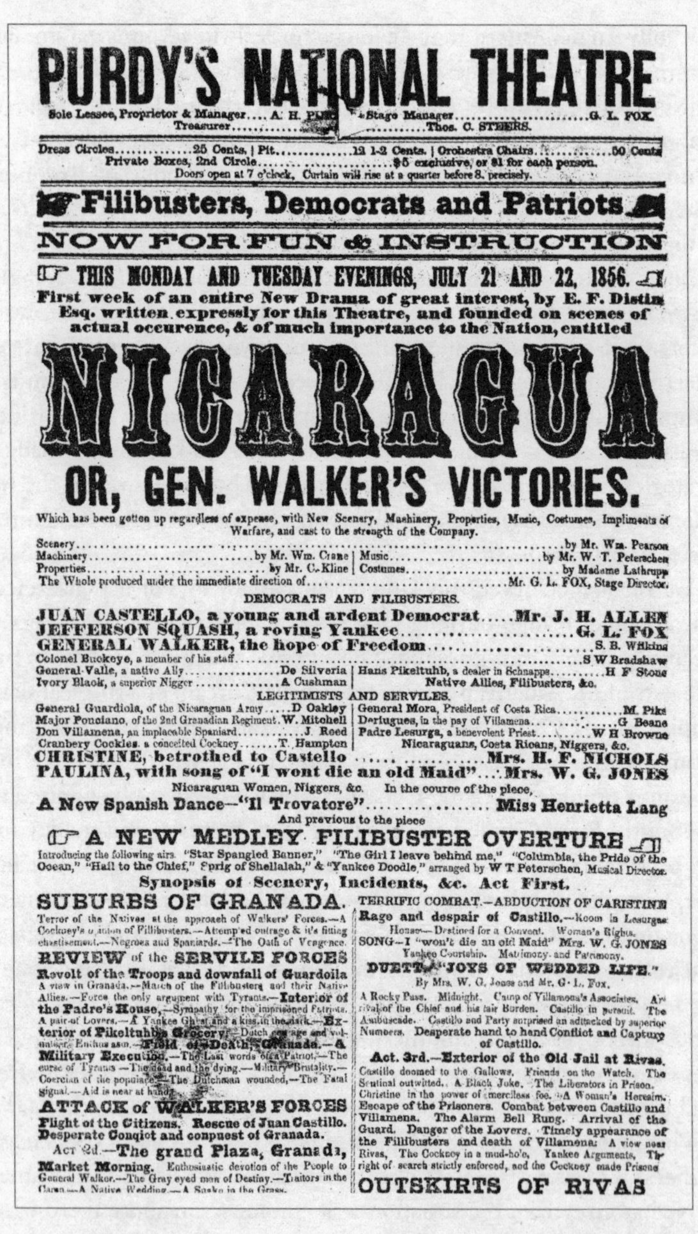

Playbill for Nicaragua, *a celebration of Walker's adventures*

the Transit Company was in the possession of the enemy; the bay was commanded by an extremely powerful British naval squadron. As the ship dropped anchor the men opened their coffin-shaped boxes and armed themselves with Minié rifles, Colt revolvers and bowie knives. The plan was to seize the steamer in a surprise night attack and head upriver to reinforce Walker.

Oliphant was put in charge of a boat for the cutting-out operation against the Costa Ricans. But before the sun set a British warship, the steam corvette HMS *Cossack*, sent over a boat to the *Texas*. Captain Cockburn boarded the American steamer and informed the filibusters that if any attack took place the guns of the *Cossack* would open fire. When Oliphant began to remonstrate he was recognised by the naval captain as a British subject and, 'being where a British subject had no right to be', was ordered off the ship and into the boat. Oliphant, dressed in the rough clothes of a filibuster, was brought before the commodore of the British squadron on his imposing new 80-gun screw-propelled ship of the line HMS *Orion*.* The commodore was astounded to recognise the young brigand immediately – they were cousins.

The filibusters' attack on the river steamboat was called off in the face of the firepower of the British squadron. The Costa Ricans stole away to the safety of their riverside fortresses. This was one of the most important moments in Walker's career. A golden opportunity had been blocked by the Royal Navy. From the decks of the *Orion* Oliphant over the next few days observed through a telescope the activities of his former companions as they prepared to fight their way through Nicaragua to Walker. The interference of the Royal Navy was not finished. With the guns of his ship trained on Walker's forces, Captain Cockburn ordered the mercenaries to form up on the beach and he went down the line picking out British subjects. By now the filibusters knew the danger of their situation; almost all the Europeans affected to be British and Cockburn affected to believe them and carried them back to the *Cossack*. From then on the British received,

* HMSs *Orion* and *Cossack* were new state-of-the-art warships that had recently seen service in the Baltic theatre of the Crimean War.

without question, deserters and returned them free of charge to the United States as a way of draining Walker's force.

Walker's most deadly enemy was not Britain or the United States, Costa Rica or Honduras; it was one of the most ruthless businessmen in history. The speculators who had funded Walker's rise to power did not do so out of the goodness of their hearts. In payment, the president had revoked the charter of the Accessory Transit Company, seized its property and gave it and exclusive rights to the transit business to his sponsors, C. K. Garrison and Charles Morgan.

In invading and controlling Nicaragua, Walker had not just trodden on the toes of the British; he had made a mortal enemy of no less than Cornelius Vanderbilt. Garrison and Morgan were on the board of Vanderbilt's Accessory Transit Company. By paying Walker to intervene in Nicaragua and propelling him to the presidency, they had orchestrated one of the most outrageous boardroom coups in history. Vanderbilt burned with rage; his company had been stolen from him by two former subordinates and a diminutive maverick from Tennessee. According to the obituary of the great tycoon in the *New York Times*, Vanderbilt wrote to Garrison and Morgan: 'Gentlemen, You have undertaken to cheat me. I won't sue, for the law is too slow. I'll ruin you. Yours truly, Cornelius Vanderbilt.'

There is no other evidence that the letter was sent. But the tenor of it certainly reflected Vanderbilt's determination to regain his company. He sent money, weapons, mercenaries and military advisers to the Central American countries to stiffen their resistance to Walker. He earned the enduring nickname 'Commodore', however, for his brilliant military strategy. The lacklustre Costa Rican troops were put under the leadership of one of his agents and ordered to capture the Transit Company's riverboats so that they could control the entirety of the San Juan River and Lake Nicaragua. In doing so the Costa Ricans, under the command of Vanderbilt and his agents, struck a devastating blow against Walker. By the time the *Texas* got to Greytown the Nicaraguan president was – although he did not yet know it – blockaded. And, as importantly, the migrant route across the Isthmus was closed for business. Vanderbilt's revenge was complete.

Morgan and Garrison were ruined; Walker's forces divided.

One half perched at Greytown, half-starved and riven with tropical diseases; the other half, including the president, was stuck at Rivas in dire need of supplies and reinforcements to hold off the Central American army.

Walker's army crumpled and the gringo president fled to the protection of the United States on 1 May 1857. Within six months he was back in Nicaragua; he was followed there by the US navy and arrested by a party of marines on arrival. In 1860 the residents of the Bay Islands invited him to carve out an English-speaking colony from Honduras. 'President' Walker (as he still styled himself) jumped at the chance. This time he was arrested by the Royal Navy at Trujillo and handed over to the Honduran authorities, who executed him in front of a firing squad. He was thirty-six.

Motivated by an unquenchable, almost psychotic lust for power rather than by money, he was a hero to some and an outrageous villain to many more. The real William Walker has remained an enigma; few penetrated beyond his icy, unsmiling exterior. As his regime in Nicaragua fell apart he began to bear himself 'like an Eastern tyrant – reserved and haughty'. He was remote from his followers, locked in his own private delusions.[41]

Not the least remarkable thing about his madcap career was the way in which many of the dominant themes of the 1850s converged upon him. His conquest of Nicaragua was a by-product of the gold rush. Walker's powerbase was California and many of his followers were drawn from the floating population of Forty-niners cursed with incurable wanderlust. Gold and the resulting torrent of settlers made Nicaragua so important to the world. The business of migration was one of the biggest, perhaps *the* biggest, businesses going in the 1850s: it was worth fighting a private war over if you were Cornelius Vanderbilt.

Nicaragua offered tempting opportunities for the future – not least the possibility that it would be the site of a canal that would connect the world's two great oceans and revolutionise global trade.* Now that

* The canal has still not been built. But there is every chance that a 178-mile ship canal across Nicaragua will soon be constructed, with Chinese backing.

was a prize worth spilling blood and treasure for. Vanderbilt and his former associates hired armies to fight for it; Britain and America went to the very brink of war. That war-that-never-happened was triggered by events in Central America, but its long-term cause was the tensions that built up between Britain and the United States during the Crimean War. When 'the Immortals' marched into Nicaragua they probably did not much heed a distant war between Britain and Russia. Why should they? But the world was more complicated than they could imagine.

And then there was slavery. Although a Southerner, William Walker did not seek to take over Central America in the name of King Cotton; he was too far gone in megalomania for that. Only when things began to unravel did he legalise slavery in a bid to galvanise support in the southern American slave states. He went from being a hero writing the American future in the jungles of Nicaragua, an adventurer who could unite his countrymen behind him, to become, in the eyes of many, the pawn of slave power and the puppet of Pierre Soulé. 'In doing this,' lamented one former supporter from New York, 'William Walker has lost his best and only chance of being a great man.' Indeed, the whole idea of American expansion had become intertwined with slavery. It all went back to 1854 and the fateful passage of the Kansas-Nebraska Act.[42]

Until then many people across the Union were prepared to support, or tacitly permit, the annexation of Cuba in the name of Manifest Destiny. Had the Pierce administration and the Democratic Party not impaled itself on Kansas it would have had the strength to seize the island. After the Act was passed the Democrats were wiped out in the Northern states, where attitudes had hardened against the expansion of slavery. The slave states chose to make a stand over Kansas, and won a pyrrhic victory: the incomparably richer prizes of Cuba and Central America were lost to them as the country split along sectional lines. Americans in the North who had been inspired by Manifest Destiny and enticed by the rich commercial potential of Cuba now looked on territorial gains in the Caribbean and Gulf with downright hostility. After the Act was passed the Pierce administration became the lamest of lame ducks.[43]

The stupendous Great Exhibition of 1851 signalled the outrageous self-confidence of the 1850s.

The rough, tough, bearded world of the frontier: gold-seekers prepare food in Victoria in 1851–2. The clothes, beards and demeanours were typical of pioneers across the West's many settler frontiers. All that is missing here are the ubiquitous revolvers, rifles and bowie knives.

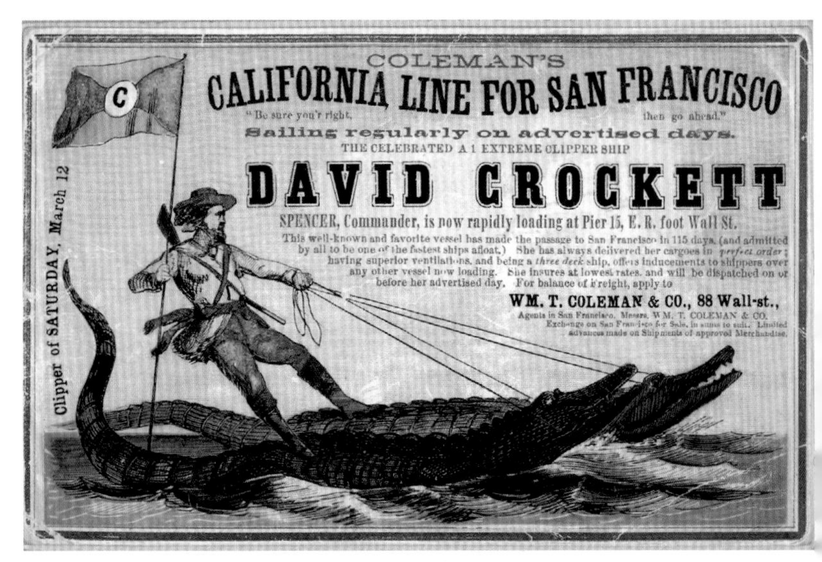

Advertising speed: Coleman's clipper line announcing the departure of the *David Crockett*.

'It is our turn to be masters now; you will be our servants yet!' – Australian gold-diggers celebrating their new fortunes with lavish weddings.

Chicago's grain elevators were at the heart of a steam-driven transportation network that opened up vast tracts of prairie land to the world market.

The western arm of global capitalism: a slave auction in New Orleans. The worldwide demand for cheap, slave-produced raw cotton helped sustain the boom of the 1850s.

The eastern arm of global capitalism: a view of an opium warehouse in India shows the stupendous scale of the narcotics trade, which was deemed essential in opening up China to world commerce.

The Bund, Shanghai: the headquarters of the West's presence in China.

Heroes and Villains

Laurence Oliphant: traveller, diplomat and spy

The 'Prophet Warrior' of the Caucasus, Imam Schamyl. A largely forgotten hero of the 1850s, he was regarded at one point as holding the fate of several global empires in his hands.

The Napoleon of Central America: William Walker

The self-proclaimed rejuvenator of Asia: Sir John Bowring

The reluctant, but devastatingly effective, imperialist: Lord Elgin photographed in Beijing in 1860 after he had sensationally occupied the city, burnt the Summer Palace, and brought the Qing dynasty to terms.

Prince Gong, photographed in Beijing after he capitulated to the British and French: 'a proceeding', commented Elgin of the photo shoot, 'which I do not think he much liked'.

The man who set out to wire the world: Cyrus West Field

Master of time: Julius Reuter

Giuseppe Garibaldi, hero of the world

John Brown: the abolitionist who saw himself as the agent of Jehovah

(left) The samurai scholar, Yoshida Shoin

(below) Ii Naosuke, dictator of Japan at its moment on the crossroads of history

The Victorian heyday in Asian eyes: the arrival of Commodore Matthew Perry's technologically advanced 'Black Ships', depicted as a sea monster by a Japanese artist.

The ship that wired the world: Brunel's *Great Eastern* arriving in Newfoundland with the Atlantic Cable. In 1866 she laid two transatlantic cables and a third in 1869, the year in which she laid the cable connecting Aden with Bombay.

The Kansas-Nebraska Act fundamentally altered the history of the United States. But it was an earthquake that reverberated beyond North America. Without it, American public opinion might have responded with united hostility to British interference in the Caribbean; an immeasurably stronger Pierce government would have been well placed to lead the country to war over Cuba and Nicaragua. And had the US prevailed, the South would have won territories that controlled some of the world's key commodities and sent senators to Washington. The course of the nineteenth century would then have been radically altered. With so much political and global economic power concentrated in the tropics, the future of slavery would have been assured long into the future.

Or at least that's how some influential voices in the South had it. After his fall from power, and before his final ill-fated foray, Walker toured the Southern states as a hero. His defeat did not end dreams of empire-building; it exacerbated them. During the late 1850s a secret society called the Knights of the Golden Circle attracted 65,000 followers, including twelve state governors and several members of the cabinet. The KGC had an eager army ready and willing to follow in Walker's footsteps and Americanise Mexico, Cuba and Central America and add these places to the land-hungry realm of King Cotton. Expansion had become an obsession in the South, a sacred cause that would determine its long-term survival and continued prosperity. Blocked by the North first in Kansas and then in the Caribbean, many were openly questioning the value of remaining in the Union. Imagine, they said, a new republic, an empire of cotton and sugar fringing the Gulf, with New Orleans as its great metropolis.[44]

For thousands in the South these were not airy dreams but an achievable goal. Walker's legacy was to awaken this lust for conquest and embolden the South to contemplate going it alone. At the Southern Convention in Montgomery in 1857 a motion was proposed stating that the 'increase of Southern commerce and Southern power [in Nicaragua] is of paramount importance to all other questions before the American people'. As one delegate put it, 'the only way in which the South could extend her territory and institutions, was by way of Central America, and from there northward towards the

United States'. At the following year's Convention held in Vicksburg another delegate declared that if the South became an independent nation it could use the threat of turning off the supply of cotton to bend the world to its will: 'crowns would bend before her; kingdoms and empires would enter the lists to win her favors . . . it would be in her option to become the bride of the world, rather than to remain as now, the miserable mistress of the North'. Then would the South be free to create its own empire and reopen the slave trade to make Mexico and Central America productive.[45]

But, as the Anglo-American crisis of 1855–6 made clear, Britain would not hesitate to intervene militarily to prevent American expansion. In the previous decade it had thought about colonising California and Oregon, but had stepped aside to allow America to take them. But in the 1850s it was taking a harder line in the Caribbean. That was a geopolitical reality that no amount of Southern fantasy and belief in its leverage over the world could overcome. Throughout the 1820s, 1830s and 1840s, the South had been able to gain tracts of productive land from decaying relics of the Spanish Empire – such as Florida and Texas – in which to grow cotton and spread slavery. Now, in the mid-1850s, it had to face up to the fact that continual expansion was being prevented, for different reasons, by the North and by Britain. The feeling that the cotton empire was being hemmed in, even at its moment of greatest vitality, gnawed at Southerners. Looked at in this way, the spat between Britain and America was more than a sideshow: it marked a crucial moment in the South's history.[46]

So many might-have-beens. The dispute between the US and Britain simmered but did not come to the boil; the slave states did not fulfil their imperial destiny. But examining this strange passage of history is like lifting a stone and seeing the unfamiliar bug life below. It reveals a world of fault lines and unexpected flashpoints in places that don't often receive much attention. Nicaragua and the Caucasus, William Walker and Imam Schamyl have this in common: remote lands and self-motivated adventurers could suddenly flare up into global significance and ignite chain reactions that detonated around the world. Because disaster was avoided at the last minute, or grandiose plans failed to materialise, such incidents are often passed over. But examine the details and a clearer picture of the era becomes apparent. The decade of progress, wealth and idealism had a darker side. The rush for riches and resources, the hunger for imperial acquisitions, made the world dangerously volatile in the 1850s. Looked at from the vantage point of Dagestan or Greytown, instead of Melbourne or San Francisco, the golden years of the mid-century appear rather different.

Gold, migration, slavery and the Crimean War: Walker blundered into these intermeshed issues and events that dominated the decade in his monomaniacal bid for personal aggrandisement. But Walker would not have achieved as much as he did if he had not spoken so powerfully to the spirit of the age. 'What is life, or what is success,' he declared, 'in comparison with the consciousness of having performed a duty, and of having co-operated, no matter how slightly, in the cause of improvement and progress?' Some of those who rallied to his banner called themselves 'regenerators'. They wanted to be part of a movement that would, as they saw it, invigorate an important part of the world with Anglo-Saxon values. Young, idealistic and intoxicated by the 'buccaneering influences' of Walker, these men could see no moral difference between colonisation by a state and colonisation by an individual. Walker's cause was, for them, the cause of civilisation in general. For them, the president embodied the reckless utopianism of the heyday.[47]

But a private individual who tries to change the world is labelled a buccaneer and terrorist. Nation states act with impunity. The aims and methods of William Walker and his small band of 'regenerators' were little different from those of the great powers in East Asia. While the Southern states were fixated on the tropics as the field for imperial expansion, those of the North looked further afield. One American journalist said that 'the eye of the nation, which has for some years been resting on the quartz mountains of California, is now bent on the ancient shores of Asia; there will doubtless be opened the next act of the drama of our republican empire'. And it was there, in East Asia, that the US would join Russia, France and Britain in competition for the continent's spoils. In the words of William H. Seward, the Pacific region would be 'the chief theatre of events in the world's great hereafter'.[48]

9

Tsunami

YOKOHAMA

The opening of commerce with Japan is demanded by reason, civilization, progress, and religion.
Democratic Review, 1852[1]

The peaceful world is now shaken up; those above quake, those below quiver.
Popular song, Edo, 1854[2]

Most of the time, the giant catfish Namazu is held under a rock by the god Kashima. But in December 1854 the deity let his guard slip and Namazu began thrashing around. The first intimations of his violent underwater contortions were felt in the Tokai region of Japan with a severe earthquake later calculated to have a magnitude of 8.4. More than 10,000 buildings in central Japan were destroyed; over 2,000 people lost their lives. An hour later an enormous tsunami surged towards the coast. It hurled great sea-going junks onto land, destroying bridges, buildings and temples; thousands of houses were submerged in the tidal wave and washed away; a pine forest was obliterated. At Suruga Bay the great wave deposited 700,000 cubic metres of sand on top of a village, leaving it for ever buried under a sand dome.

Namazu was not finished. The very next day another submarine megathrust hit the same region, with yet more sickening losses of homes, buildings and people in Osaka and elsewhere. The resulting tsunami also impacted on Shanghai and was observed at San Francisco and Lima. On 11 November the next year the third and last of the Ansei Great Earthquakes had its epicentre in the megalopolis of Edo (modern-day Tokyo). With a population in excess of a million it was the third largest city in the world after London and Beijing. In all, 50,000 houses and fifty temples were shaken to the ground or consumed by fire that day or during the eighty aftershocks; more than 7,000 people were killed.

The devastating earthquakes and tsunamis that battered Japan in 1854–5 sparked an unprecedented popular reaction. In the weeks that followed the Edo quake over 400 separate prints, known as *namazu-e*, depicting anthropomorphised catfish in all sorts of situations appeared on the streets. Some are expressions of anger and vengeance – the crazed sea monster is viciously chastised by Kashima and put back under control, sometimes with the help of the lower classes of Edo. The woodblock prints were displayed to ward off further earthquakes. They were also a means of shaming those who profited from the destruction – landlords and builders, fast-food sellers and physicians rush to the aid of their friend Namazu to save him from the retribution inflicted by Kashima and the mob; sometimes construction workers are shown worshipping the fish. The wreckage of an earthquake is an opportunity for dramatic change: in a few broadsides Namazu inspires radical wealth redistribution. For although Namazu unleashes fire and carnage, destructive events are pregnant with revolutionary change. Namazu is Janus-faced: in one sense he brings misery; in another he offers 'world rectification', albeit by brutal means.

Weird, mysterious and darkly humorous, the *namazu-e* of 1855 have myriad interpretations. Above all they reveal an anxious, apprehensive society. Japan had been rocked by serious earthquakes since 1847. Other ominous signs had plagued Japan since then: there were devastating cholera outbreaks and food shortages; the country was convulsed with popular unrest. In 1854 doomsayers were spurred on by yet another portentous sign: the imperial palace at Kyoto and thousands of houses in the city were destroyed in a mighty conflagration; a month later Kyoto was shaken by the most powerful earthquake ever felt there.

Victims of the Ansei earthquakes take revenge on Namazu while those who benefited from the disaster rush to the catfish's aid

The *namazu-e* reflect this tumultuous age of fire, earthquake and tsunami. Were the gods punishing Japan? Did it herald epochal transformation? And if so, was Namazu a destructive force or the harbinger of 'world rectification'?[3]

The grinding of tectonic plates was just one cause of upheaval. The peace and stability of Japan was being rocked by forces gathering outside the empire. Back in the sixteenth century Japan had been inundated by Westerners who brought the destructive gifts of gunpowder and Christianity. Japan was racked with internal discord. In the 1630s the shogun – Tokugawa Iemitsu – expelled all foreigners from the sacred soil and forbade Japanese from leaving the country. Trade was permitted with Korea and China, but the only Westerners allowed to do business with Japan were representatives of the Vereenigde

Oostindische Compagnie (the Dutch East India Company), who were strictly confined to the tiny manmade island of Dejima in Nagasaki harbour.

The emperors during this time resided in Kyoto with only ceremonial functions. Real power was exercised by the shoguns and their military government – the *bakufu* – in Edo. It was the special responsibility of the shogun – the hereditary military ruler – to keep Japan and its divine emperor pure from the contamination of foreigners. Intercourse with the outside world did not come to an end under *sakoku* – the seclusion laws – but it was strictly regulated. For over two centuries Japan remained closed, a mystery to the West.

But then the British came in force to the region. The First Opium War (1839–42) shocked Japan. China was overwhelmed by the awesome modern military hardware deployed by the Royal Navy and, with the imposition of the Treaty of Nanking, utterly humiliated. Monstrous steam gunboats wrested trading rights from China; Britain gained Hong Kong. As a result the great Asian empire fell into internal discord. 'How can we know whether the mist gathering over China will not come down as frost on Japan?' asked one Japanese scholar in 1847.[4]

That was the year the earthquakes began. Since then the Japanese watched events in Asia with alarm. Britain consolidated its power in India and annexed parts of Burma. China descended into vicious civil war with the rise of the Taiping Rebellion. Japan was surely next on the British hit list. The warships of the 'outer barbarians' – the United States, Russia and Great Britain – were active in east Asian waters, making no secret of their desires to wrest concessions from Japan – by force if necessary – and drag her into the global economy.

Ships from France, the United States, Russia and Britain attempting to enter Japanese waters and open treaty negotiations were turned away in the 1840s. But then in July 1853 four steam-powered warships flying the Stars and Stripes powered into Edo Bay. Japanese guardships crowded around the monstrous barbarian vessels; one displayed a large sign in French reading 'Depart immediately and dare not anchor!' But the squadron, commanded by the unruffled Commodore Matthew Perry, outgunned all the cannon positioned in Edo Bay; the ships dropped anchor amidst the shoals of Japanese vessels.

Previous expeditions had met with procrastination and dealt with minor representatives of the *bakufu*. Perry acted the part of a grand dignitary, remaining in his cabin and refusing to deal with any but the highest-ranking officials. His ships began sounding the channel to Edo; the Japanese were told that he was charting the bay in case imperial officers refused to receive a letter written by President Millard Fillmore and the Americans had to fight their way into the capital. No other expedition to Japan had been so determined or uncompromising: this was a potent manifestation of gunboat diplomacy. The *bakufu* had no option but to send important dignitaries to meet Perry in person at Kurihama.

On the day appointed, Perry's immense flagship USS *Susquehanna* fired a salute of thirteen guns and the commodore boarded his barge. Under a profusion of immense banners streaming in the breeze, thousands of Japanese samurai lined the entire shore of the bay armed with swords, lances and matchlock firearms; bodies of cavalry stood behind immense screens of decorated canvas stretched out along the beach. The American ships were anchored so that their broadsides faced the ranks of warriors.

Meanwhile Perry's boat made for an improvised jetty built of sandbags. When he stepped ashore his officers formed a double line and fell in order behind him. They were preceded to the house of reception by 300 marines and sailors dressed in pristine uniforms marching to bands that played 'Hail Columbia'. The Stars and Stripes and the broad pennant came next, carried by the most athletic-looking sailors. They were followed by two ship's boys who bore President Fillmore's letters written on vellum and contained in rosewood boxes with gold embellishments. Then came Perry in full dress uniform, flanked by two enormous African-American bodyguards. Both sides attempted to outdo each other in pageantry and spectacle.

Inside the prefabricated pine reception pavilion Perry and his suite entered a chamber richly decorated with violet silk hangings with the imperial coat of arms embroidered in white. Perry was conducted to a priest's throne on a raised dais. He sat opposite two of the highest officials of the *bakufu*, who were dressed in heavy silk brocade interwoven with gold and silver figures. These men 'assumed an air of statuesque formality which they preserved during the whole interview'; they never said a word or changed expression, only standing up briefly when Perry entered. For some minutes no word was spoken in the chamber. The uneasy silence was broken by one of the interpreters. Then the boys and the African-Americans presented the boxes containing the letters. In return the American party received a letter of receipt. And with that the ceremony was over. Foreign military had forced their way onto Japanese soil without resistance from the shogun. It was one of the most significant moments in the history of the world in the 1850s.[5]

Perry promised to return with even more ships in a few months' time to receive the emperor's reply. Just after he left, Admiral Vasilevich Putiatin of Russia entered Nagasaki harbour.

Perry returned to Japan in 1854 and got his treaty after threatening war: American ships were allowed to procure stores at Shimoda and Hakodate, and a consulate was established at the former port. By then the Crimean War had begun and the conflict between Russia and Britain swirled into Japanese waters. In October, Admiral Sir James Stirling of the Royal Navy negotiated the use of Nagasaki and Hakodate for his ships as bases for the war against Russia in the Pacific. A little later his enemy, Admiral Putiatin, did one better than Perry or Stirling, securing Nagasaki, Shimoda and Hakodate for his fleet.

That year was a turning point for the whole of Asia, from the walls of the Caucasus Mountains to the eastern shores of Japan. Nature seemed to respond to the upheaval in human affairs. Putiatin was still in port when the first of the three Ansei Great Earthquakes shook the seas.

It was little wonder that the artists of the *namazu-e* and their audiences caught this sense of imminence. Menacing and full of foreboding, Perry's steam warships were dubbed the 'Black Ships'. Many artists made the connection between the American incursion and the Edo earthquake. In one image Namazu is partially submerged and resembles one of Perry's Black Ships lurking in Edo Bay. The benevolent-looking giant catfish spurts out gold coins, while onshore people implore it closer.

In another print Perry and the catfish fight a neck-to-neck

tug-of-war. 'You stupid Americans have been making fun of us Japanese for the past two or three years,' Namazu declares to Perry during their contest. '. . . Stop this useless talk of trade; we don't need it . . . Fix your rudder and sail away at once.' Perry then boasts that the United States is a benevolent democracy. But the catfish is having none of it. 'Shut up Perry,' he snaps as the umpire indicates that he has won the tug-of-war. The Americans have come as pirates, bent on despoiling Japan, despite their fine words. 'Knowing this, the gods of our country have gathered together and have caused a divine wind [*kamikaze*] to blow and sink your ships and those of the Russians. For sure in the eleventh month . . . the gods struck out against your rudeness.'[6]

Namazu may not have got Perry, but he got the Russians. At Shimoda, Putiatin's Black Ship, the steam frigate *Diana*, was spun around on its anchor over forty times in the tsunami, badly damaged and finally sunk in a storm. Perhaps the earthquakes and the deadly tsunamis were aimed not just at waking Japan from its lethargy. Perhaps Namazu was shaking the seas to smash the Black Ships of the intruders.

—

The *namazu-e* reflect the doubts and divisions that were tearing Japan apart in the wake of Perry's mission. There was unanimous agreement that Western intrusion would bring untold harm; the pride and purity of the empire would be destroyed by contamination. Just look at the once-mighty empires of India and China. But there unanimity broke down. Japan was facing an agonising dilemma. By the 1850s it was clear that Western technologies – particularly those of war – could not be matched by non-Western countries any more. The question, then, was how far a country could go in accepting modernisation and free trade while preserving its identity and uniqueness.

Many bitterly resented the *bakufu* for caving in to the Americans. The shogun had failed in his sacred duty of keeping out the barbarians. Tokugawa Nariaki, the daimyo (feudal lord) of Mito, led the argument that Japan should resist all encroachments by the maritime powers. In a paper entitled 'Japan, Reject the Westerners' he recalled the time when Japan was strong enough to keep foreign ships away and warned

Perry wrestles with Namazu, from a print of 1855

his fellow daimyo of the delusions of Western modernity: 'At first they will give us philosophical instruments, machinery and other curiosities . . . [but] trade being their chief object, will manage bit by bit to impoverish the country; after which they will treat us just as they like . . . and end by swallowing up Japan.' He concluded by saying that Japan had one opportunity to drive away the barbarians: once they gained a foothold, however precarious it seemed at first, they would never, ever be shaken off. One of Nariaki's followers declared that 'we must now make our islands into a castle and think of the ocean as a moat'.[7]

Ii Naosuke, the powerful daimyo of Hikone, retorted that 'when one is besieged in a castle, to raise the drawbridge is to imprison oneself and make it impossible to hold out indefinitely'. Naosuke did not want trade any more than Nariaki. But he knew that Japan was helpless against modern warships. The answer was to appease the barbarians by offering them minimal concessions, including limited trade, thereby buying time and foreign currency so that Japan could acquire the military technology necessary to defend itself in the mid-nineteenth century. Japan would then not need a drawbridge or a moat: she would be equal in power and status to the West, able to dictate terms to them.[8]

The *bakufu*'s ruling council and the daimyo sided with Naosuke's view. Japan must strengthen herself; but she could not fight a war. Perry was appeased. In part the hope was that by dealing with the US, the most dangerous enemy – Great Britain – would be held at arm's length. One daimyo advised that Japan should court the United States, the Netherlands and Russia to keep out Britain. In other words, 'to use foreign countries to control foreign countries'.[9]

The ruling classes were divided by the incursion of the West. But whether Japan responded by closing itself off or by offering temporary concessions, there was confidence that the elite could save Japan. The self-assurance was not shared lower down the scale. People were aware that a 'precipice in time' had suddenly opened up. Edo was flooded with woodblock prints depicting confused Japanese officials, smoke-belching ships, strange red-haired foreigners and maps of the world. They reflected the mixture of fear and curiosity felt on the city's streets during these strange years. Here were the terrifying ships that had disturbed Japan's long seclusion. But there is also pride: the maps put Japan squarely at the centre of the world, and some prints depicted a sumo match that Perry and his men attended on their second visit – the weedy Europeans are contrasted with the monumental wrestlers. The images also reveal millennial apprehensions in these years and the loss of faith with the nation's protectors. The Great Earthquakes and tsunamis that tore Japan apart in the wake of the intrusion were a sign that the shogun had lost the support of the gods. 'The peaceful world is now shaken up,' went one popular song at the time, 'those above quake, those below quiver.'[10]

It was clear that the world had changed overnight; the arrival of Perry heralded the collapse of the old order and the dawn of a new era, whatever that might be. The simmering discontent and millennial apprehensions were not just felt on the streets. Fencing academies throughout Japan were inundated as never before with samurai eager to perfect their *kenjutsu* (swordsmanship) skills and relearn the traditional martial arts (lost in the soft days of peace) for the coming wars against the Western powers. The daimyo were required, in alternate years, to reside in the shogunal capital of Edo. They brought with them their retainers, numbering in the thousands. That meant that in Edo young samurai trained and learnt in the company of their contemporaries from all over Japan. They also saw, at first hand, the ominous Black Ships in Edo Bay.

Idealistic, fiercely patriotic and eager to fight, these young samurai studied, lived and debated together at a fundamental time for Japan. The academies buzzed with passionately held anti-foreign views and seethed with discontent at the old ruling classes. They looked to Tokugawa Nariaki of Mito as their guide through the chaos of the 1850s and despised the daimyo's enemies – those great lords who had dishonoured Japan by daring to make concessions to the foreigners. The youthful samurai spoiled to avenge Japan in the imminent war. 'Since foreign ships have come to several places,' wrote one of those zealots, Sakamoto Ryoma, in a letter to his father, 'I think there will be a war soon. If it comes to that you can be sure I will cut off a foreign head before coming home.'[11]

And their fury was not directed just at Perry, or even at those in the elite who shamefully favoured temporising the barbarians. These young warriors were influenced by Shinto teaching. Look at modern Japan, they said, and the cause of the present discontents would become clear. During centuries of peace and tranquillity Japan had seen rapid urbanisation and the creation of a commercialised market economy grafted onto a strictly graduated feudal hierarchy. The upper ranks of samurai had left the land, hung up their ancestral arms and decamped to the towns to serve their daimyo as sedentary bureaucrats, administrators and courtiers.

At the same time economic changes had given rise to wealthy merchants, artisans and townsmen. Although they were of lowly status in the Confucian ordering of society, the warrior elites – and even the daimyo – had become dependent on them for credit and services. In their biennial visits to Edo the daimyo and their armies of followers plunged into a metropolis teeming with tempting luxuries and over 6,000 restaurants, dozens of theatres and hundreds of bookshops, bedazzling fashions and no less alluring courtesans. The shogunal

A samurai warrior

capital was an expensive, intoxicating megacity, rife with consumerism and dissipation. It had the power to corrupt the nation's warrior elite. Even those who did not accompany their lords to Edo every other year found modern Japan a confusing, expensive place, one that had apparently deviated from its values and virtues. By the 1850s there was a large class of lower-ranking samurai living in reduced circumstances, increasingly alienated, indebted to the moneyed men, and with no significant military role.[12]

It was these samurai who flooded into the fencing academies in the 1850s and prepared themselves to defend the sacred soil of Japan from the intruders. Their anger was directed at the *bakufu*. Two and a half centuries of shogunal rule had left Japan bloated and weak. Honourable samurai had slipped into disregard or had become addicted to the enervating luxuries of urban life. The great rice merchants were ruining society, controlling prices and keeping them artificially high to the detriment of ordinary Japanese. Against this the shogunate was powerless; discontent was gnawing at the vitals of the nation. The younger samurai crowded together in the academies yearned for a return to a purer, traditional Japan, one that prized agrarian values over commercialism and celebrated the warrior ethos. They had this in common with the consumers of popular prints in Edo: the arrival of Perry and Namazu marked the beginning of revolution – 'world rectification'.[13]

That was why traditional martial arts became so important. It was also why these younger samurai turned away from the shogun and began to look towards the imperial court at Kyoto. This meant that their growing antipathy to the shogun and their feudal betters did not seem disloyal. They were going over their heads and fixing on a higher power. The emperor represented the ancient, uncontaminated Japan after which they hankered. The arrival of Commodore Perry was heaven-sent: Japan was about to awake from a long slumber, rediscovering its ancestral virtue and restoring its emperor to his celestial throne as it did so. Virulent xenophobia and adherence to the emperor were inseparable in the febrile atmosphere of 1853–4: the slogan of the nascent movement was *sonno joi*: 'revere the Emperor, expel the barbarians'.

*

Those Americans who set foot on Japan's sacred soil in 1853–4 were tantalised by the thought of discovering an Oriental medieval society preserved in aspic. As one American writer had it: 'the laws and customs of Japan are obstinately unalterable . . . All must proceed as it has done for centuries; progress is rendered impossible.' Its doors so long barred to outsiders, this civilisation was one of the last mysteries remaining on the planet to be unravelled and pawed over. And they were equally excited by the reception they would get from the people of the land where time stood still. They would be like visitors from outer space, representatives of the mid-nineteenth century equipped with the full dazzling panoply of Western technology and science.[14]

When Matthew Perry landed at the small village of Yokohama in 1854 to sign the treaty of friendship with Japan, he did so with the same pomp as he had on his first visit, and with twice as many warships. This time he brought gifts. He had boxes of Colt's revolvers and the latest rifles; a telescope and a camera; books and charts. But the centrepiece of the gift-giving ceremony was an exhibition of the defining technologies of the mid-nineteenth century: a small working locomotive with tender and passenger car, and two telegraph machines connected to each other over the distance of a mile by gutta-percha-insulated wires suspended on poles. The locomotive whisked passengers round its track at twenty miles an hour, and the telegraph communicated messages in Japanese, Dutch and English instantaneously.

Modern Western civilisation, Perry was saying, was embodied in communications technologies. Nothing else was so ground-shaking in the 1850s and nothing did more to establish Western superiority. In the West, China and Japan were bywords for once-great civilisations that had slipped into torpor and lassitude, gifting the future to the more energetic and enterprising European races. In this way the Eastern empires were a necessary mirror for the West, one that justified its ambitions for global domination: 'after all,' wrote one American, 'is it not inevitable that sooner or later those besotted Oriental nations must come out of their barbarous seclusion, and wheel into the ranks of our civilization?' According to this writer, Britain would rejuvenate India and China, America the Pacific islands and Japan; the two powers would meet at Shanghai: 'The Anglo-Saxons,' he asserted, 'are the masters of the world.'[15]

In some accounts of the ceremony the Japanese dignitaries and spectators are depicted as children, amazed by the nature-defying abilities of the modern machines. That was how people in the West wanted the Japanese to respond in this epoch-making clash of civilisations. Their excitement and awe would reconfirm the leadership of the West by highlighting the primitive state of non-white civilisations and justify America's effort to drag Japan into the modern world.[16]

But other accounts give the exact opposite impression. 'The Japanese were more interested in [the magnetic telegraph] than anything else,' said one eyewitness, 'but never manifested any wonder.' They examined the military hardware and other miracles of modernity, and had their photographs taken 'without expressing the slightest astonishment'. Perry's officers were amazed – and not a little disgruntled – that their Japanese counterparts were, among other things, au fait with the Panama Railroad (then under construction) and the workings of steam engines. It was rather an anti-climax for the intrepid explorers.[17]

For high-ranking Japanese officials the display of modern machines was impressive enough; yet it was not quite the bolt out of the blue many had assumed. Japan may have secluded itself from the rest of the world, but it had not closed itself off. That was a distinction that people in the West were slow to grasp. The shogun's court subscribed to the *Illustrated London News*, for example, and the *bakufu* had acquired books and papers detailing global politics and scientific discoveries through their Dutch and Chinese trading partners. This knowledge was strictly regulated, but the seeds of scientific enlightenment were diffused in small numbers across the archipelago. Perry did not know it – and nor did many Japanese – but his telegraph was not the first on Japanese soil.

In 1851 a telegraph constructed from home-made batteries, copper wires and porcelain insulators began tapping out messages in the grounds of the castle of the daimyo of Shinano Province. Its inventor was the samurai scholar Sakuma Shozan and, remarkably, he achieved his feat from knowledge gleaned solely from an entry in a Dutch encyclopaedia.

The Opium War opened Sakuma's eyes to the 'violent extravagance and wickedness' of the British and 'their long-cherished desire to conquer Japan'. This threat could only be countered by modernising Japan very rapidly. He not only experimented with electricity and telegraphy, but also magnetism, chemistry, photography and the manufacture of modern weapons. In the mid-1850s he ran an academy in Edo that attracted large numbers of students – 5,000 it was claimed – from all over Japan. 'This Sakuma is an extraordinary man of really heroic proportions,' wrote one of his pupils, Yoshida Shoin, '. . . Those who enter his school to study gunnery he compels also to study the Chinese classics, and those who enter to study the Chinese classics he compels also to study gunnery.'[18]

1851 was propitious in Japan's modernisation. One of the foremost promoters of technological innovation, Shimazu Nariakira, succeeded to the daimyo of Satsuma that year. He had been fascinated by Western technology from a young age and took delight in new inventions; in 1848 he imported a daguerreotype camera, the first of its kind in Japan. But he was also alive to the threat from Britain – Satsuma is on the far south-west of Japan, the front line of Western intrusion. 'Now,' he wrote, 'Japan's situation should be called extremely dangerous, our moment of crisis. Such being the case, military preparations should come first, otherwise our national prestige shall suffer and we shall finally become the minions of other countries.'[19]

As soon as Nariakira became daimyo, work began on a cluster of factories just outside Kagoshima, his capital. There was a blast furnace to produce pig iron and a reverberatory furnace to refine it. In the neighbouring domain of Saga – which included Nagasaki, Japan's only gateway to the world – the daimyo was also building furnaces and had founded a chemistry research institute. Out of these factories came cannon, rifles and ammunition; later they produced modern agricultural machinery. At the same time and nearby, Nariakira's engineers began work on a model paddle steamer and sailing warships in contravention of the seclusion laws that banned the construction of ocean-going vessels. In 1854 Nariakira launched the *Iroha Maru*, a three-mast cannon-carrying ship, and the *Shohei Maru*, the first of five Western-style warships constructed in his yards.

So much for the notion of Japan as a fossilised, unprogressive society. True enough, experiments with electricity, photography, chemical processes and industrial development were sporadic and thinly scattered. It was also a fact that the *bakufu* punished those who dabbled in foreign sciences, labelling them a threat to the social order. When one of Sakuma Shozan's pupils, Yoshida Shoin, attempted to board Perry's ship in the dead of night in order, Prometheus-like, to travel to the ends of the planet and uncover and bring back the secrets of Western knowledge to Japan, he was locked in a cage, sent to jail in Edo and then sentenced to house arrest. The tight control and conservatism of the *bakufu* stood against change.

But the Japanese were among the most educated people in the world, with high literacy rates. Throughout the first half of the nineteenth century it was not just the elite – the more energetic daimyo and the scholarly samurai – who innovated with technology; at a lower level farmers, sake brewers and artisans were developing mechanised processes. There was a reservoir of skill and enterprise in Japanese society. Take for example Tanaka Hisashige, a craftsman skilled at making astronomical instruments and clocks and experimenting in pneumatics and hydraulics. He was famous for his lifelike mechanical robots and, most spectacularly, for his year clock, completed in 1851. Made from over 1,000 intricate parts, it runs for a year without needing to be rewound and has six faces showing the Western clock, the day of the week, the month, the phase of the moon, the Chinese year and the Japanese time system. On top of the complex and beautiful mechanism is an astronomical model with the sun and the moon rotating round a globe decorated with the map of Japan. (In 2004 the Japanese government sponsored an attempt to make a working replica of the clock; it took 100 engineers using twenty-first-century technologies six months to do so.) Following this achievement Tanaka, now in his fifties, moved to Saga – the cutting edge of Japanese modernisation – where he made a small locomotive that puffed around a circular track and working models of screw-propelled steamships. He then turned his mind to armaments and telegraphy. The electrical engineering firm Tanaka founded was one of the companies that later merged to form Toshiba.[20]

These were the ingredients for explosive industrial take-off in the second half of the nineteenth century, something that the Western expeditionaries were ignorant of when they arrived bearing their high-tech gifts.

Above all there was a determination that Japan should never enter into a Faustian pact with the West: that, unlike other underdeveloped countries, it should refuse to pay the price for modern technology by becoming mired in debt and dependence on foreign powers and financiers. In the 1850s peoples who had held out against the West for time immemorial were flattened by the onslaught of modernity. Japan was prepared to be a bastion against the tidal wave. The race to catch up with the West brought with it tensions. The samurai who were perfecting their martial arts saw any attempts to modernise as a covert way of Westernising Japan. The great Sakuma Shozan was killed by the single stroke of a sword wielded by the samurai Kawakami Gensai, one of the most notorious assassins of the time and a member of *Ishin-shishi* – a secret society of 'men of high purpose' dedicated to eradicating all trace of foreign contamination from Japan by acts of terrorism.

But the modernisers and the radical activists of the *sonno joi* movement were united by the determination that Japan should preserve its integrity and not become a colony or dependant of the maritime powers. Defending Japanese customs at the same time as keeping the foreigners at bay with modern techniques was bound to be a disturbing experience. Sakuma Shozan came up with the slogan that encapsulated this attempt to preserve timeless Japanese culture while hurtling into the future: 'Western science, Eastern ethics.'

10

The Civilising Mission

HONG KONG

The English barbarians are born and grow up in wicked and noxious villages beyond the pale of civilisation, have wolfish hearts and brutish faces, the looks of the tiger and the suspicion of the fox.

Placard displayed in Canton[1]

The best time to see Hong Kong was from the sea at night. One writer said that the view of the moonlit colony was 'one of the grandest and most beautiful that can be imagined'. The boomerang-shaped outline of the city huddled under the mountains was evident from thousands of lanterns; an armada of international vessels were decked in light as they bobbed on the glowing phosphorescent water.[2]

Ships were continually arriving and departing from all over the world: whalers that traversed the globe, cargo vessels in search of freight, East India merchantmen, Yankee tea clippers, opium dispatch vessels, paddle steamers, P&O mail ships and, at any time, hundreds of Chinese junks and lorchas that ceaselessly plied between mainland China, Singapore and Hong Kong. The island commands the mouth of the Pearl River Delta. Today the region is one of the world's most productive and densely populated, a startling megalopolis in the estuary with 60 million living in its contiguous cities. Hong Kong is on the southern tip of this urban network and Guangzhou its hub. In the 1850s Guangzhou, then known as Canton, was one of the largest cities in the world, behind only London, Beijing, Edo and Paris. Back then the journey between Hong Kong and Canton took just six hours along the delta, with its endless winding creeks and myriad islands. Canton was the entrepôt of Chinese trade and, as such, the traffic between it and Hong Kong was ceaseless.

The seedy atmosphere that permeated Hong Kong was exacerbated by the huge floating population that incessantly washed through its cramped streets and into its opium dens, gambling houses and brothels. And just as impressive as this crowded harbour were the wharves, bamboo jetties, warehouses and factories that massed on the water's edge. In these vast godowns (dockside warehouses) were stored the commerce of the world – commodities from Europe, the Americas, India, Australia and China. Alongside this orgy of trade grew specialised businesses to serve it: banks and law firms, insurance companies and shipbrokers, financiers and journalists.

The skyline was dominated by the great craggy mountains scattered with huge boulders, forests and ferns and often veiled in mist. They overhang the city of Victoria, a crescent of buildings that was expanding at incredible velocity, conquering and obliterating for ever the intoxicating beauty of the 'Fragrant Harbour'.

For Western observers it was 'a sort of oasis of English houses and customs in the midst of semi-barbarism'. It was exotic and European at the same time. One travel writer captured this sense as he strolled past the all-important racecourse: 'We walked on with fine cacti making hedges, and belts of mango and lychee trees. Here and there, water rushed down over blocks of granite, with a charming Chamonix sound about it . . . Ferns grew plentifully about, and the foliage everywhere was delicious. As we turned to come back, we met several equestrians and people in carriages, out for their afternoon ride. The whole place was so tranquil, and pretty, and homelike, that I got very low-spirited . . .'.[3]

Other parts of Hong Kong were not so sublimely tranquil. A Chinese population of 4,000 in 1842 when the British took possession of the island had swelled to 80,000 by the mid-1850s. Many of these immigrants shared tiny living quarters on boats in the harbour. Business and pleasure were conducted in the Chinese part of Victoria. For many Western tourists who dropped into Hong Kong this was China,

Hong Kong in the 1850s

or what they thought was China. Here in the narrow jostling streets they took in the sights, sounds and smells of this young, frenetic city. Noses were assailed with the smells of food sizzling over open fires and incense emanating from joss houses and temples. Houses, shops, tea houses, apothecaries and restaurants competed for space, their profusion of rectangular signboards suspended just above the heads of pedestrians.

Next stop was the market. Exotic and beautiful fruit and vegetables filled the stalls; pigeons and fowls, still alive, were tied by their feet and suspended in the air; great mounds of delicious fishes freshly caught in the South China Sea lay on the fishmongers' slabs.

The city was at once a construction site and a food market: among the carpenters banging nails into wooden frames, street-food vendors busily chopped up meat, children hawked sweetmeats and artisans sat on the floor making their wares. One minute tourists found themselves caught up in a religious festival, the next they saw the melancholy sight of a chain gang being marched off to labour on the roads at the point of an Indian sepoy's bayonet. Out of the windows of grog shops came the din of carousing sailors from Britain, America, France, Germany and everywhere else. The Chinese population predominated; but on a morning stroll you would see among the crowd peoples who had been drawn to the commercial emporium of Asia by the skeins of global trade: Jews and Bengalis, Parsees and Persians, Malaysians and, of course, Europeans in black hats and black frock coats or military uniforms. Coolies toiled under prodigious weights; expressionless men carrying sedan chairs parted the swell of humanity for a brief moment as they barged through at top speed.

Meanwhile the Western inhabitants of Hong Kong relaxed in their clubs playing endless games of billiards, or they went to the American bowling alley. The white settlers in Hong Kong – and the five Chinese treaty ports opened up for trade – had a reputation in Britain for their loucheness. A whiff of scandal always hung around Hong Kong, with rumours that the European elite never scrupled to do lucrative deals with local gangsters or get rich on shady government contracts. Great fortunes were made by the merchant princes; the money was often dissipated on luxuries imported at incredible prices and in the island's infamous red light district. Certainly, the Europeans and Americans in Hong Kong and Shanghai lived like kings on the rim of the Orient. Most of the immigrant Chinese population crammed into boats in the harbour, where they dwelt in poverty. This was a city of fantastic contrasts and frenetic progress.

Some 700 miles by sea to the north a similar scene was being enacted. Whereas Hong Kong was a British colony, the foreign settlement outside Shanghai was a frontier town. Soon after the city was opened to trade in 1842 British merchants began moving into a strip of land alongside the Huangpu River outside the walled city. Like Melbourne or Bendigo or a thousand American frontier cities, the foreign settlement at Shanghai boomed. But, unlike a makeshift colonial or Western boom town, it was famous for its beauty and regularity. 'A fine quay extending for upwards of a mile fronting the river, is lined by the separate and handsome dwellings of the merchants, surrounded by broad verandas and balconies, and generally enclosed with beautiful gardens.'[4]

The Bund – the waterfront promenade – with its palatial houses, consulates and commercial headquarters, was by now one of the most famous addresses on the planet. Modern Shanghai grew and formed itself around the international settlement; the Bund, , is a magnet for tourists today. Already in the 1850s Shanghai was an essential part of the tourist's itinerary; few other cities in the world were as celebrated in the middle part of the century.

Above all Shanghai stood for something important. The colonial grandeur of the settlement and the madly opulent lifestyle of the expatriate merchants were intended to broadcast that the Western presence was permanent and highly lucrative. The West was not going away. Soon, it was assumed, other cities in China would resemble Shanghai, with impressive modern businesses, cavernous godowns groaning with the trade of the world, and the ceaseless bustle of commercial activity on the roads, rivers and waterfronts. The Western concession at Shanghai was colonisation nineteenth-century-style on turbo-charge, and proud of it.

But Hong Kong and Shanghai stood for something else. The ostentatious Western architecture expressed as stone churches, business headquarters, warehouses, clubs and consulates represented China's humiliation and powerlessness. The intrusion had been forced on China by gunboats and unequal treaties. In Shanghai and the four other open ports, Westerners had extra-territorial privileges – in other words, they were exempt from the jurisdiction of local law. The foreign settlements became self-governing and self-policing city states, glaring affronts to Chinese sovereignty.

'Jesus Christ is free trade and free trade is Jesus Christ.' This startling declaration was uttered by Dr John Bowring, one of the most extraordinary Britons of the nineteenth century. In 1854 Bowring arrived in Hong Kong as governor. His aim was to bring 'commerce, peace and civilization' to one-fifth of the inhabited world and one-third of the human race, rescuing this region and its people from, as he saw it, the dead weight of superstition, slavery, tyranny and stale custom. Just look at Hong Kong, he said. A barren, unhealthy island with 'every impediment which would clog its progress' had been turned into one of the world's most dynamic cities in a few short years because of 'the elasticity and potency of unrestricted commerce'.[5]

Sir John Bowring did not look particularly like the spiritual renovator of the world, a man of action in Asian waters. In his early sixties, he resembled a rather crabby and otherworldly professor, with wild white hair cascading down to his shoulders from a protruding bald forehead. But Bowring was a force of energy, a polyglot who published over forty books, edited the utilitarian *Westminster Review* and sat as the Radical MP for Bolton. A deeply religious man, he penned several hymns, the most famous being 'In the Cross of Christ I Glory'.

Radical politics and Christianity intersected in an unshakeable faith in the regenerative effects of free trade. In the late 1830s Bowring was one of the founding members of the Anti-Corn Law League, Britain's first mass political pressure group. Dubbed the country's 'commercial knight errant', throughout his career he led a peripatetic life, proselytising for free trade in a succession of European and Middle Eastern cities with evangelical fervour. He hectored a bewildered Pope on the moral urgency of free trade. 'You have no idea,' he wrote to his friend, the Earl of Clarendon, 'what a *mighty* influence is awakened and how the holy spirit is spreading.'[6]

'Communication,' wrote Bowring, 'is civilization in activity . . . He who can communicate cheaply and rapidly with all his fellows must be elevated by the very fact of that communication.' Free trade facilitated global communication by creating extensive – and inordinately complex – networks that criss-crossed the globe, linking diverse peoples and flattening natural and artificial barriers. It was a revolution: the old authorities that held mankind's moral, intellectual and material progress in check for millennia would be swept away in the deluge. 'The power which Rome once wielded – the power of the world – has been replaced by a mightier power: the power of commerce.'[7]

Ironically, it was his defeat on the dog-eat-dog battlefield of capitalism that propelled Bowring from armchair intellectual to man of action. When his business failed, he turned to his powerful patrons, Lords Clarendon and Palmerston, for employment. They found him the job of British Consul in Canton, a position he held from 1848 to 1853.

It was a dispiriting experience. It was clear to him that the Chinese walls of resistance to free trade were strong enough to hold back even the 'power of commerce'. And it was equally clear that his superiors lacked the vision and resolve to use military means to secure liberal ends.

But then, in 1854, he was elevated to governor with plenipotentiary powers and superintendent of trade in East Asia. An intellectual transported to a position of authority, with military force at his or her disposal, is rarely a pretty sight. Bowring was a radical firebrand who was given the power – military power – towards the end of his life to put theory into practice, with the whole of East Asia as his laboratory.

Hong Kong provided a suitable home for a free-trade evangelist. More than just a convenient base, it was Britain's beacon of unrestrained capitalism shining out to the world. One of the earliest historians of the colony declared with mystic awe that it was Hong Kong's 'mission in the history of the universe' (no less) to bring Asia into harmony with 'a divine tendency' – the 'genius of British free trade and political liberty'. In the heady years of the 1850s it symbolised the unbridled confidence and magical thinking of the age as surely as a frontier boom town, a long-distance railway or a submarine telegraph cable. Infected with the fast progress and utopianism of the time, Bowring wanted to import that dynamism to the whole of Asia from his vantage point in the colony.[8]

Hong Kong had become one of the key hubs of global trade and finance, a free-trade emporium on the margins of an enormous, disintegrating empire. This fact shaped the colony's early history. It was the headquarters of a buccaneering form of capitalism, a place of boundless possibilities, irresistible temptations and pervasive criminality.

During his voyage to Hong Kong at this time, the American businessman George Train described the 'sickening smell' throughout the ship. 'Go where you would you could not escape its stupefying influence, down below or on deck, in your stateroom or at the dinner table, the continuing nauseating smell . . . gave you the headache and the blues, to say nothing of keeping your eyes half closed when you did not care to sleep.'[9]

It was the sweet, sickly smell of opium, and it followed Train everywhere he went in Chinese waters. Just outside Hong Kong and the five treaty ports sat enormous hulks that received the drug for transshipment to mainland China. When he arrived at Wusong, Train saw six beautifully constructed and fully armed British merchant ships that had brought opium from India.

The world wanted a lot from China; it consumed its tea, wore its silk and displayed its porcelain; Western businesses slavered to access 400 million potential customers. But China did not want much to do with the rest of world. For centuries foreign traders had been confined to a narrow plot of land in Canton, where they purchased commodities from a select body of merchants known as the Hong. But the Chinese did not need or want foreign imports. Silver coin raised in revenue in India or purchased in Mexico paid for the tea and silk.

This imbalance of trade was unacceptable to Western traders and governments. The only solution was to import something the Chinese might want. During the first decades of the nineteenth century illegal imports of opium soared and the outflow of silver abated, then reversed, as the Chinese began paying for the drug in cash as well as tea. Alarmed at the corrosive moral effects, not to say the drain of precious metals, the Qing dynasty determined to stamp out the illicit narcotic trade. In 1839 the Governor of Canton confiscated 20,000 chests of opium worth £2 million.

Outraged at the slight to national honour and the destruction of its citizens' merchandise, the British government went to war. British steam warships blasted away Chinese defences and penetrated the Yangtze as far as Nanking. The war was brought to a close with the Treaty of Nanking (1842), the first of a series of 'unequal treaties' with Asian powers. It opened up four Chinese ports to Western traders in addition to Canton – Xiamen (Amoy), Fuzhou, Ningbo and Shanghai – and ceded Hong Kong to Britain.

Within a few years, it was asserted, the opium trade would wither away as manufactures, technologies and business services conquered China. By the 1850s it was clear that that dream was as far from being fulfilled as ever. The largest market in the world remained closed to the West. For one thing, China had well-developed local industries and crafts; the demand for British products was disappointingly low. And there was the political barrier. Xenophobia increased with the accession of the militantly anti-Western nineteen-year-old Xianfeng emperor in 1850. Chinese officials became more obstructive in dealing with their Western counterparts.

By the 1850s, then, far from melting in the heat of free trade, narcotics had become one of the planet's key commodities. From the City of London to Hong Kong and Shanghai, via intricate networks of trade and finance, opium was the moving force of exchange and the foundation of mighty fortunes. It set in train plenty of legitimate business, but left violence, lawlessness and political chaos in its churning wake.

As was the case with slave-grown cotton, people in Britain did not like to think of opium as one of the pillars of global capitalism. William Gladstone, British Chancellor of the Exchequer, savaged Hong Kong as a nest of pirates and smugglers, the sleazy home of 'the worst, the most pernicious, demoralising, and destructive of all the contraband trades that are carried on upon the surface of the globe'. But without the drug, what would Hong Kong be? The whole pattern of global trade and commerce would have been radically different.[10]

If slavery was an essential part of world capitalism, its global impact was relatively confined compared to opium. The lives of hundreds of millions of Asian people from India to China were bound up with it, some directly but many more indirectly. Like petroleum in our age, opium dictated geopolitics. Not only had Britain fought China to gain a market for the drug, it had embarked on smaller opium wars in the drug-growing regions of Sindh and in the central Indian kingdom of Gwalior to maintain its monopoly on production. According to Richard Cobden, many of the world's greatest multinationals owed their fortunes to war and smuggling, yet they were insured against the resulting anarchy by the protection of the Royal Navy. 'The [opium] trade,' he said, 'is founded on . . . licence and lawlessness; it flourishes in times of disorders and commotion, and anything which plunges the East into anarchy and confusion, is promoting the interest of these merchants and serving their unholy gains.'[11]

The most vicious example of the 'licence and lawlessness . . . anarchy and confusion' ushered in by opium was occurring in southern China. In 1836 Hong Xiuquan entered Canton to take his civil service examinations. He failed to pass, but while he was in the city he heard a Christian missionary preach. The next year, after he failed the exam again, he suffered a mental breakdown. During several nights of hallucinatory dreams he was visited by a fatherly figure and an elder brother, who instructed him to rid China of demons.

As the years passed, Hong began reading evangelical Christian tracts. In doing so, he came to understand his dreams. They meant that Hong was the Chinese son of God, the younger brother of Jesus. Hong's conviction grew as he studied the Old Testament with an American Southern Baptist missionary in Canton. He began preaching in the remote southern district of Guangxi, winning over thousands of converts.

By early 1851 Hong Xiuquan had gathered a sizeable, disciplined army of followers. As the Qing dynasty began to launch a campaign to suppress the movement, he proclaimed the Heavenly Kingdom of Transcendent Peace (the Taiping Kingdom) and declared himself *T'en Wang*, the Heavenly King and ruler of the world. On 30 March, at an extraordinary public ceremony in Guangxi, Jesus Christ spoke through his mouthpiece, a Hakka peasant, to his faithful followers. 'Trust completely in Your Heavenly Father and Heavenly Brother,' Jesus said; 'they will take charge of everything, so you need not worry or be nervous . . . Those who betray God won't be able to escape Heavenly Father's and Heavenly Brother's punishment. If we wish to have you live, you will live; if we want you to perish, you will die.' After Jesus finished uttering these chilling words, an opium smoker was publicly tried, beaten with a thousand blows, and then executed. This ritual of divine revelation and savage retribution would be repeated through the year.[12]

The austere puritanism and theocratic absolutism of the Taipings combined with radical ideas such as land redistribution, communism and the emancipation of women. It spoke powerfully to those whose lives had been turned upside down by the intrusion of the West. God had spoken to Hong Xiuquan; the Taipings were His Chosen People and they would return China to greatness by creating a heavenly order on earth. Throughout 1851 the Taipings defended themselves from attack. In June the Heavenly Army suffered defeat. Jesus spoke again, telling his followers to set aside their fears and focus on the coming 'Earthly Paradise' where they 'would receive rewards beyond their expectations'. By September the Taipings had captured a city, the first of many as they established a vast rebel empire in south-west China. Later in the year, Hong's disciples burnt their houses and began their 1,000-mile exodus to the Promised Land in Nanking.[13]

The Taipings regarded opium with special fury, savagely executing anyone caught abusing it. They also detested their rulers. The Manchu were the demons that God had instructed Hong to eradicate. According to the Taipings the effete ruling class had corrupted Chinese values and the empire had declined from the pinnacle of world power to abject weakness. They had let in drugs and barbarians. 'Each year [the Manchus] transform tens of millions of China's gold and silver into opium and extract several millions from the fat and marrow of the Chinese people and turn it into rouge and powder,' said one Taiping leader. 'How could the rich not become poor? How could the poor abide by the law?'[14]

By 1853 the Taiping army – now numbering 750,000 disciplined soldiers – had established their Heavenly Capital in Nanking. Millions fell under the rule of this fanatical, brutal theocratic insurgency as the Taipings secured control over cities along the fertile Yangtze valley and thrust north towards Beijing.

The coastal regions of China were on fire; Manchu authority collapsed and anarchy took over. Canton was threatened; in 1853 Xiamen was taken over by Triad secret societies allied to the Taipings who slaughtered the Manchus. From the church tower in the Settlement at Shanghai Westerners watched the countryside around the city degenerate into disorder as imperial forces and Taipings fought it out for control of the city.

The lagoons and islands of the Chinese coast were infested with pirates, drawn to the China Sea by the upsurge in trade and the breakdown of authority. On the Pearl River Delta opium clippers were waylaid and despoiled, merchant junks towed off to remote lagoons and ransomed, cargo ships plundered and their crews killed or mutilated. So dangerous were the waters and so great the threat of hijacking that P&O steamers stopped carrying passengers, and on American ships

Chinese were locked in iron cages or confined in the hold. *The Times*' special correspondent counted over 200 armed junks in Hong Kong harbour one day. Of these at least a quarter were professional pirates using the coasting trade as cover for their murderous activities and another quarter were part-timers. Hong Kong itself was where they fitted out for pillaging expeditions and where they fenced the stolen goods, often with the connivance of the colonial authorities.[15]

The Hong Kong that lurked behind the smart colonial façade was revealed with the arrest of the baby-faced buccaneer Eli Boggs. Notwithstanding his delicate white hands, coy smile and carefully parted black hair the American was considered one of the 'boldest and bloodiest' pirates operating in the China Sea, with a fleet of fifty armed junks under his command. On one occasion he was alleged to have single-handedly boarded a ship, killed fifteen people and driven the rest overboard. He also ran a protection racket in Hong Kong. When a Chinese merchant refused to pay, Boggs had him cut up and sent pieces of the body to other merchants on the island as a warning.[16]

When Eli Boggs was put on trial he claimed to have been framed by Ma-chow Wong, the leaseholder of the fish stalls in Hong Kong's Central Market. This position gave Wong enormous influence over the fishing community and, by extension, insider knowledge about the movement and activities of pirates on the Pearl River Delta. A respectable member of the community, of course, Ma-chow Wong offered his prized knowledge to the Hong Kong police and the Royal Navy. To the British he was the best informer going; gossip fresh from the fish stalls went direct to the police. To the criminal underworld, Wong was a dangerous man. And that gave him power, power he was happy to exploit. Hong Kong's pirates fell under his sway. Those who did not do his bidding were shopped to the British; people – pirates or not – who offended him were offered up to the authorities on fabricated evidence. The pirates' booty had to be sold in his shops and their profits shared with him. He became 'the orbit round which the satellites of plunder revolved', an untouchable underworld boss on a truly impressive scale.

Wong's gangster paradise fell apart when he was named by Boggs, who also made allegations that Wong operated in partnership with the British assistant commissioner of police. A few days later a police search revealed property belonging to a recently plundered merchant junk in one of Wong's shops. And it was proved that he was not just extorting money and favours from the pirates – he was a pirate chief himself. Evidence showing that Wong had organised the attack on the junk, in which the owner's wives and son were killed, was also produced. The assistant commissioner of police was accused of sharing the profits of piracy and running brothels in the city. Neither Eli Boggs nor Ma-chow Wong were severely punished. They were sentenced to exile, left Hong Kong and faded from history.[17]

'I carried out the principles of Free Trade to their fullest possible extent,' Sir John Bowring remembered with pride of his time as Governor of Hong Kong. '. . . Vessels came from every quarter and every nation. They entered, they departed, and no official interfered.' The results were plain to see. Free trade had made Hong Kong a nest of pirates, drug smugglers and people-traffickers. Here were the results of a fatal collision between Britain's freebooting style of capitalism, on one side, and on the other an imperial government that was set against barbarian intrusion but powerless to enforce its decrees in the face of civil war.[18]

The problem, as far as Bowring saw it, was not that free trade had been pushed too far, but that it had not gone far enough. If China opened itself up to the world, opium-smuggling and piracy would be driven out by legitimate business. The Chinese, Bowring believed, only had themselves to blame.

The time had come, Bowring said, to 'strike boldly at the head – the heart – and not trifle with the extremities of these great empires' of China and Japan. 'We have been trifled with – tantalised too long.' In his first year as Governor, Bowring went with warships to Bangkok and compelled the King of Siam to sign a treaty opening up his country's markets. When the Siamese court said it was going to leave some trade restrictions in place Bowring counter-threatened to leave HMS *Rattler* anchored in Bangkok for as long as it took for it to capitulate. The Siamese princes backed down and Bowring boasted that he had 'emancipated [Siam] from her intolerable yoke – and dragged her from her miseries into the bright fields of hope and peaceful commerce'. Japan, Korea and Cochin China (Vietnam) were next on his list. In June 1854 Bowring was at Shanghai, ready to go to Japan 'properly accompanied by Ships of War' to impose a treaty on the shogun that would go far beyond the limited concessions granted to Perry.[19]

This activity would be the prelude to his great work – finally and irreversibly opening up China to the miracle of free trade. The Permanent Under-Secretary for Foreign Affairs warned the Foreign Secretary, Lord Clarendon, of Bowring's 'excess of vitality and zeal': 'He would probably be over the Great Wall before we had time to look around us.'[20]

But these dreams of extending free trade throughout the region came up against cold reality. In 1854 the problem of piracy was so bad that the Royal Navy was entirely taken up with safeguarding what remained of trade in the region. And then the Crimean War spilt into East Asia.

On the banks of the Amur River there was a 30,000-strong Russian army under General Nikolai Muravev. There was also a squadron of warships based at Petropavlovsk on the tip of the Kamchatka Peninsula. The enemy, British officers suspected, was poised to establish a naval base that would give them control over the Sea of Japan and access to the Pacific. In 1854, when the British fleet went north to search for the Russian ships, Hong Kong entered panic mode: tens of thousands of Taiping pirates were about to descend, some said; others fretted that Russian frigates would enter the harbour and wreak carnage. In Sydney and Melbourne people were on alert against a surprise attack. 'We are no alarmists,' stated the *Argus*, 'but it would awaken in us no surprise to learn that a Russian fleet had issued from the Amur,' and the first they would know about it would be the opening salvo from the Russian ships as they bombarded Melbourne.[21]

British commanders in the Pacific expected the Crimean War to explode into East Asia. But it never did. The Russians were too weak to mount seaborne assaults on Japan or Hong Kong, let alone Melbourne. The British and French launched ineffectual attacks on Petropavlovsk. Apart from those there were few hostilities in the Pacific during the Crimean War.

But despite the absence of fighting, the Crimean War had a massive impact on East Asia. China and Japan were drawn into the imperial face-off between Russia and Britain and the wider competition between the maritime powers, with profound consequences for both.

Japan was caught in the crossfire of the Crimean War. Before the war the British had been lukewarm about Japan, preferring to leave it to the Americans. But the thought that Russia might establish naval bases there for commerce-raiding changed all that. Of all the maritime powers, the Japanese feared Britain the most. Britain alone of the great maritime nations was an Asian and Pacific power, with a string of bases from the Middle East and southern Africa through to the Indian subcontinent, Singapore, Australia and Valparaiso in Chile; and there was, of course, Hong Kong and Shanghai. The flagship of the China Squadron, HMS *Monarch*, was, with 84 guns, the most powerful warship between the east African and the American west coasts.[22]

Japan was braced for the full storm of British naval power in 1854. The Dutch informed the *bakufu* that Bowring was preparing to cross from Shanghai with the Royal Navy. In August Admiral Sir James Stirling, commander of the China squadron, went to Japan with a force greater than Perry's. He came again in September 1855, this time with the entire China squadron and two French warships. Notably absent was Sir John Bowring. He had been on the point of leaving Shanghai for Japan in June 1854 when news of the outbreak of the Crimean War reached him; he could not desert his post with the Russian threat hanging over China.[23]

Stirling reported that 'Japan was as useful to us as a British colony in that location'. He did not seek the trading rights Bowring was so keen on. His aims were purely military: British interest in Japan was determined by the immediate needs of the Crimean War – the use of Japanese ports as forward supply bases against the Russian navy in the Sea of Okhotsk.

The tsunami that the Japanese thought would break over their country did not happen, not at once. The Crimean War gave them time to come to terms with the shock of intrusion: there was no Japanese Shanghai or Hong Kong in the 1850s, no influx of cheap manufactures, missionaries or rapacious opium merchants. The world was aflame; few people had their eyes focused on Japan.

If the Crimean War hastened and shaped Britain's encounter with Japan, it did the same for Russian relations with China. For years General Nikolai Muravev had been advocating that Russia should seize the Amur valley and Sakhalin before the British inevitably did. 'It is an indisputable fact,' he fumed, 'that for their complete and entire control of trade in China the British need the mouth of the Amur and navigation on the river.'[24]

Muravev had been ignored and criticised for attempting to stir up trouble with the Chinese and the British. Nicholas I and his ministers were set against conquest in northern China. But the Crimean War gave Muravev a once-in-a-lifetime chance to fulfil his dreams. He knew that the Qing regime despised the British; using the Crimean War as cover, he told Chinese officials that he was moving his troops down the Amur to prevent British forces from seizing land in Manchuria. The Chinese were happy to help their enemy's enemy. The result was a spectacular land-grab on the part of Muravev, an event that went largely unnoticed at the time and, it has to be said, by most traditional narratives of the Crimean War.[25]

By the time Russia was defeated in the faraway Crimea, Muravev's forces were occupying a cool 400,000 square miles of Chinese territory north of the Amur River. Colonists were being encouraged to this new frontier to establish settlements and trading stations; Muravev was having steamboats constructed in Belgium that would ply the Amur and open it up as a major artery of global trade. With his army came botanists, ethnographers, geologists and explorers. A San Francisco newspaper talked of profits of 600 per cent to be made by Californian entrepreneurs supplying Russia's remote colonists with food, alcohol and equipment. Muravev had achieved this extraordinary feat virtually unnoticed and without bloodshed – the British had been so preoccupied by the naval threat from Russia in the Pacific that they failed to spot Muravev's real intentions.[26]

The British *Quarterly Review* pointed out that as a result of all the bloodshed in the Crimea, Russia might have been made to 'recede a few steps in Europe' but since then had 'made one of her giant leaps in Asia'. The Tsar now had the chance of securing extensive seaboard

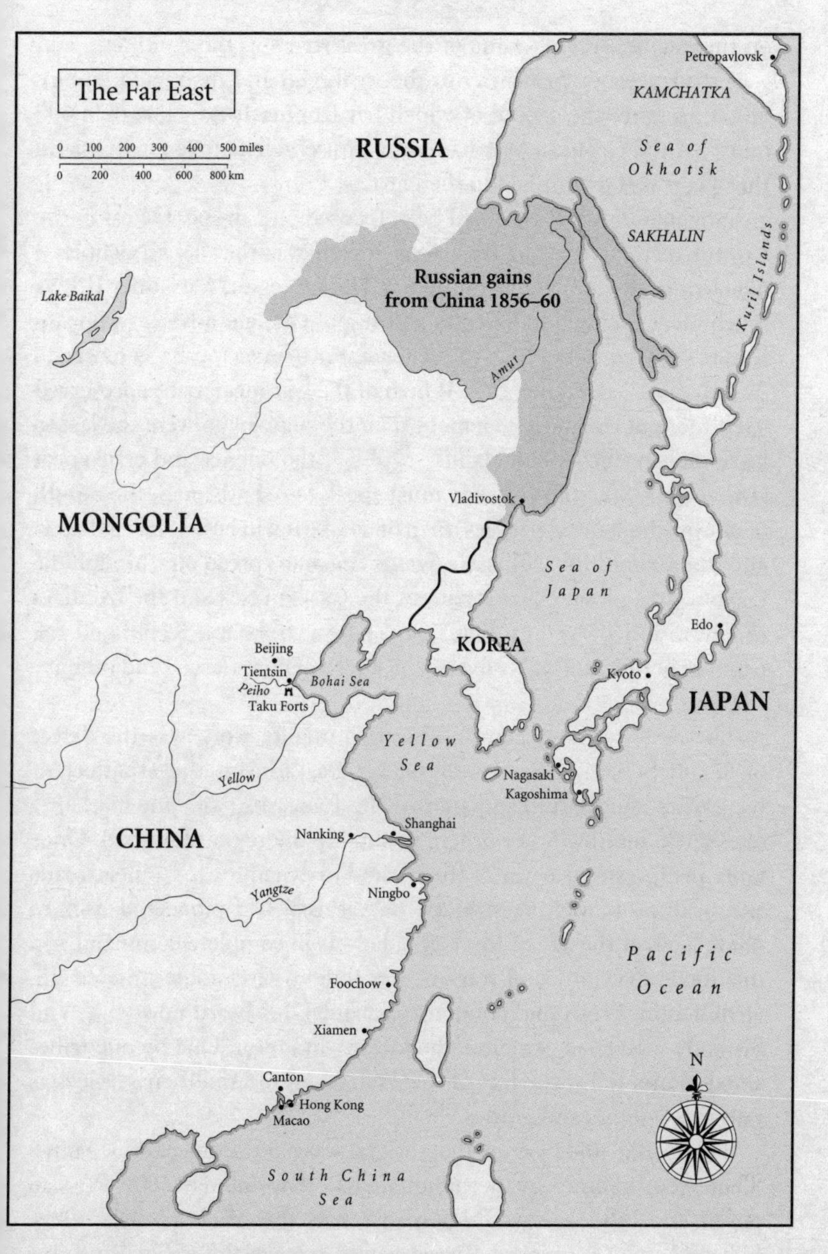
The Far East
0 100 200 300 400 500 miles
0 200 400 600 800 km
RUSSIA
Petropavlovsk
KAMCHATKA
Sea of Okhotsk
SAKHALIN
Kuril Islands
Lake Baikal
Russian gains from China 1856–60
Amur
Vladivostok
MONGOLIA
Sea of Japan
Edo
KOREA
Beijing
Tientsin
Peiho
Taku Forts
Bohai Sea
Kyoto
JAPAN
Yellow Sea
Yellow
Nagasaki
Kagoshima
CHINA
Nanking
Shanghai
Yangtze
Ningbo
Pacific Ocean
Foochow
Xiamen
N
Canton
Hong Kong
Macao
South China Sea

on the Pacific, access to one of the great rivers in the continent, and was in 'dangerous proximity to the weakened and distracted empire of China, from the capital of which her frontier is now less than 600 miles'. If the Crimean War had been a miserable failure for Russia in Europe, it had triumphed in the Far East.[27]

For generations, Russia had been focused on Europe. Defeat in the Crimea made it turn to Asia. One reason was that Tsar Nicholas I died before the conclusion of the war. His successor, Alexander II, was much more inclined to listen to a group of younger advisers who saw Russia's divine mission as the regenerator of Asia. 'Who is closer to Asia than us?' asked one. '. . . Which of the European tribes preserved itself more of the Asiatic element than the Slavs, who were the last to leave their primeval homeland? . . . Yes, if the science and civic spirit [*grazhdanstvennost*] of Europe must speak to Asia through the mouth of one of [the West's] peoples, then of course it will be us.' Immediately after the Crimean War Russian agents began to spread out through the gigantic area of land that separates the Caspian Sea and the frontiers of China and India – to Khurasan, Khiva, Bukhara, Kabul and the Chinese border city of Kashgar – in search of commercial and military treaties.[28]

The first step in accomplishing this mighty work was the defeat of Imam Schamyl. Prince Alexander Baryatinsky, the architect of the earlier successful campaign in the Caucasus, was put in charge of 300,000 men with the orders to subdue the region for good. Once more the forests gave way to the axe and to dynamite, to tunnels, roads and bridges. It was the strategy Baryatinsky had pioneered back in 1851. Now, at the age of forty-two, he was in complete command and this total environmental warfare was fought with single-minded determination. From the mountains Schamyl, his beard now white and his body weakened, watched the forests disappear. One by one tribes allied to him fell at the feet of the Tsar. Many offered their services as guerrilla fighters and guides.

Baryatinsky told his men that the end was not the defeat of Schamyl. They were fighting in the mountains so that one day soon Russian power and influence would spread across the whole of Asia. Once Schamyl was vanquished Russia would remake the world. Imagine, he told his soldiers, Russian railways pushing into Central Asia and Persia; imagine the great empire that would be created, with the wealth of Asia and the glory of conquest lying at the Tsar's feet; imagine Russia at the frontier of India. The Caucasus was the pivot upon which the future of Russia would turn. Schamyl's final defeat did not occur until 1859. But long before then, he was effectively neutralised.

Russia's sudden lunge into Asia on two fronts – in the Caucasus and Manchuria – was motivated in part by the need to expunge the humiliation of defeat in the Crimea but mainly by the fear that the British would engross the whole of Asia before it could get there. The *Friend of India* newspaper boasted that 'the whole of Asia, from the Indus to the Sea of Okhotsk, is destined to become the patrimony' of the British. A long-time American in India noted that 'it now appears likely that at no distant day, the greater part of Asia will be divided between Russia and England, each power extending its border till they meet'. No one could say when or where this confrontation would occur or 'what Asiatic nations will receive their laws from London, and what nations will receive them from St Petersburg'.[29]

British warships were in the Sea of Okhotsk keeping an eye on the Russians and charting the coastlines in preparation for future hostilities. When British steamers were apprehended bringing weapons and money to Schamyl, a furious Alexander demanded 'categorical explanations' from the British government. Answer came there none. Alexander fumed at this 'unmentionable infamy': the British were up to their old tricks, encroaching on Russia's turbulent frontier.

A report circulated in St Petersburg titled 'On the Possibility of an Armed Clash Between Russia and England in Central Asia'. Prince Baryatinsky, sensitive as ever to British aggression, believed that British secret agents were at work on the southern shores of the Caspian Sea. He feared the sudden 'appearance of the English navy in the Caspian Sea', a situation that would be 'a fatal blow not only to our influence in the East, not only to our foreign trade, but also to the political existence of the Empire'. The solution was to draw the unstable khanates of Central Asia under Russian control, securing trading monopolies and thus erecting a barrier against the British.[30]

As far as the British were concerned, Russian expansion was a dire threat to *their* empire. As one Russian wrote: 'Now that Russia has consolidated her power in the Caucasus, she will possess an impregnable basis to extend her conquests on the one side into Asia Minor, and on the other into Persia and India.' Almost as soon as the Crimean War was concluded an Anglo-Indian army invaded Persia. As the British guns roared out on helpless Bushehr, one of the gunners turned to an officer and said, 'That's one in the eye for the Russians, sir!' The British were in Persia to force the Shah to return the disputed city of Herat to Afghanistan. The Russians had the right to station consuls wherever they liked in the Persian Empire; the British feared that if Persia secured Herat it would become 'the advanced guard of Russia' in a future attack on India. Dost Mohammad, the ruler of Afghanistan, needed every support to prevent him siding with Russia. It was as if the Crimean War had never really ended.[31]

But it was in China that the British had most to fear. Thanks to General Muravev's solo conquest, Russia's sphere of control now extended into Manchuria to the Sea of Japan and the border with Korea; it was set to be a Pacific power and had earned the right to dictate, with the other major Western states, the destiny of China. But Russia was seen as having even greater secret influence. All Western countries were forbidden from dealing directly with the Qing government in Beijing; they had to speak through the imperial commissioner at Canton. That was with the exception of Russia, which, thanks to its long-standing overland trade links with China, was allowed to retain the *E-lo-ssu* – the Russian Hostel in Beijing. It was a small perch in Beijing, to be sure, but the fact that the Russians were there at all was alarming to the Western powers. It meant that the Tsar could open up channels to the Qing emperor and intrigue against them.

To some observers it was apparent that Russia was going to exploit the Taiping Rebellion and do to China what the British had done to India. The urgency, therefore, of reforming China was never greater. Admiral Stirling warned that 'if China be not electrified and organised by British energy and management; or brought under the influence which a more extended Commerce will give us, she will soon fall within the Dominion of Russia'.[32]

Britain, France and the United States believed that many of their problems in China would be solved if they could talk directly to the imperial government in Beijing. The Crimean War had delayed efforts to force diplomatic relations on the Chinese emperor. But in 1856 the time had arrived, according to Lord Clarendon, when 'the Emperor . . . must be made to adopt a more liberal system in his intercourse with foreign nations' and the humiliating position that foreign diplomats were placed in completely changed. His second demand was that 'the vast resources of that vast Empire [must be] opened up to the industrial enterprise of other nations'. When that happened the smuggling and drug-trafficking would cease, replaced by the benign effects of free trade. As soon as the Crimean War ended Sir John Bowring was badgering the Foreign Office to be allowed to go to Beijing in force with British, American and French naval escorts. His moment of destiny had arrived.[33]

But before Bowring could muster the strength to knock on China's front door, another opportunity presented itself thanks to the chaos of piracy. In October 1856 a lorcha named the *Arrow* arrived at Canton carrying a cargo of rice from Macao. While it was riding at anchor near a fort called the Dutch Folly, it was approached and boarded by Chinese police officers. They suspected the *Arrow* of being engaged in smuggling, and three of its crew were alleged to be notorious pirates; the twelve Chinese sailors were taken off and the vessel impounded.

The British Consul in Canton was Harry Parkes, a twenty-eight-year-old man whose life and career had been spent in China. As soon as he learnt what had happened he demanded that the Imperial Commissioner, Yeh Mingchen, return the crew and apologise. What had angered him so much was that the Chinese officers had hauled down the Union Jack from the *Arrow*. It was an unforgiveable insult to the honour of Great Britain.

Why, the world might have asked, were the British so het up about Chinese officials arresting Chinese pirates on a Chinese-built and owned boat in a Chinese port? What was the Union Jack doing on a pirate ship in the first place?

It all went back to the previous year when Sir John Bowring passed a law allowing Chinese residents of Hong Kong to register their ships as British. This was a fresh outrage to China. Suddenly all kinds of Chinese-owned vessels were sailing about, using the protection and privilege of the British flag to evade Chinese customs officers. Smugglers and pirates now, it seemed, had the backing of the Royal Navy. Chinese officials were unable to bring to justice their own criminals in their own ports. It was as serious an invasion of national sovereignty as could be imagined. Many pirates had forged registry documents and flew the flag illegally. Yeh claimed that the *Arrow* was one of these illicit ships. In fact the *Arrow* had been legitimately registered in Hong Kong by a Chinese resident of the colony; but its licence had lapsed ten days before. Yeh would not apologise, but he agreed to release nine of the twelve prisoners.

Parkes refused to accept the release of the nine crewmen because they were not returned on the *Arrow* in his presence, as he had demanded. He was more than happy to escalate the incident. Later he wrote that God's hand was 'clearly traceable in the whole affair'. And his view was shared by Bowring; the *Arrow* incident would be 'the stepping stone from which with good management we may move on to important sequences'.[34]

And what might those sequences be? The opening-up of the vast city of Canton perhaps? And where would that lead on to? Canton was the gateway to China. Once Western privileges were conceded there, there was no stopping the chain of events. Next stop Beijing. Fate had gifted Bowring and Parkes the opportunity to change the world. The *Arrow* incident was not, for Harry Parkes, about piracy or the honour of the British flag: 'it was the cause of the West against the East, of Paganism against Christendom'.[35]

In short, Bowring now had the excuse – long sought for – to rain shells down on Canton until his demands, and the demands of France and the USA, were met. British warships were dispatched to Canton. It made no difference to Yeh. Then the Royal Navy captured the forts defending the approach to the port. Bowring made his demands: access to the city. Yeh tried to return the prisoners and bring the matter of the *Arrow* to an end. But events had gone far beyond that. On 27 October the Royal Navy opened fire from the Dutch Folly and breached Canton's walls. Two days later marines forced their way through the damaged wall and made straight for Yeh's *yamen*, his official headquarters. Harry Parkes followed them in and read a proclamation announcing the free entry of foreigners to Canton. The *yamen* was then looted. 'We are so strong and so right,' trilled Bowring. 'We must write a bright page in our history.'[36]

But still Yeh would not give in. The Royal Navy commenced shelling Canton from long range in the first week of November and more forts were captured and blown up. This display of the firepower and dominance of the Royal Navy was to no avail. Rather magnificently, Yeh sat in the ruins of his *yamen* bedecked in the gorgeous finery of his viceregal robes, defiantly refusing to yield. He offered a reward of 100 Mexican dollars for anyone who killed an enemy. Gathered round him were 20,000 of his 'braves', fanatically anti-Western militia fighters.

In December the whole Western factory area – containing the massive *godowns* that until recently held the wealth of Chinese trade, the consular buildings, the banks, financial houses and residences – was burnt to the ground. Despite the firepower concentrated on Canton and the destruction being wrought on the city, Yeh knew how precarious the British position really was. There were only 1,000 troops; they were a long way from reinforcements; and, with so many warships in the Pearl River Delta, Hong Kong was left almost defenceless against pirate attack.

That was the Western position in China in microcosm: however powerful the invaders were, however badly the peripheries of the Chinese empire were mauled, the Chinese could afford to sit out the tempest until it blew itself out.

Yeh was correct. Eventually the British withdrew to launch commando raids on Chinese war junks that had hidden away in the maze of shallow creeks in the delta. The cutting-out operations against these junks were relished by the men on the ground and newspaper reporters: here was the Royal Navy in fine form, and there were plenty of tales of derring-do. By the end the entire Chinese navy had been destroyed. But it was a minor diversion and a hollow victory. In Canton triumphal arches were raised to Yeh, the victor. The bombarding ships had gone; more deliciously, the foreign presence in Canton, so glaring and degrading, was now an abandoned, charred ruin.

*

Yeh and the people of Canton might have been even more cheered by what was happening on the far side of the world, in the Houses of Parliament in London. Sir John Bowring was attacked as a filibuster and warmonger. The Earl of Derby, leader of the opposition Conservatives, told the House of Lords that the *Arrow* affair was the 'most despicable cause of war that has ever occurred'. 'I am an advocate in a cause which I believe to be that of policy, of justice, and of humanity,' he declared. 'I am an advocate for weakness against power, for perplexed and bewildered barbarism against the arrogant demands of overweening, self-styled civilisation. I am an advocate for the feeble defencelessness of China against the overpowering might of Great Britain.' Lord Lyndhurst called Bowring's bombardment of Canton 'the consequence of the mischievous policy of one of the most mischievous men I know'.[37]

After several nights of acrimonious debate in the Commons on the legality of the war, Palmerston's government lost a vote of censure. Parliament dissolved and a general election was called.

Britain at the beginning of 1857 was in a belligerent mood. The country had almost come to blows with the United States and at the time of the election British troops, marines and sailors were fighting in Persia and China. The disappointing showing during the war against Russia, and the unsatisfactory peace settlement, had challenged the British world view and their sense of mission; that faith in the British century would be vindicated and restored in the Iranian deserts and the markets of China. The public did not share the pacific view of their MPs. Palmerston was swept back into power in the so-called 'Chinese Election' with a majority of eighty-three, the largest for any government in over two decades.

So confident was Palmerston that he was more in tune with the national mood than his political rivals that even while the election contest was ongoing he put together a major military and diplomatic expedition that would force open the locked doors of China and Japan for good.

Having engineered the crisis with China, Bowring was discredited and had to take a back seat in the great work. The man chosen to lead this all-important mission of vengeance was the doyen of imperial administrators, James Bruce, 8th Earl of Elgin. Within a few weeks of the general election he was bound for Hong Kong in the company of his trusted private secretary and confidant, Laurence Oliphant, fresh from his failed bid to become a Nicaraguan potentate.

Elgin was a very different man to Sir John Bowring. Where the free-trade zealot was impulsive and trigger-happy, Elgin was every bit the disinterested and detached nobleman. A somewhat plump forty-six-year-old with snowy hair, he concealed his innermost thoughts and passions behind an impenetrable aristocratic veneer and rigorous self-control. As a natural conciliator he detested violence, in particular by the strong against the weak; as a high-minded man of culture he deplored the way that Britain was about to despoil a venerable civilisation. Bowring's piratical response to the *Arrow* affair was to him deeply distasteful; 'nothing could be more contemptible than the origin of our existing quarrel,' he said. '. . . That wretched question of the *Arrow* [was] a scandal to us.' He hated the rapacious merchants of Hong Kong and Shanghai who had brought so much harm to China.[38]

Yet Elgin was prepared to achieve Palmerston's aims by any means, including military intervention. He was, after all, the duty-bound servant of Crown and Empire, and he would not back down from hard choices or actions that he privately felt repugnant. Neither he nor Oliphant had been to China before. Before they even got there to accomplish the mission, another Asian region exploded into anarchy and confusion.

11
Retribution

LUCKNOW

. . . the people of this country [Britain] will never consent to leave India . . . and will go on playing Double or Quits until they win.
Sir George Lewis[1]

Whose work are we engaged in, when we burst . . . with hideous violence and brutal energy into these darkest and most mysterious recesses of the traditions of the past?
Elgin[2]

'We have no kisses for cowards!' cried the prostitutes in the Meerut bazaar on the evening of 9 May 1857 as the men of the 3rd Native Light Cavalry sloped towards the brothels. That very day Lord Elgin and Laurence Oliphant, bound to exact vengeance on China, were crossing Suez on the first train to make the journey.[3]

Among the prostitutes was 'Mees Dolly', the British widow who had been expelled from the cantonment for theft and had found a home and a means of life in the brothels. That day eighty-five elite skirmishers from the regiment had been disgraced in front of their comrades on the parade ground: their uniforms were stripped from them and they were marched off to gaol with their highly polished boots in their hands, their ankles fettered, to begin a sentence of ten years' hard labour.

Their crime was to have refused to accept rifle cartridges at a drill parade back on 24 April. The new Pattern 1853 Enfield rifle required that the cartridge be bitten before its contents were poured down the muzzle. According to rumours sweeping through the Bengal army, the cartridge was greased with pork and beef fat, respectively forbidden to Muslims and Hindus. The elite troopers had made a stand against the sacrilegious practice. The regiment's commander was determined to degrade the refuseniks in front of two infantry regiments and their comrades in the 3rd Cavalry.

After the tense, humiliating performance on the parade ground, when the men reached the brothels Mees Dolly and her fellow prostitutes began taunting the troopers. Were they really men if they allowed their brave comrades to be marched off with anklets of iron? 'Go and rescue them . . . before coming to us for kisses.'[4]

The next evening Mrs Elizabeth Muter, wife of a captain in the King's Royal Rifles, drove to church to await her husband and his men as they paraded for Sunday service. 'The sun was sinking in a blaze of fiery heat that rose hazy and glowing from the baked plain.' Meerut, a major military base about forty-three miles north-east of Delhi, was the headquarters of the Bengal Artillery, with three Indian and two British regiments. The British station was considered the prettiest and most pleasant in India. Mrs Muter waited for the marching bands of the church parade as she always did of a Sunday evening. But instead she heard what she thought was holiday-making in the Meerut bazaar. Her husband and the troops had not turned up and the service was cancelled. As she drove towards home it seemed that the horizon was on fire, 'as if the whole cantonment were in flames'.

The tranquillity of Sunday had been transformed, in the space of an hour, into mayhem. Elizabeth Muter encountered huge crowds of rampaging men on the road that led to the bazaar. Then she saw two British artillerymen pelting down the road, pursued by a pack of rioters hurling stones at them. Mrs Muter used this diversion to get back to her bungalow. There she found her servants gathered together 'in a flutter of alarm'. The *khansamah* (house steward) said he could not be responsible for the safety of the Muters' property. 'At the same time he advised me to conceal myself; a proposal which, from the indignant refusal on my face, he saw I regarded as an insult.' 'To conceal myself in my home,' Mrs Muter thought to herself, 'in the lines held by a regiment that had reckoned up a century of renown! And from what?'

Then she heard what. There was 'a Babel of voices nearly drowning the ceaseless rattle of musketry'; next came the heavy tramp of an English battalion on the march. Captain Muter sent word to his wife that the native regiments had mutinied and had broken open the gaol; they were shooting any European, man, woman or child they could find and setting fire to the garrison. Elizabeth was escorted to the Quarter Guard, where the wives and children of the regiment were seeking shelter. She heard 'through the roar of countless voices . . . the boom of heavy guns: one, two, three – then silence'. By midnight the British lines were defended by European troops all armed with the deadly Enfield rifle and Mrs Muter returned home.

For the rest of the night she paced her veranda and watched flames devour nearby bungalows. The fires appeared to get closer and closer to the British lines. When dawn broke the sun revealed a scene of complete devastation. Elizabeth Muter knew that she was witnessing the beginning of a revolution. Meerut was the centre of a cyclone that would sweep away British rule in India. The taunts of Mees Dolly and the prostitutes had come to this.[5]

The first night of the Rebellion saw property burnt and forty-one Europeans dead, including eight women and eight children. During that night of carnage and destruction an eighteen-year-old named Kate Moore had become worried that her aunt, who was on a visit to Agra, might find it hard to get home in the chaos. She sent a telegram: 'The cavalry has risen, setting fire to their own houses and several officers' houses, besides, having killed and wounded all European soldiers and officers they could find near the lines; if aunt intends starting tomorrow evening, please detain her from doing so as the van has been prevented from leaving the station.'[6]

This sentence, a mixture of the nightmarish and the mundane, announced the beginning of the Rebellion. Shortly afterwards the line was cut. But because the telegram came from an eighteen-year-old, and a girl at that, no action was taken. Another message was picked up at the telegraph station at Delhi warning of discontent. Then the line went dead. The telegraph operatives shut up shop for the night. The next day Charles Todd, the boss, set out to see where the fault in the line to Meerut had occurred.[7]

News of the mutiny at Meerut was unknown to them at this point. But during the night of 10–11 May mutinous troopers of the 3rd Light Cavalry galloped the forty miles from Meerut and across the bridge of boats into Delhi. They rode into the open-air audience hall at the Red Fort with swords drawn. They proclaimed Bahadur Shah, King of Delhi and heir to the fabled Mogul Empire, ruler of India. Then the same kind of violence and destruction that had engulfed Meerut the night before was re-enacted in Delhi. European men and women were shot on sight or taken prisoner by mobs, as were Indian converts to Christianity. European houses and buildings, churches and cemeteries were destroyed as the civilian population and the sepoys turned on their despised masters.

Survivors of the massacres made it to Flagstaff Tower on the ridge to the north of Delhi. As they crammed into the small fortress the enormity of what was happening became clear. It was certain that Charles Todd, who had set out to repair what he thought was an electrical fault, had fallen into the hands of the mutineers. A teenage Eurasian telegraph signaller named William Brendish managed to send a message to the military garrison at Ambala reporting what was happening.

It was a 'very meagre' dispatch with hazy details of what had occurred in Meerut and Delhi; it concluded with the words 'we are off. Mr C Todd is dead, I think. He went out this morning and has not yet returned. We learned that nine Europeans are killed.' Brendish then fled for Flagstaff Tower. No sooner had he arrived, however, than he was dispatched back through the mayhem to the telegraph station. This time he reported to Ambala that mutineers from Meerut had captured Delhi and that the city's troops were refusing to fight. The station at Ambala received nothing else thereafter except for some mysterious and garbled signals; the operator assumed someone who knew nothing of telegraphy, a mutineer, was playing with the equipment in a deserted telegraph office over the dead body of William Brendish.* Then the line went dead.[8]

'The cutting of our telegraph-wires and the interruption of our posts,' commented the civil servant John Kaye, 'were among the first hostile acts of the insurgents in all parts of the country.' As regiment after regiment mutinied and large parts of northern and central India began to fall to the rebels the British, deprived of communication and surrounded by hostile forces, could only imagine the full enormity of the situation across the subcontinent.[9]

First the telegraph cables were severed. Then the technology was used against the British. Cable was refashioned into bullets; tubular iron telegraph poles found new life as basic cannon. At Allahabad mutineers turned their cannon on railway engines and destroyed the track and sheds. 'There seemed,' said Kaye, 'to be an especial rage against the railway and the telegraph.'[10]

The rebels attacked them not out of fear of modern technology, but for strategic reasons. They also symbolised a new form of scientific imperialism that was transforming India.

There were 200 million human beings in India, or 16 per cent of the world's population. Just how this mass of humanity was to be governed baffled policymakers and, frankly, bored the public in Britain. Part of the problem was the haphazard way in which British rule had extended over the subcontinent and the way that power was exercised. The affairs of the subcontinent were managed by a business, the East India Company, which was itself supervised by the British government in London and its imperial officials in India.

Over the centuries the Company's position in India had evolved from economic predominance to political, territorial and military power. From its footholds in Bombay, Madras and Calcutta the Company had, since the seventeenth century, expanded its territory in a series of military conquests. Robert Clive's victories extended Company rule over Bengal. The Maratha Empire, which ruled most

* In fact, Brendish made his escape, lived to a ripe old age and received various accolades as one of the saviours of India. The story that he sacrificed his life for British India, however, was believed and became one of the myths of 1857.

of India in the eighteenth century, was finally defeated in 1818 after a series of wars. After that turning point in Indian history a third of the subcontinent was under the direct rule of the Company, and the Indian rulers in the remaining two-thirds entered into subordinate alliances with the British. Above all the Company possessed a military monopoly. Of its three armies the Bengal Presidency was the largest with 135,000 men, the Madras Presidency came next with 50,000 and the Bombay Presidency had 45,000. In addition there were 39,000 British soldiers.

What was Britain's justification in ruling India? Was its job to maintain order and leave Indian society more or less to itself? Or did Britain have a higher calling to 'regenerate' India? Attempts by utilitarian administrators to transform India by rational law-making, land reform, Christianity and education failed to make much impression on the subcontinent.

But then came a reforming governor-general. James Ramsay, 1st Marquess of Dalhousie, was the supreme example of a technocratic administrator. Just thirty-five when he assumed the reins of power in 1848, he was a short, stocky workaholic with a forceful personality. In India he woke at 6 a.m. to read a chapter of the Bible. Before he sat down to digest the newspapers over breakfast he had already read his official papers. He was in his office by 9.30, worked through lunch, and left at 5.30 to participate in the rounds of official entertainments he loathed.

Dalhousie's considerable energies were directed at revolutionising India. Just before he left office in 1856 he claimed credit for unleashing in India the three 'great engines of social improvement, which the sagacity and science of recent times had previously given to Western nations – I mean Railways, uniform Postage, and the Electric Telegraph'.[11]

Here was scientific imperialism being tested on the Indian subcontinent. Technology – specifically communications technology – would do the work of a thousand laws and administrative novelties without the need for meddling and tinkering with the labyrinthine, impenetrable social and religious customs of India. Peoples could be transformed, societies regenerated, superstitions cast aside by the magical properties of modern science without direct, heavy-handed human intercession. The modernisation and material prosperity of India became the moral justification for the British Raj. Rigid caste barriers, bad governance, 'barbaric' customs and unproductive practices would be blow-torched by the white heat of the technological revolution.

Commercial telegraph lines opened in India in 1851 with eighty-two miles of cable; within four years the system had expanded to 4,000 miles. They stretched from Calcutta to Agra, from Agra to Peshawar and Bombay, and from Bombay to Madras; gutta-percha-insulated cables went under the great rivers. A few years later the network expanded to over 11,000 miles.

Dalhousie's aim was to tie India into the forces of modernisation that were transforming the West in the 1850s. He envisaged a rail system that would bind together the vast extent of the subcontinent, delivering 'commercial and social advantages . . . beyond all calculation'. He foresaw 'great tracks teeming with produce' connecting remote parts of India to the global market. A comprehensive rail network woven for tens of thousands of miles into the Indian soil, he said, would give rise to 'the same encouragement of enterprise, the same multiplication of produce, the same discovery of latent forces, the same increase of national wealth' that was revolutionising the 'western world'.[12]

'In a few weeks,' declared the *Railway Times* in 1853, '. . . the iron road that is probably destined to change the habits, manners, customs, and religion of Hindoo, Parsee, and Mussulman, will commence its work on the Indian Peninsula.' The first passenger train in the history of Asia left Bombay for Thane on 16 April 1853, its epochal twenty-one-mile journey heralded by a twenty-one-gun salute. Aware of the significance, tens of thousands came to witness it. The onlookers were not made up exclusively of the inhabitants of Bombay, or even India; they came from Sind, Afghanistan, Central Asia, Persia, Arabia and East Africa. The railway was hailed by a paper – *The Overland Telegraph and Courier* of Bombay, appropriately enough – as 'a triumph, to which, in comparison, all our victories in the East seem tame and commonplace'.[13]

Karl Marx informed readers of the *New York Daily Tribune* that the British were establishing a 'net of railroads over India' and said the 'results must be inappreciable'. India's production of the world's most important industrial staple – raw cotton – matched that of the American South. But, unlike the American slave states, India lacked the infrastructure to export it. Almost all the cotton was spun in India for domestic production. Dalhousie's rail revolution was intended to connect even the remotest parts of the subcontinent to the global economy. In time, surely, Manchester would break its dependence on America; the Southern producers would lose their key market. The task was the most challenging and extensive railway endeavour on the planet. By 1857, 20,000 labourers were breaking their backs and risking their lives digging tunnels and constructing viaducts in the forbidding rocky slopes of the Bhor Ghat ridge near Bombay. India had 400 miles of railway track in use at that time and a further 3,600 in the pipeline.

There was, however, a major block on the railway whisking India to the future, as Dalhousie saw it. And that was the part of India not under full and formal British control. These 'native' states the governor-general eyed with distaste and impatience: how much better for everyone when they were governed on a rational footing, with railways, telegraphs and roads. Governed, in other words, by the British.

According to the 'doctrine of lapse' the territory of any Indian ruler who died without heirs or who ruled in a way that was 'manifestly incompetent' could be annexed by the Company. Hindu law permitted adopted sons to inherit land and royal titles; Dalhousie did not. He applied strict interpretations to the doctrine of lapse and in so doing the Company devoured significant new territories during Dalhousie's period in office: Satara in 1848; Jaipur and Sambalpur in 1849; large parts of Burma in 1852; Nagpur and Jhansi in 1854; Tanjore and Arcot in 1855; Udaipur and Oudh in 1856. Dalhousie also removed territory from the Nizam of Hyderabad – the largest of the princely states – in lieu of debts to the Company. This territory – Berar – contained some of the richest cotton fields in India.

Dalhousie rigidly applied the doctrine of lapse in order to consolidate British rule and to increase the revenues of the Company. Nagpur was a case in point. It was large (80,000 square miles), strategically located in central India on the route between Calcutta and Bombay, abundant in cotton fields, and exceptionally rich: 'too good a plum,' Dalhousie candidly said, 'not to pick out of this Christmas pie'. He wanted it to increase Britain's military strength, 'to enlarge our commercial resources' and to 'consolidate our power'. The annexation of Nagpur was of doubtful legality. As one contemporary commentator had it, 'the real law by which Nagpur was added to our dominions was the law of the strongest'.[14]

But the kingdom Dalhousie eyed most greedily was Oudh, the wealthiest of all Indian states. Greed is the right word, for Dalhousie used epicurean language with regard to his choicest morsel. It was the 'cherry which . . . has long been ripening'. At another time he said that 'Lately I have been busy trussing up the Kingdom of Oudh preparatory to putting it on the spit.'[15]

The great *Times* journalist William Howard Russell described the view of Lucknow, capital of Oudh: 'A vision of palaces, minars, domes azure and golden, cupolas, colonnade, long facades of fair perspective in pillar and column, terraced roofs – all rising up amid a calm still ocean of the brightest verdure . . . Spires of gold glitter in the sun. Turrets and gilded spheres shine like constellations . . . There is a city more vast than Paris, as it seems, and more brilliant, lying before us. Is this a city in Oudh? Is this the capital of a semi-barbarous race, erected by a corrupt, effete, and degraded dynasty? I confess I felt inclined to rub my eyes again and again.'[16]

The fairy-tale magnificence of royal Lucknow visually clarified the issue. For years the British had moaned about the misgovernment of the kingdom and the scarcely believable extravagance of its kings and court. The recently completed Qaisarbagh, the king's dazzling palace so vividly described by Russell, symbolised for Dalhousie the wasteful government of Oudh. Rather than depose the ineffectual Wajid Ali, Nawab (king) of Oudh, and replace him with a more suitable Indian successor as previous governors had done with other fun-loving rulers, Dalhousie turfed him out on the grounds of misgovernment and the East India Company took direct control over the kingdom. Another brake on modernisation had been removed by India's technocrat.

*

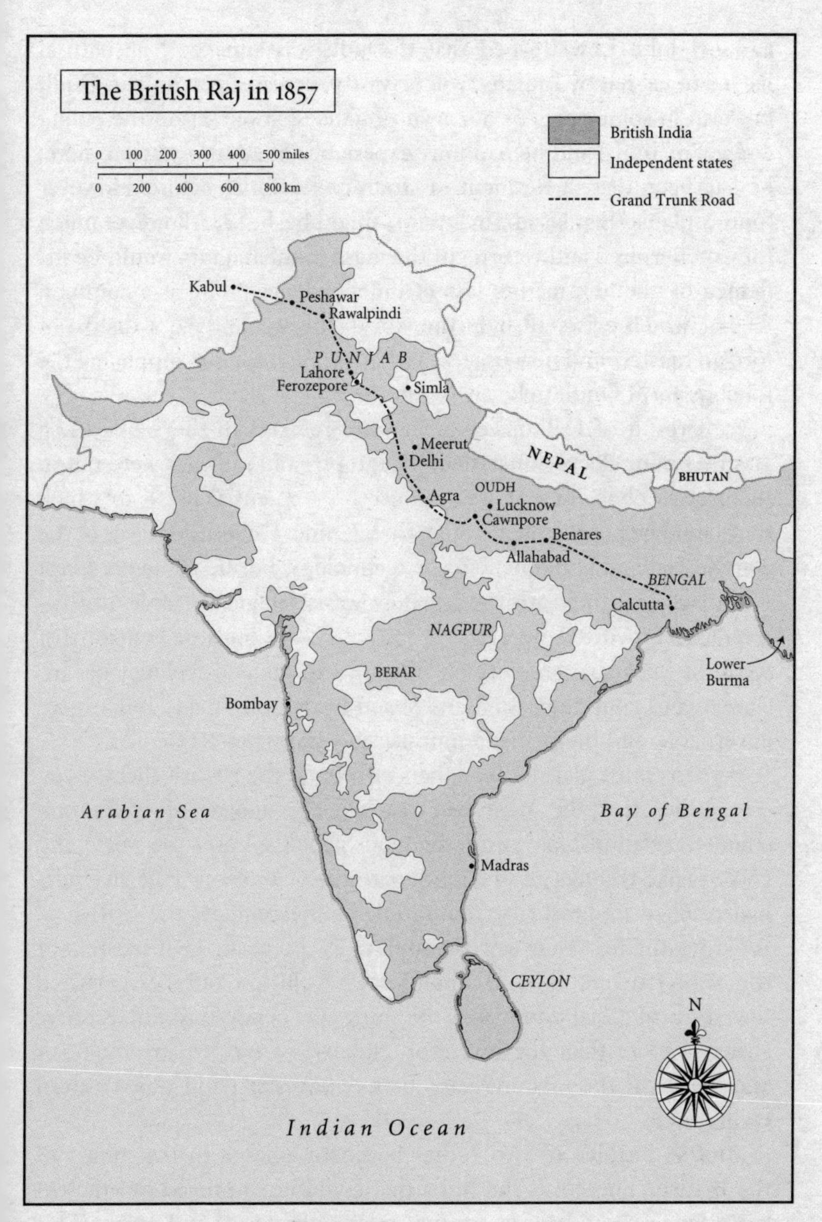

The British Raj in 1857
0 100 200 300 400 500 miles
0 200 400 600 800 km
British India
Independent states
Grand Trunk Road
Kabul
Peshawar
Rawalpindi
PUNJAB
Lahore
Ferozepore
Simla
Meerut
Delhi
NEPAL
BHUTAN
OUDH
Agra
Lucknow
Cawnpore
Benares
Allahabad
BENGAL
Calcutta
NAGPUR
BERAR
Lower Burma
Bombay
Arabian Sea
Bay of Bengal
Madras
CEYLON
N
Indian Ocean

Colonel John Low warned that if Oudh was annexed 'the natural hatred of us' felt by Indians 'will be vastly increased not only in Oudh but also in some parts of our own provinces'. Low sat on the ruling council of India, and he had long experience of administration there; he had been British Resident in Oudh two decades before. However 'pure and just' British administrators might be, he said, however much they modernised and reformed, the mass of inhabitants would be inflamed by the humiliating loss of independence. 'This is a common feeling which exists all over the world,' he wrote, 'viz: a dislike of foreign masters and new usages, and I see no reason for supposing the inhabitants of Oudh to be an exception.'[17]

As a result of Dalhousie's whirlwind reforms in the 1850s India was becoming deeply unsettled. Telegraphs and railways were not in themselves obnoxious; it was the use – or potential use – to which they could be put that made them threatening. Dalhousie talked of the commercial and social benefits of technology, but he made no secret of the fact that their primary purpose was strategic. As far as Indians were concerned, telegraphs and railways were tools of control that were the exclusive preserve of Western experts and technicians. Instantaneous communication and steam locomotion made India more governable, and the British immeasurably more powerful.

Perhaps most glaring were their effects on the British themselves. The belief that the West had reached the apogee of civilisation coloured relationships with indigenous people all over the world and encouraged feelings of innate superiority. Company rule in India had reached its height by the middle of the century; the British in India for the first time saw themselves as the permanent masters of the subcontinent. The technological revolution only exacerbated that sense of racial dominance. Nothing was believed to differentiate societies more than the invention and use of modern technologies and highlight the yawning gaps between Western and non-Western civilisations.

And so an idea of two Indias began to emerge in the minds of the British. There was the India that could be improved by efficient governance, the rule of law, trains, telegraphs, roads and canals. This India could be managed from above, by European officials, officers and technicians without entanglement with the other India – the India that was, in their eyes, chaotic, impervious to change and incorrigibly barbarous. In May 1857 the naturalist Brian Hodgson noted that things were 'strangely altered' since the time he arrived in India forty years before. Back then 'knowledge of and respect for' Indians and Indian customs were 'exacted and enforced'; '*now*, one hears ordinarily and from the mouths of decent folks nothing but contemptuous phrases (nigger &c) applied to the people' and 'manifestations in ordinary social intercourse of . . . haughty contempt of inferior races'. Many other writers attested to the recent prevalence of that 'hateful word' in the 1850s.[18]

Many British commentators made the shocking effects of technology on Indian society a boast. 'A privileged race of men, who had been held in veneration as the depositaries of all human knowledge, were suddenly shown to be as feeble and impotent as babes and sucklings,' wrote John Kaye with evident glee of Hindu teachers. '. . . There was no means of contradicting or explaining away the railway cars, which travelled, without horses or bullocks, at the rate of thirty miles an hour, or the electric wires, which in a few minutes carried a message across the breadth of a whole province.'

Kaye believed that communications technology taught Indians 'the great truth that Time is Money', and that the expected gains in material prosperity would undermine the authority of 'their spiritual guides'. 'That the fire-carriage on the iron road was a heavy blow to the Brahmanical Priesthood is not to be doubted,' he crowed.[19]

If the British were powerful and arrogant enough to take states as large and potent as Nagpur and Oudh, what could they *not* do? With the full force of technology in their arsenal, now they could extend their control even further and deeper. Perhaps that meant an all-out assault on Indian religions and the caste system. Perhaps it meant the forcible conversion of Muslims and Hindus to Christianity. In Bengal a group of missionaries circulated an address that made this horrendous possibility seem likely: 'The time appears to have come,' it declared, 'when earnest consideration should be given to the question, whether or not all men should embrace the same system of religion. Railways, steam-vessels and the electric telegraph are rapidly uniting all the nations of the earth.' Nothing seemed impossible in this topsy-turvy decade.[20]

Religious fears exacerbated already existing economic grievances. Even before Dalhousie's spurt of modernisation, the Indian economy had been knocked backwards by its increasing interaction with global trade. India's manufacturing and commerce, once dominant in Asia, had been hit by competition with the industrialising West, most notably its textile industries. Free trade had seen the decline in commodity prices. The result was extensive de-urbanisation as former artisans and workers sought survival in the countryside, already hard pressed by increasing revenue demands and land reforms. They joined large numbers of demobbed soldiers who had lost their jobs when the British dismantled the armies of conquered kingdoms. Railways would perhaps not enrich the subcontinent or make it more modern, but ruin its remaining industries as it became a supplier of primary produce to Western industry and opium to China. Where would that leave the Indian population? Perhaps as a vast rural proletariat servicing the needs of the West without sharing the profits. And where India led, the rest of Asia would surely follow.

Religious and economic anxieties generated by change were felt acutely in the greatest of the Company's armies, that of Bengal. For generations becoming a sepoy in the service of the Company meant honour, status and profit. The British had extended to sepoys a number of caste privileges that they would not necessarily have been able to enjoy in their home villages. Serving overseas would break their caste, so they were spared the obligation. The soldiers were jealous of their hard-won status; it made up for the meagre pay and appalling living conditions.

But in an age of modernisation and efficiency such concerns for caste were becoming costly for the British. Lower-caste recruits were drawn into the Bengal army, as were men from frontier societies such as Ghurkhas, Sikhs and Punjabi Muslims. The General Service Enlistment Act of 1856 meant that new recruits, whatever their caste, would have to fight in Persia, Burma, China or wherever else they were required to act as the iron fist of British imperial ambitions in Asia. There was a feeling that 'the English had done with the old

Bengal Army, and were about to substitute it for another that would go anywhere and do anything, like coolies and pariahs'.[21]

The annexation of Oudh marked an important moment for the army. Low pay had been compensated with plunder on wars of conquest. Now the British Indian Empire had reached the limits of its expansion, reducing sepoys from proud warriors to armed policemen. Many of the soldiers came from Oudh and they enjoyed the status of being part of an independent Indian kingdom. Now, the last vestige of Indian independence had been removed and the British were unquestionably masters of the whole subcontinent.

Anxiety ate away at the army, exacerbated by heightened fears about religious contamination. Much of it was unknown to the British. 'I have lived to see great changes in the *sahibs*' attitude to us,' wrote the veteran Bengali sepoy Sita Ram Pande with sadness. He recalled his early days as a young sepoy when the *sahibs* joined their men in watching *nautch* girls perform their famous erotic dances and attended the sepoys' sporting events. By the later 1850s European officers did not want to have anything to do with their men and rarely talked to them, according to Sita Ram. 'One *sahib* told us that he never knew what to say to us,' he said. 'The *sahibs* always knew what to say, and how to say it, when I was a young soldier.'[22]

If the majority of British officers were oblivious to the disquiet of the men they commanded and the anxieties gripping the Bengal army, there were plenty in India who were not. If the army buttressed the Raj, so too did the rulers of the independent states. After Dalhousie's spate of annexations, however, there were plenty of dispossessed rulers and princes who suspected the Company of plotting more land-grabs. Seizing Oudh in 1856 proved to be the last straw; if Wajid Ali could be deposed, who was safe? Rumours and conspiracy theories began to swirl through the army at the end of the year and into 1857, instigated by agents of some of the Indian princes.

'All kinds of news, both true and false, are discussed in the bazaars of large military stations,' wrote Sita Ram Pande, 'and anything injurious to the fortune of the government is listened to with the keenest interest.' The news buzzing about the bazaars was that Britain had suffered a serious defeat at the hands of Russia in the Crimea; that the

British army had been devastated and all the warships sunk: British power was an illusion. The gossips also spread the rumour that the Europeans were grinding down pig and cow bones and mixing them with the sepoys' rations. A prophecy did the rounds stating that British rule would be limited to a century. The Battle of Plassey, which confirmed British hegemony, had been fought on 23 June 1757.[23]

What the plotters needed was an issue so explosive that it would unite all the diverse grievances nursed within the army. And another legacy of the Crimean War provided the solution. The P53 Enfield rifle-musket had more than proved itself during the war with Russia. At the beginning of 1857 thousands of these rifles had arrived in Bengal as part of the army's modernisation programme. Word flared through the army that the cartridges were greased with animal fats. When a Hindu sepoy bit into one of these cartridges and ingested the fat of a cow, he would be polluted and his caste broken for ever; a Muslim sepoy would be defiled by the pig fat. British officers went out of their way to assure their men that no animal fats were used in the grease. Then they let the men prepare their own grease. But still the rumours persisted. If it wasn't the grease, then it was the cartridge paper that was impregnated with offensive fats.

Trust between British officers and the sepoys had completely broken down. The early months of 1857 saw growing restlessness and insubordination as conspiracy theories that the British were going to forcibly convert India to Christianity took hold. One staff officer better informed than most felt strong intimations of menace: 'I can detect the near approach of the storm, I can hear the moaning of the hurricane, but I can't say how, when, or where it will break forth.'[24]

Even so, when the Rebellion broke forth at Meerut in May, the British were taken by surprise. Many a colonel refused to believe that his loyal men would mutiny and refused to take precautions. The Rebellion spread from Meerut to Delhi and from garrison to garrison across central India. When it became clear that British power was on its last legs, the Rebellion gripped towns and countryside as well.

European refugees made their way on foot across lands that were aflame with rebellion to find shelter in the surviving outposts of the

Raj. Agra Fort filled with 6,000 women and children; they watched from the walls as the rebel army massed. At Lucknow 3,000 civilians, British soldiers and loyal sepoys found shelter from the vengeance of the people of Oudh behind the thick walls of the Residency. For ninety days they endured continual artillery bombardment and sniping. The situation was even more forlorn at Cawnpore, where 1,000 non-combatants and 300 troops under the command of Sir Hugh Wheeler had to defend themselves behind feeble entrenchments against 15,000 rebel soldiers.

The wider world held its breath in anticipation. Events in India had the possibility of transforming the politics and commerce of the entire planet. China beheld the annihilation of its greatest enemy, and followed with glee the success of the rebels. For Karl Marx, writing in the *New York Daily Tribune*, the Rebellion 'coincided with a general dissatisfaction exhibited against English supremacy on the part of the great Asiatic nations' and was 'intimately connected with the Persian and Chinese wars'. A contributor to the *Democratic Review* looked with satisfaction at the 'beginning of the end' of British power in Asia. According to the Paris correspondent for an American newspaper, the Rebellion was regarded by Europe as 'much, very much more than a military revolt'. If Britain lost its Indian Empire it would plummet to 'a fourth-rate power', leading inevitably to the reordering of Europe. Americans had much to fear from the Rebellion. Although 'it matters little to us who rules India', said the *New York Herald*, the conflict 'is of the highest moment to this country'. The British would hold on to India at all costs, reported the *Herald*, but in doing so it would require vast armies and the expenditure of unfathomable sums of money so that the 'immense effort of the Russian war will seem like a pastime, and the effort to crush the elder Napoleon mere trifling in comparison'. The paper foresaw the suppression of the Rebellion disordering the commerce and money markets of the world and, most alarmingly, diverting 'British capital from foreign investments'. The nightmare scenario was painted in vivid colours: the liquidation of US railroad stocks and bonds, investment draining from America, the Bank of England raising its lending rate, and the bankrupting of America's best customer. In the first shock of the news from Meerut, Delhi,

Cawnpore and Lucknow the *Herald*'s language was apocalyptic: 'there has been no precedent for such a contest as this one, since the Goths and Vandals overthrew the Roman Empire'.[25]

No wonder the world was absorbed by the Indian Rebellion. The twists and turns of the uprising dominated newspapers around the globe, filling column after column in Berlin, Vienna, Paris and New York. With the British reeling with shock, the rebels had their best chance to bring the Raj to an end in the summer of 1857. In these months the British came within an ace of losing the subcontinent for ever. There was no Imam Schamyl, however, in the India of 1857 to swallow up the British in a brutal guerrilla war. And there was no Garibaldi either, no nationalist hero who could spread the Rebellion into southern and western India and transform localised rebellions into a war of Indian independence. People rebelled that year for myriad reasons. All that united them was a deep loathing of the British and the desire to see them gone for ever. Rajas tried to realise their local ambitions. Rural insurgents took advantage of the collapse of authority and got what they could in the anarchy. There was no united front against the British and never could be in such a vast, diverse country. Instead of carrying the Rebellion across India, tens of thousands of sepoys concentrated on Delhi, Lucknow and Cawnpore.[26]

The British, in contrast to the rebels, kept open their lines of communication and, in doing so, developed a grand strategy. In the newly conquered Punjab there were 36,000 Bengali troops primed for mutiny. At Lahore four regiments that were on the point of seizing the ammunition magazine at Ferozepur were disarmed before they could act. The British had William Brendish to thank, the boy who had volunteered to return to the telegraph office at Delhi to send word of the mutinies. A few years later the famous British officer Herbert Edwardes told a meeting in Liverpool: 'Just look at the courage and sense of duty which made that little boy, with shot and cannon all around him, [send] that message which I do not hesitate to say was the means of the salvation of the Punjab.'[27]

Brendish's telegraph saved the magazine at Ferozepur. It reached the British commander-in-chief, who was summering in the Himalayan foothills at Simla, and he began organising forces to retake Delhi. On the North-West Frontier, at Peshawar, a group of young officers acted decisively when the boy's telegram arrived. They knew that unless the mutiny was crushed at once it would metastasise with unstoppable speed. They created the Moveable Column, a rapid-response force drawn from the Pathan hill warriors as well as Sikh and Punjabi volunteers. This force lived up to its name, moving relentlessly round the region that is now Pakistan in the hottest month of the year under the command of the ruthless, charismatic young officer John Nicholson, restoring order, disarming Bengali regiments and rounding up mutineers and executing them. 'The Electric Telegraph has saved India,' declared one official. It was true to this extent: had they lost the Punjab the British position in India would have been untenable.[28]

But control of the Punjab by no means guaranteed the survival of the Raj. When the seriousness of what was happening was known, messages began to ping back and forth across the wires. But when they were sabotaged at Cawnpore by the rebels, cables sent from Peshawar and Bombay could no longer reach Governor-General Canning (Dalhousie's recently arrived successor) at Calcutta. One of the great unsung heroes of the Rebellion (on the British side) was a twenty-five-year-old lieutenant named Patrick Stewart. When the mutinies broke out he hastened back to Calcutta from Ceylon. On the way he stopped at Madras and found that telegraph messages sent to Canning had been diverted there. Stewart took the messages with him by ship and as soon as he arrived in the capital of British India he commenced work on a telegraph line linking Madras and Calcutta. By this means messages sent from the Punjab and Bombay could reach Calcutta in a great loop bypassing rebel-held areas.[29]

Canning reassured London that he was able to communicate freely and instantaneously throughout India and beyond. By use of the telegraph he could cable Madras for reinforcements and summon troops and artillery freshly returned to Bombay from the war in Persia. And telegraphic communication was used in conjunction with fast steamships. Imperial troops answered the call from Ceylon, Burma, Mauritius, Cape Colony and Malta.

By great good fortune Lord Elgin's army, bound for war against China, was in transit in the Indian Ocean at that very moment.

Canning telegraphed Madras to dispatch a fast steamer to intercept Elgin. It missed the expeditionary force off Ceylon, but Canning had taken the precaution of sending a second; it overtook the British army passing through the Sunda Strait, the water that separates Sumatra and Java. Elgin re-routed his entire force to India in early June.[30]

'Time is everything,' Canning reported to London. 'Delhi once crushed, and a terrible example made, we shall have no more difficulties.' The governor-general cabled the commander-in-chief, advising him 'to make as short work as possible of the rebels, who have cooped themselves up [in Delhi], and whom you cannot crush too remorselessly'.[31]

Easy to cable from Calcutta: much harder to fulfil grandiose orders. The telegraph sparked quick action in key locations; but speed of communication could not make up for the chronic lack of manpower. In the 750 miles that separate Calcutta and Delhi there were just five British regiments. The build-up of forces took longer than Canning, at the end of a cable 900 miles from the action, expected. British, Afghani, Sikh, Ghurkha, Pathan and Punjabi forces were summoned to Delhi from the north-west; the Corps of Guides, an elite frontier strike force, marched 580 miles from Hoti Mardan to the city in an astonishing twenty-two days through summer heat in the month of Ramadan. Gradually the forces on the ridge above Delhi built up. Outnumbered and without heavy artillery, they were unable to make the decisive blow Canning wanted. But by quick dispatch, however, they had gained an important position, and they refused to be dislodged.

Elsewhere in India, the British made up for their deficiencies in numbers with technology and speed. Within a month of the Rebellion beginning, Colonel James Neill was fighting his way from Calcutta up the Grand Trunk Road towards the beleaguered British survivors at Benares, Allahabad and Cawnpore with a small force of 2,000 men. Neill's contingent was accompanied by Pat Stewart who, having established his emergency cable between Madras and Calcutta, was providing the army with mobile field telegraphs.[32]

The military field telegraph had been used in the Crimean War, but not in the way that Stewart deployed it in India. Stewart and his men established lines in advance of the British army, under fire and harried by cavalry behind enemy lines. Headquarters was put in immediate contact not only with Calcutta and other Indian cities, but with the front line. Elsewhere in India, Nicholson's Moveable Column was accompanied by a telegraphic signaller who connected it to HQ at Rawalpindi. Post and wire marched as quickly as the army's vanguard.[33]

Colonel Neill relied on something other than speed and communications technology in his thrust up the Grand Trunk Road. A God-fearing zealot, he adopted the maxim that 'The Word of God gives no authority to the modern tenderness for human life.' Not only were hundreds of suspected rebels rounded up and hanged in rows, but women, children and the elderly suffered the same fate. Villages were burnt in an orgy of indiscriminate revenge and terrorism.[34]

Lieutenant Stewart was with Brigadier-General Henry Havelock, who, having arrived from the Anglo-Persian War, raced up to Allahabad with a modest column of 1,403 Europeans, 560 loyal sepoys and eight guns. Havelock's 'Army of Retribution' left Allahabad on 7 July. Nine days, 126 miles and four victories later his exhausted and hungry men were bivouacked on the outskirts of Cawnpore. The rebel army was taken completely by surprise by a remarkable lightning campaign. Havelock's men carried the infamous Enfield rifle. Its superior range and accuracy devastated and demoralised larger armies of sepoys armed with their antiquated muskets.

Even before he reached Cawnpore Havelock had learnt of what had happened to the British garrison there. During three weeks of siege the feeble entrenchment had come under deadly attack from cannon and sniper fire. Wheeler, his 300 soldiers and the civilians held out for three desperate weeks in their rudimentary fort against 12,000–15,000 rebel troops. Water ran low and, when the bodies of the dead could not be buried, dysentery and cholera broke out. The rebel artillery managed to destroy the British hospital and with it all the medical supplies and medicine. Rebel attacks were held at bay by volleys of canister shot, but the British took fearful casualties. After three weeks a third of the British soldiers were dead and many were seriously ill or wounded.

The only outcome to the stalemate was the slow death by disease and starvation of the beleaguered British. On 25 June the rebel leader, Nana Sahib, offered Wheeler and the survivors safe passage by boat on the Ganges to Allahabad if they surrendered. Wheeler had little choice but to agree. Two days later the British left the battered entrenchment and were escorted to boats waiting at the Satichaura Ghat on the banks of the Ganges. As they boarded the boats the Indian boatmen abandoned the vessels. Then the massed ranks of rebels opened fire on the British soldiers. Only a handful of troops escaped the massacre. About 125 survivors, all women and children, were taken to a villa named Bibighar in Cawnpore.

When Havelock and his men entered Cawnpore on 17 July they discovered the remains of the British entrenchment; a ruined make-shift fort stood in the midst of roofless, battered buildings and a mess of broken furniture, abandoned shoes, torn books, empty bottles and everything else that was mute testament to the horrors of the three-week siege. Then they entered the Bibighar villa. The first officers who arrived in the house found the rooms ankle-deep in blood, torn hair, children's shoes, ladies' bonnets and bloody handprints on the walls. Even more horrific was the well in the courtyard in which there was 'a ghastly tangle of naked limbs'. These were the remains of 200 prisoners, mainly women and children, who had been taken from the Satichaura Ghat. They had been slaughtered as Havelock neared Cawnpore.

Over the succeeding days many Europeans saw the shambles at the Bibighar. None could fail to be moved and sickened by what they saw. Descriptions of the bloody rooms and the well spread through India. It filled the British with murderous rage. The enemy was dehumanised in their eyes; the sepoy was a fiend who must be brutally punished and humiliated before he was eliminated from the face of the earth. Every rebel, and suspected accomplice, would have to pay the price for what happened at Bibighar. Stories, most of them fiction, spread through the British army of unimaginable tortures and massacres.

From that moment on the British were not just fighting to restore order; they were avenging the women not only of Bibighar but all across India whom they believed had been raped. 'Remember Cawnpore': that was the slogan that rallied British soldiers as they fought through India in the following months. 'Since they had butchered our defenceless women and children,' wrote a colonel, 'we would have been more than human, we would have been less than men, if we had not exterminated them as men kill snakes wherever they meet them.'[35]

While Havelock stormed up the road to relieve Lucknow, James Neill was left in Cawnpore to restore order and punish rebels. This he did with relish. Mutineers were given a brief trial before being sentenced to death. But before they were hanged they were degraded and defiled. Hindus were force-fed beef and Muslims pork. Those Neill judged to be rebel ringleaders had to clean up the blood in the Bibighar, some with their tongues. 'To touch blood is most abhorrent to high-caste natives, they think that by doing so they doom their souls to perdition,' Neill gloated. 'Let them think so.' Defilement, humiliation and then death became the pattern throughout India.[36]

Three hundred miles to the north-west of Cawnpore, the British celebrated the recapture of Delhi on 20 September with a raucous drunken party in the Jama Masjid, the great mosque built by Shah Jahan. Delhi was captured after a fearsome artillery bombardment and vicious street fighting. The city became a scene of carnage and wanton destruction as British troops and their Indian allies rampaged through the streets destroying property and hunting down wealthy inhabitants. A young British officer recorded 'heaps of dead bodies scattered throughout the place and every house broken into and sacked'. Throughout the 'pacification' of north-east and central India homes, shops, bazaars, temples and mosques were systematically despoiled and the loot carted off.[37]

In Canton, Imperial Commissioner Yeh and the population at large eagerly received exaggerated accounts of Britain's losses in India. Elgin's expedition against China had been diverted by the Rebellion, and Canning was unwilling to return any troops until the situation was firmly in hand. This infuriated British officials and merchants in Hong Kong and Shanghai. Laurence Oliphant spoke for them when he wrote: 'it would never do for the world to suppose we cannot manage two Eastern difficulties at the same time'.[38]

The best chance the rebels had of eliminating the Raj for ever had passed when Delhi fell. But the Rebellion was yet to be defeated.

Havelock's small force fought its way to Lucknow, uncoiling telegraph wire as it went. But it was not strong enough to extricate the beleaguered defenders and had to join the garrison in the confines of the Residency. Only in November did an army led by Sir Colin Campbell evacuate them. Oudh remained in rebel hands, as did large parts of central India. By the time Delhi fell, the idea that British commercial and military power was about to collapse, and drag down the global financial system with it, was long gone.

By the end of 1857 the British were resurgent. The mood was for vengeance to be meted out against not only the Indians but also the Chinese, whose debts of honour were still unpaid to the British Empire after the *Arrow* affair. Strong enough to regain India, troops were resupplied to Elgin and came under pressure to inflict a blow on China. The attitude he observed in colonies and bases from Calcutta through to the South China Sea nauseated him. 'I have seldom from man or woman since I came to the East heard a sentence which was reconcilable with the hypothesis that Christianity ever came into the world,' he informed his wife. 'Detestation, contempt, ferocity, vengeance, whether Chinamen or Indians be the object.'[39]

The earl did not possess the military strength to proceed to Beijing; instead he reluctantly conceded to the demands of Sir John Bowring and the naval commanders to attack Canton. He wrote to his wife as he made his way to wreak havoc on the city in late December 1857: 'When we steamed up to Canton, and saw the rich alluvial banks covered with the luxuriant evidences of unrivalled industry and natural fertility combined; beyond them, barren uplands, sprinkled with a soil of reddish tint, which gave them the appearance of heather slopes in the [Scottish] Highlands; and beyond these again, the white cloud mountain range, standing out bold and blue in the clear sunshine – I thought bitterly of those who, for the most selfish objects, are trampling underfoot this ancient civilisation.' The commodore in command of the naval forces noted the earl's melancholy countenance. 'Yes,' Elgin replied, 'I am sad, because when I look at that town, I feel I am earning for myself a place in the litany, immediately after plague, pestilence and famine.'[40]

Canton came under bombardment, 'slow and continuous, with a

sombre monotony', from the Royal Navy's gunboats. The attack was co-ordinated by field telegraph. 'I hate the whole thing so much,' Elgin opined. He watched the shelling from atop a hill with Laurence Oliphant, who recorded: '200 feet below lay the city, mapped out before us; a vast expanse of roofs, a labyrinth of intricate lanes . . . a pagoda here, there a many storied temple, or the successive roofs of a *yamen* embowered in luxurious foliage, above which towered a pair of mandarin poles – beyond all, the tapering masts of our ships'. The most striking thing was the silence and lack of movement 'on the part of a population of a million and a half, that lay as though entombed within the city walls'.[41]

On the last day of the year the joint British, Indian and French force scaled the walls and entered a city that had denied access to Westerners for centuries. There was very little opposition to the storming of China's second city. As in India at the same time, the response among the soldiers, marines and sailors to this momentous event was to sack fabled Canton, 'the most important and flourishing mercantile emporia in the empire', and carry away as much of its wealth as possible. 'My difficulty has been . . . to keep our own people in order, and to prevent the wretched Chinese from being plundered and bullied,' a pained Elgin noted in his diary. 'There is a [Hindi] word called "loot" which gives, unfortunately, a venial character to what would, in common English, be styled robbery.'[42]

This disapproving attitude did not stop Elgin taking fifty-two massive boxes of silver coin and sixty-eight packets of silver ingots from Canton's treasury. For him the haul represented 'reparations', not plunder. Commissioner Yeh, the official unfairly blamed for starting the whole thing, was captured and packed off to a villa outside Calcutta. 'The imperturbable Chinamen go on just as though nothing had happened,' reported *The Times*' special correspondent. The Anglo-French army of occupation was 'puzzled by the tenacious, childlike, helpless obstinacy – the passive resistance of their enemy'. There was something unsettling about the indifference of the Chinese to the invasion of their country.[43]

Canton thus degraded, Elgin set off to fulfil his primary task: to enter the Peiho River, seal off the gateway to Beijing and force the emperor to give in to Western demands. His squadron of navy steamers rendezvoused with a Russian warship carrying Admiral Poutiatine, a French warship and the USS *Minnesota*. Four plenipotentiaries from the world's four major powers were hovering in the Bohai Sea, close to the imperial capital, ready to press their demands.

The allies made heavy weather of crossing the bar at the mouth of the Peiho. Foreign gunboats entering the 'river of the north' for the first time was clearly a warlike act and they expected stiff opposition from the forts at Taku guarding the entrance to the river. But, as at Canton, there was only token resistance. The flotilla proceeded upriver to the metropolis of Tientsin (Tianjin) ninety miles south of Beijing. China had never been trespassed to this extent. At the negotiations with the Chinese imperial commissioners Elgin acted the role of 'uncontrollably fierce barbarian', something he predictably found 'disgusting'. Several days were spent by the Western ambassadors 'fighting and bullying, and getting the poor Commissioners to concede one point after another'.[44]

Under military duress the unfortunate commissioners capitulated and agreed to the Treaty of Tientsin: protection for Christians in China, the opening of five more ports, and the right of Westerners to travel into the interior and trade along the Yangtze. In addition, the British got over £1 million in compensation for their losses at Canton and in war reparations.

All this was bad enough for the Qing dynasty. But the demand that offended them most, which trenched most seriously on Chinese dignity, was that of allowing foreign representatives to reside in Beijing. For the British this was the single most important goal, something that trumped trading rights. Elgin believed that official intercourse between Western and Chinese officials at the highest levels would prevent the misunderstandings that were ruining the Empire and sullying Britain's reputation. When China's relations were put on a proper footing, when it joined the family of nations, problems of recurrent war, piracy and drug smuggling would be resolved. Full diplomatic relations would also mean that Britain could give military assistance against the Taipings. 'Though I have been forced to act almost brutally,' Elgin wrote to his wife, 'I am China's friend in this.'[45]

But Elgin could not appreciate just how much the issue represented a fundamental and irreconcilable collision of world views between China and the West. For time immemorial the tributary system had governed Chinese relations with the world. Ambassadors of barbarian states came to Beijing at the specific invitation of the emperor and as his vassals. That notion of national prestige was important, but it was not the full story. Almost in tears the imperial commissioners begged and pleaded with Elgin to desist from the demands for permanent representation in Beijing: 'this point is absolutely impossible to concede. If we dare to authorise it at our own discretion, the emperor will certainly punish us severely.'[46] Ambassadors could reside almost anywhere else and deal with China on a footing of equality and in a manner in accordance with Western protocol, they said. Just not in Beijing. Not under any circumstances.

If the Manchu conceded the right of ambassadors to reside in Beijing, implying equality between the rulers of the world, the Qing emperor would be discredited in the eyes of all his subjects. He might even lose his throne. By the Mandate of Heaven, the emperor ruled all mankind; if barbarians denied this, then why not the Chinese people? The issue of ambassadors threatened to unravel the Confucian concept of the world. The result would be complete anarchy.

And there was a fundamental reason why the emperor should be insulated from barbarian chiefs. As one memorial put it, people on the streets were discussing the demand and 'there is no one who does not think with trembling and anxiety that the barbarians' residence at the capital would threaten the safety of the imperial ancestral temple and of the altars of the deities of the soil and the grain'. This was not just about national prestige: it was about pollution. And that was something that Western diplomats, soldiers and traders could not begin to comprehend.[47]

Elgin set out for the Far East to get retribution for China's 'insult' to Great Britain. Using this pretext, he overawed China with the West's military superiority. At last the great empire had been prised open.

Immediately after the signing of the Treaty, Elgin gave orders to be taken to Japan.

Was the Scottish earl about to treat Japan as he had treated China? The US Consul in Shimoda, Townshend Harris, had been in Japan for two frustrating years, in which the *bakufu* (the military government that ruled Japan on behalf of the shogun) had blocked his efforts to get a commercial treaty; he realised that 'no negotiations could be carried on with them unless the plenipotentiary was backed by a fleet, and offered them cannon balls for arguments'. British aggression in Canton and Tientsin and the imminent arrival of Elgin did the job of cannon-balls. The captain of the warship escorting Elgin said that 'the hour had arrived for Japan to yield to reason, or to be prepared to suffer as the Court of Peking had done, for its obstinacy'. The nightmare that had haunted Japan for almost two decades – the arrival of a British fleet – was about to be realised. Townshend put pressure on the shogun's council to sign a treaty granting extensive trading and diplomatic rights. It was necessary to agree to this treaty, said Townshend, to set a precedent and prevent Elgin demanding even more.[48]

But Elgin was not quite the figure of fear that the Japanese had been led to expect. For a start he did not have a clear idea of what he hoped to achieve, but he was sick of China and its climate. Japan offered the respite of cooler weather. It also offered contrasts of a different sort. The British on the Elgin Mission learnt to hate China and the Chinese, regarding the shabbiness of the cities and the poverty of the people as a sorry let-down, almost as a deliberate personal insult to them. In Japan they were utterly seduced by the neatness of the cities, the politeness of the people and the pretty scenery. 'Charming' was the word that Elgin frequently used as he fell deeply in love with Japan. It put his mission in context. Lord Elgin was clear-sighted enough to realise that, unlike China or India, which had been embittered and degraded by one-sided contact with the West, centuries of seclusion had left the Japanese apparently happy, peaceful and prosperous. 'God grant,' he prayed, 'that in opening their country to the West, we may not be bringing upon them misery and ruin.'[49]

Witnessing Japan in the period after the first shock of Western intrusion but before the onset of rapid modernisation was a unique treat. Oliphant was overcome with curiosity, his senses so surfeited with new impressions that he was left dazzled. Here was a society in a 'high state of civilisation' that had developed in isolation from the rest of the world. Everything was new and different; it was an explorer's paradise. The British party sat on the floor and tasted sushi. They examined Japanese artefacts and crafts with delight, visited gardens and temples, and perambulated through the clean, regular streets of Nagasaki and Edo. They might not have understood much of what they saw, but they at least respected and admired the unfamiliar culture they encountered.

Absent were assumptions of racial superiority and triumphalism. As Oliphant told readers of *The Times*, Japan had no equal in the world for its climate, fertility, 'picturesque beauty' and social peace and prosperity. It was no wonder, he said, that the Japanese had little need for the rest of the world, and he agreed that contact with the West would be destructive. On the other hand, he noted with some astuteness, the Japanese were better equipped to master Western technology and science than any other Asian society. Japan was, in this view, a *tabula rasa*, a laboratory in which Western modernisation could be tested.[50]

The Elgin visit had more of the character of a sightseeing holiday than armed intimidation. Elgin arrived with scant force and very little time. But fear of the Royal Navy still held sway and the ongoing British aggression in China and India was well known; Elgin found he was pushing at an open door. The *bakufu* was prepared to give the British as much as – and a little more than – it had just conceded to the Americans. Ports were open to them, tariffs on a number of manufactures were set at the low rate of 5 per cent, and consular representation established in Edo and the open ports. The Treaty, and the painless way in which Elgin negotiated it, was hailed as a triumph and a matter of historical significance back in Britain. In her speech opening Parliament Queen Victoria said that the Treaty 'opens a fresh field for commercial enterprise in a populous, and highly civilised country'.[51]

Beneath the placid waters of Japanese society things were not as tranquil as the British mission had been seduced to believe. Hotta Masayoshi, the leading member of the *bakufu*, had drawn up the Treaty of Amity with Townshend Harris. Meanwhile the staunch isolationist and advocate of armed resistance Tokugawa Nariaki had taken the daring step of sending his vassals to the imperial court at Kyoto to gain the emperor's ear. When Hotta asked Emperor Komei to agree to the treaties with the US and Great Britain he found that the emperor was resolutely opposed. He had little choice but to resign.

This demonstration of imperial will was unknown in modern Japan. For centuries emperors had kept out of politics. Now in the crisis of the late 1850s the two loci of power in Japan, the symbolic as represented by the Emperor and the real as exercised by the shogun, were on a collision course. The *bakufu* was tottering, its supremacy under unprecedented challenge. Its new leader, Ii Naosuke, was determined to regain control. And in the pivotal year 1858 there was an opportunity to do so. On the day Elgin approached Edo the shogun died (not that the British were ever told). There were three candidates from the three Tokugawa branch families eligible to succeed him. The race was between two of them, the twelve-year-old Tokugawa Yoshitomi or the twenty-one-year-old Hitotsubashi Yoshinobu. The older of the two happened to be Nariaki's seventh son, who had been adopted by the Hitotsubashi-Tokugawa family to put him in the line of succession.

If Nariaki got his way and his son became shogun then Japan would reject any agreement with the West. The *bakufu*, as it was presently constituted, would be discredited. But Ii Naosuke and the *bakufu* backed the boy candidate. This was a naked struggle for power in Japan. Once again the emperor became involved when Nariaki lobbied on behalf of his son; several other *daimyo* did likewise. On the succession crisis hung the future of Japan. Would it slam shut the doors to the West just after it had opened them? Would the emperor enlarge his long-dormant powers at the expense of the shogun? It was Nariaki, leader of the radical *sonno joi* ('Revere the Emperor, Expel the Barbarian') movement, versus Ii, head of the traditional *bakufu*.

In the end the *bakufu* won. Ii secured the majority of the *daimyo* and the Tokugawa boy became the fourteenth Shogun of Japan. Ii was all-powerful and he made everyone aware of the significance of his victory. Nariaki was put under house arrest. Courtiers were imprisoned and lesser-ranking samurai who had travelled to Kyoto to lobby on behalf of their masters were rounded up and carted back to Edo in cages.

The Ansei Purge, as it is known, was a move to crush all opposition to the commercial treaties. The most zealous critics of the decision to open Japan were beheaded as the *bakufu* ruthlessly consolidated its power. Japan may have looked peaceful and content to Elgin and Oliphant. But the façade of politeness concealed the brutal assertion of power and deep fissures that were shattering society.[52]

'We are . . . quitting,' Elgin mused as he steamed back to Shanghai, 'the only place which I have left with any feeling of regret since I reached the abominable East – abominable, not so much in itself, as because it is strewed all over with the records of our violence and fraud, and disregard of right.'[53]

Those 'records' could be read across Asia in 1858. While Elgin was busy forcing China and Japan to capitulate, British forces were snuffing out the last embers of the Indian Rebellion. *The Times* special correspondent William Howard Russell was with Lieutenant Patrick Stewart on a day in February on the banks of the Ganges. Stewart took aim with his rifle and fired at the end of a fuse placed on a rock. The weapon was known as 'Jacob's Rifle', invented by Brigadier John Jacob to fire an exploding shell. Stewart's aim was true; the bullet hit the rock, exploded and ignited the fuse. It was a showy and ingenious way of setting off a series of mines. The resulting explosion blew to pieces Hindu temples on the banks of the Ganges. 'Alas! dirty fakirs and Brahmins,' commented Russell, 'your triumph was but short.'[54]

Here was modern technology harnessed in the name of vengeance – the hyper-modern twinned with the primeval desire for retribution.

Pat Stewart's efforts gave a modern sheen to the British forces as they regained control of India. But modern civilisation could not tame vengeance. The mutineers may have behaved appallingly in the maelstrom of the Rebellion, but the revenge meted out by the British and their allies was quantitatively much worse and scarcely less restrained.

William Russell was with Stewart again as British troops stormed the Qaisarbagh, the stunningly opulent complex of palaces and mosques in Lucknow, on 14 March. The British and their Indian allies numbered over 30,000, and they had fought their way into Oudh to smash the Rebellion.

Russell asked his readers to imagine a scene reminiscent of Samuel Taylor Coleridge's *Kubla Khan*: gardens and courtyards with orange groves, statues, rows of lamp posts, aqueducts and pavilions with 'burnished domes of metal'. When Russell entered these enchanted grounds battle was raging between redcoated British soldiers and rebel sepoys; bullets hissed through the air, dead Indian soldiers lay among the orange groves and the statues were reddened with blood. Surrounding the gardens were magnificent colonnaded buildings with richly gilt roofs and domes. Even as the battle was fought round them British soldiers battered down the doors with the stocks of their rifles or shot them off their hinges. From inside the buildings Russell heard shouts and smashing glass.

Then out of the doors poured a torrent of British soldiers laden with plunder: 'Shawls, rich tapestry, gold and silver brocade, caskets of jewels, arms, splendid dresses. The men are wild with fury and lust of gold – literally drunk with plunder.' Some men came out with fine Chinese porcelain or mirrors, which they smashed to pieces on the ground before running back in for more. Others emerged attired in regal costumes encrusted with precious metal and gems. Out of the windows flew the riches of the royal house of Oudh considered worthless by the marauders: broken furniture, ripped-up paintings, gorgeous clothing, musical instruments, books and everything else that could be found in a palace. These piles of discarded plunder were turned into bonfires. Those who walked along the marble pavements in the courtyards crunched through three inches of broken fragments of glass from chandeliers and mirrors.

Russell proceeded through the labyrinthine palace complex, from court to court, stepping past dead sepoys, 'their clothes smouldering on their flesh', and looking with wonder at the chests stuffed with china, goblets, cups of the finest jade and the caskets brimming with emeralds, diamonds and unbelievably enormous pearls. Everywhere were 'banditti' of the Queen's army, 'faces black with powder; cross-belts specked with blood; coats stuffed out with all sorts of valuables'. Russell had heard the expression 'drunk with plunder' many a time; but he had never witnessed the terrors of the uncontrollable orgy of blind destruction that lay behind the phrase. Nonetheless,

Russell, like the officers and civil officials, pocketed his share of the loot.[55]

After the capture of Lucknow, the conquest of Oudh and a subsequent campaign across central India all resistance in India was suppressed by the beginning of 1859. The whirlwind transformation of Asia in 1857 and 1858 showed, for Victorians, that military force and modern technology were inseparable in the civilising process. All the burning and looting and despoiling of Oriental pleasure domes – be they in Delhi, Lucknow or Canton – had a meaning that went beyond lust for instant riches. It was the enactment not only of vengeance for crimes and insults committed against Western civilisation, but the rebirth of 'backward' societies so long in thrall to sensuous illusions and surfeiting excess. It was not just soldiers armed with clubs who were prepared to vandalise, it was the writers who set up visions of Oriental splendour with their pens only to belittle them as infantile and then revel in their immolation.

From the destruction of barbaric gewgaws and the levelling of absurd enchanted palaces, it was held, would come a more rational world of progress, free trade and technology. When Patrick Stewart, the whizz-kid of telegraphy, detonated ancient Hindu temples with one shot of a state-of-the-art exploding bullet he was making vivid the total technological and military supremacy of the West over benighted and backward savages. And when the British sacked Lucknow they were levelling the barriers that appeared to them to stand in the way of modernity. In building the palaces at Lucknow and stuffing them full of riches, the kings of Oudh had outraged the British and demonstrated their unfitness to rule. The annexation of Oudh helped spark the Rebellion; the vandalism of its palaces signalled the vanquishing not only of the revolt but of the old, hidebound, unreformable India.

British power had proved instrumental in opening up the great and formerly impregnable empires of China and Japan in the later 1850s and dragging them into the modern world. Technology was the symbol of nineteenth-century civilisation and the weapon that battered down all resistance. The trusty Enfield rifle was the new imperial enforcer, its range and accuracy giving those armed with it advantages over numerically superior forces.

Many years later, in 1902, a twenty-foot grey granite obelisk was raised in Delhi to commemorate the actions of William Brendish and others in warning British India about the mutinies by telegraph. Time has rendered much of the inscription on this memorial unreadable, but it is still possible to make out the words stating that 'The electrical telegraph has saved India' (saved India, that is, for the British). The inevitable triumph of modernity was at the heart of stories and myths about the Rebellion. Five years after the obelisk was unveiled the journalist Louis Tracy penned a novel about the events of 1857 called *The Red Year*. One scene dramatically reveals not only contemporary beliefs about the inevitable triumph of Western science, but also modernity's close connection with barbarism. In it a captured sepoy facing execution looks up at the cable on the telegraph pole from which he is about to be hanged by the British. 'Ah,' he says, 'you are able to hang me now because that cursed wire strangled all of us in our sleep.'[56]

PART III

News of the World: The Age of Bronze

12

Empire of News

FLEET STREET

This war . . . has knocked the delusion out of the imagination of the world . . . and proved that England still is, and still deserves to be, the paramount empire of the globe. In this war we have no allies but our own spirit and our own energy . . .

Illustrated London News[1]

It was one of the most forbidding and lonely places in the world: 'a huge black rock lifts its head up out of the deep water . . . the eternal wash of the Atlantic has worn deep hollows, and in some cases masses of rock stand out isolated from the great granite wall that breaks the ever-restless ocean that thunders against it'. Yet the name of this forsaken spot had suddenly become 'as well known and . . . as familiar as that of New York or Boston'.[2]

For centuries Cape Race, on the south-eastern tip of Newfoundland, spelt danger to fishermen; in storms and frequent dense fogs the only warning of the proximity of the murderous rocks was the noise of the waves exploding against them with the fury of heavy artillery. It was a place unknown to most and avoided by those who had no choice but to be in the vicinity – that was, until the invention of the telegraph.

From 1858 it would be impossible to pick up a newspaper in the United States without seeing 'Cape Race' printed in large, bold, prominent type. There was one simple reason for this: Cape Race was the first bit of land sighted by ships bound for Boston, Portland or New York from Europe. Liners passing the cape would let off a rocket. The sight prompted several small fishing boats to race out into the plunging waves to search for a small, airtight metal canister that was weighted at one end and had a small metallic flag at the other. At night it was lit up by a phosphorescent flare. The bobbing canister was scooped in a net and taken back to land. It contained one of the most valuable commodities of the 1850s: the news of the world, fresh from London. The headline news was telegraphed from Cape Race to the New York Associated Press two or three days before the liner reached Boston or New York.

American newspapers had a stop-press section labelled 'Via Cape Race' that boasted news from around the world two or three days earlier than by ordinary mail. From late 1857 Associated Press kept its own news yacht at the cape to intercept the liners as they approached. It was an expensive business, but the American public was hungry for news from around the planet. To take a random newspaper at a random date, the *Genesee County Herald* of upstate New York of 19 June 1858 printed world news with the dateline 'St John's, NF' that was just nine days old. The steamer *Vanderbilt* had received a news telegram from London as it was about to depart Southampton; it dropped off the signal for the AP news yacht at Cape Race, and the headlines were telegraphed nationwide as soon as it was received on land.

It was, boasted the *Genesee County Herald*, unprecedentedly quick. Leading the headlines was 'Important News from India'. The American public, like that of the leading European states and Russia, were fixated on the Indian Rebellion in 1857–8, seeing it as an event of global significance. Next up were reports of a 'fearful eruption' of Mount Vesuvius, and then news from the Treaty negotiations in China that had come by ship to Malta, where it was telegraphed to London. It also printed the closing price on the London Stock Exchange and all-important news of American railroad securities. The *Troy Daily Whig*, another New York state journal, printed the same news, but also reported price movements on the Liverpool cotton and corn exchanges. Its digest of the British provisions market read: 'Beef quiet but steady. Pork quiet and firm. Lard heavy.'[3]

In the recent past news from overseas was read first in the eastern port cities and then transmitted nationwide; now, thanks to the telegraph, world news was shared simultaneously across the United States (with the exception of Oregon, California and the territories west of the Missouri that were not connected to the telegraphic network). This shared experience gave the news a greater value. The story of how the news arrived from distant parts of the world – the sequence of the baton race of steamers, overland mail, AP intercept yacht and telegraphs, the names of the ships, their arrival and departure dates – became a news story in its own right, preceding the headlines.

The internationalised economy of the 1850s feasted on information. With so much cash loaned in advance of productivity across expansive networks, and commodities such as cotton and wheat purchased before the seeds were even sown, businesses needed fast and accurate knowledge about anything that might affect prices: weather, international events, rumours of war, internal politics, elections, revolutions, ships under construction and anything and everything that occurred or was likely to occur. As the Midwest became more dependent on exports to Liverpool, American newspapers watched the weather in the British Isles like a hawk: unseasonal storms or frosts could betoken a bonanza, fair skies and bountiful harvests a calamitous dip in prices. In the same way, Liverpool cotton brokers scanned the data on rainfall in the southern American states to anticipate the next season's yield of cotton. There could not be enough information, and it could not arrive soon enough. Time was money, and power. Merchants and banks already realised the need to construct networks of information exchange – correspondents dotted around the world who fed back to HQ a never-ceasing stream of intelligence.

But in the 1850s new technology meant that information was readily available. In a world already being transformed out of recognition by migration, economic expansion and the opening of Asia, the news revolution stands as one of the most significant developments. The appetite of publics across the globe for the latest news became a marked feature of the age. As *The Times* of London commented, there was no prize for second place in the modern world: 'If Puck takes 40 minutes to put a girdle round about the earth, Ariel will beat him out of the field if he does it in two throbs of a pulse.'[4]

*

The British public was getting its news faster and faster in the 1850s. And it was *The Times* that was in the position of Ariel, scooping ponderous Puck. At daunting expense it sped news from around the world to Fleet Street by means of numerous foreign correspondents, express mail contracts, railways, telegraphs and its own steamboats. No other paper on the planet could afford a news operation on this scale, and it made as its primary boast the extraordinary speed at which it delivered worldwide news. By 1854 *The Times* was the most important and powerful newspaper in the world, reporting graphic, gripping first-hand accounts from the front line of the Crimean War sent by its corps of war correspondents. Its international news provided the mainstay of foreign reports in the US, Europe and beyond. Its editorials were taken as barometers of British public opinion.

Under the Stamp Act *The Times* could post its weighty pages anywhere in the country for just one penny. It therefore offered a substantial read, including its unbeatable foreign coverage, at a cost that matched or undercut local papers. But it was a dangerous eminence to occupy. In an age of free trade, power concentrated in this way was seen as wrong. When it came to reporting the Crimean War – its greatest achievement – this dominance had unwelcome consequences. When the dust settled and the country's military performance improved, *The Times* was blamed for giving a partial view of public opinion and an exaggerated account of the mismanagement of the war to the world, marginalising other accounts, and tarnishing Britain's global reputation. Monopoly of opinion should be replaced by diversity.

As it reached its zenith, *The Times*' power was broken for good. In 1853 advertising duty was repealed; two years later the Stamp Act was abolished. *The Times* had to raise its price to cover the postage. Provincial papers now could compete with the Paper of Record on price. And, more importantly, from 1855, thanks to the telegraph, they could compete on content as well.

Headline news was received in London by the two main telegraph companies – the Electric and the Magnetic. These nuggets of information were transmitted not just to *The Times* but to every national and local newspaper in the country at exactly the same time. That meant that the *Essex Standard* in Colchester, or the *Derby Mercury*, or the *Aberdeen Journal* could print a digest of world news before copies of *The Times* arrived by rail.

The telegraph provided headlines or breathless stop-press columns labelled 'By Submarine Telegraph'. Fuller news stories came later, delivered by mail steamer a few days after the brief telegraphed messages. *The Times* retained its primacy in providing detailed news, but local papers now had access to breaking news. These provincial journals simply lifted foreign stories from the Paper of Record as soon as it arrived by rail and sold it for a penny, compared to *The Times*' price of five and a half pence.

The telegraph therefore made news cheap. According to *The Times*, it cheapened it. Mowbray Morris, manager of the paper, sniffed, 'I do not confide much in the telegraph, and I would it had never been invented.' The only foreign news he trusted was that sent by the paper's own handpicked correspondents; news down the wire, by contrast, he considered unreliable and founded on little more than rumour.[5]

He was right to worry. The *Daily Telegraph* was founded in 1855, the year that the popular one-penny press exploded, specifically to cater for 'the million' – the mass of middle-class men and women who had never taken a daily paper before. It was priced at one penny. It was after an enormous untapped market and the *Telegraph* ran under the slogan 'the largest, best, and cheapest newspaper in the world'. As its name promised, its reportage was based on hot news sent every day along the wires from around the country and the world. 'What we want is a human note,' said its proprietor. The *Daily Telegraph* was a prominent example of the 'new journalism' so deplored by *The Times* that emerged from 1855 – sharp, brash, irreverent and sensationalist, much like the New York daily press.[6]

The newspaper revolution of the 1850s significantly altered Britain's relationship with the wider world. In the competition to exploit the vast market for news, nationals such as the *Daily Telegraph*, provincial penny dailies and the populist working-class Sundays – *Lloyd's Weekly*, *Reynold's News* and the *News of the World* – hit on a rich seam. The deeds of heroic men on Britain's colonial frontiers and 'small wars' in faraway places offered the lower-middle- and working-class readers a banquet of patriotic adventure stories set in exotic locations. Millions of people in Britain who had never picked up a newspaper before became deeply involved in the unfolding drama of empire. The 'new journalism' helped foster a new imperialism – and an imperial ethos that would reshape Britishness.[7]

A news story from the interior of India, say, was telegraphed to Bombay, and the message transcribed and put on a mail steamer to Suez. From there it was cabled to Alexandria, transcribed once more, and put to sea again, this time propelled by steam to Trieste, where it was immediately telegraphed to London. Such a news item was brief, merely enough for a headline or a few short sentences prefixed in bold with the words 'By Submarine Telegraph'. It took a few more days for the story to catch up with the headline, when the full report arrived by mail, courtesy of steamers and railways.

News of the Indian Rebellion came this way. The official telegraph dispatches were received in London on 26 June 1857, five weeks after it broke out at Meerut. They were followed days later by the mails. These were frustratingly fragmentary and sketchy. On 13 July the next batch of telegrams pulsed in with their stark bare-bones details of atrocities that sparked sensationalist headlines and feverish speculation, followed a few days later by reports from special correspondents, official dispatches and some 20,000 private letters purporting to reveal what was going on. It was a rhythm of news that inevitably generated the maximum excitement.

The rollercoaster of emotion continued throughout the Rebellion: weeks of suspense followed by an explosion of headlines, then a glut of information as the mass of unsorted news was digested. The public was kept in a state of constant anticipation and excitement from the summer of 1857 through the autumn and into the winter as tales of horrific massacres and counter-attacks, of sudden setbacks and amazing victories poured in. The Indian Rebellion was a unique news event in the history of journalism. A month between event and report was short enough for the news to seem fresh and immediate. But it was long enough for the press to construct a narrative and chronology of events that was replete with horror and adventure, but more often than not at variance with what was actually happening.

News, for example, of the Cawnpore massacres at the Satichaura Ghat and the Bibighar arrived at exactly the same time as reports of General Henry Havelock's lightning march up the Grand Trunk Road and his succession of unexpected victories. A proper sense of chronology was distorted when events that in real-time took two weeks were compressed into a single story. Shocking news of murder was twinned with stories of heroism and retribution. This was news as melodrama, told by instalments with the breathlessness and cliffhangers familiar from a serialised novel by Charles Dickens or Wilkie Collins. There were no reporters on the ground in these crucial months; the press in Britain relied on first-hand accounts of soldiers who were in the thick of events. Journalists were free to transform these reports into vivid and, it has to be said, imaginative prose. Havelock was totally unknown in Britain. Suddenly he was propelled to a national hero, the embodiment of 'Anglo Saxon pluck'.[8]

According to *Lloyd's Weekly*, 'In the midst of blood, and outrage, and disaster – in the worst darkness of the storm, towering as a giant above all other fighting men, stands gallant Havelock! His cheering voice, his keen sword, his dashing generalship, make up, for us, a figure of radiant hope. His name becomes a "household word".' And for *The Times* he embodied the racial superiority that gave Britain the right to rule India: 'General Havelock's march is the very expression and type of our position in Hindustan. He advances, he fights, he conquers – everything goes down before him.'[9]

The glorious deeds of Havelock, as they were reported in September, were magnified out of all proportion by reports of unmitigated repulsiveness. Stories of the most lurid and sensationalist kind filled the press. European children had been force-fed the still-quivering flesh of their parents before being tossed into the air and impaled on bayonets. The last sight many men had was of their wives and daughters being raped. In Delhi, the public was told, 'Delicate women, mothers and daughters, were stripped of their clothing, violated, turned naked into the streets, beaten with canes, pelted with filth, and abandoned to the beastly lusts of the blood-stained rabble, until death or madness deprived them of all consciousness of their unutterable misery.'[10]

Little wonder, then, that the public erupted into a howl of vengeance. The Rebellion was the first major news story to be reported simultaneously throughout the country, with identical reports in *The Times*, the *Northern Echo*, the *Belfast News Letter* and many other local papers. Little wonder also that the Indian Rebellion marked a turning point in the nineteenth century. It was not the news that was significant so much as the *way* that it was reported. The melodramatic and suspenseful nature of the story heightened emotions: it was at once immediate and distant. People identified strongly with the fate of their countrymen, and at the same time felt powerless. The news was unprecedentedly fast, but agonisingly slow; overly detailed, but dangerously patchy. There was a lot of unverified rumour of terrible crimes and hardly any balanced reporting. The way that the national and local press packaged the story as a serialised melodrama stirred dark emotions and changed the country in profound ways.

Charles Dickens spoke for many at this time when he expressed his feelings in a private letter. 'I wish I were Commander-in-Chief in India. The first thing I would do to strike that Oriental race with amazement . . . should be to proclaim to them . . . that I should do my utmost to exterminate the Race upon whom the stain of the late cruelties rested . . . and was now proceeding, with all convenient dispatch and merciful swiftness of execution, to blot it out of mankind and raze it off the face of the Earth.' Millions shared his genocidal fantasies.[11]

There were, in fact, strikingly few cases of rape or torture committed by mutinous Indian sepoys. William Russell was 'moved to the inner soul by the narratives which came to us by every mail' in the autumn of 1857; when he got to India in 1858 to report for *The Times* he heard numerous stories of rape and mutilation, but never any proof. Official inquiries also affirmed that rumoured atrocities did not stand up to scrutiny. But at the crucial time, in the summer and autumn, internal communications in India had broken down, and the press in Britain reported rumours sent from Bombay or Calcutta, hundreds of miles from the action, as gospel. In the second half of 1857 the British public was absorbed by two very simple and compelling narratives

Miss Wheeler defending herself against attacking Sepoys during the Siege of Cawnpore, June 1857

– one of unparalleled barbarities inflicted on women and children, and the other of heroic British soldiers suffering privations on their single-minded crusade to save them.[12]

One of the most famous images of the Rebellion as it was reported in Britain was that of Margaret Wheeler, known as Ulrica, the eighteen-year-old daughter of General Sir Hugh Wheeler, the defender of the makeshift entrenchment at Cawnpore. There were several versions of what happened. One states that she was abducted from the boats at the Satichaura Ghat by a Muslim soldier, but threw herself into a well after killing her kidnapper and four others with a sword. The most famous says that, rather than sacrifice her honour, she grabbed a revolver and shot dead five would-be rapists before blowing out her brains with the remaining bullet. The legend was rounded off when her hair was discovered by a group of British soldiers. The horrified men divided the clump between them and made a solemn vow to kill a mutineer for each strand of hair.[13]

*

Inflamed by such stories and countless unverified tales of sickening brutality, the British needed heroic men to avenge the crimes committed against helpless women and children. And cometh the hour, cometh the men.

'It is said,' declared Lord Palmerston, 'that India is a land fertile in heroes; but the fact is that India is a land fertile in those events which give to British subjects the opportunity of displaying those great and heroic qualities which abound so much among the inhabitants of that nation to which we have the happiness to belong.' The Rebellion bequeathed the public a pantheon of freshly minted heroes. The character of the conflict gave many soldiers the opportunity to use their initiative and distinguish themselves in martial feats. Few were elevated as high as Henry Havelock.[14]

Despite the fact that the general is commemorated by a statue in Trafalgar Square, very few people who pass by it today know who he was, let alone revere him. But to the Victorians he was the embodiment of national virtue who ranked with Nelson and Wellington. The public was eager to find out more about him; the profusion of hagiographical newspaper profiles and books did not disappoint. During four decades of diligent, dull, unglamorous service in foreign lands, hard-up Havelock watched as his well-heeled contemporaries and juniors secured promotion by purchase, leapfrogging over him to senior command. During the Persian War he was an aged brigade commander, described unflatteringly in army circles as 'fussy and tiresome' with a 'little, old figure'.[15]

These were the ingredients for a true hero of the 1850s. A diligent middle-class professional, without the benefit of aristocratic connections, he displayed in his long years of frustration and neglect the virtues of self-help and moral fortitude so beloved by Victorians. He progressed by merit alone through hardships and unfairness to achieve a stunning apotheosis at the very close of his career. Religious faith rounded off the picture: he was a zealous Baptist who preached, distributed Bibles among his men, held prayer meetings and formed a regimental temperance society. Here was a Christian soldier at the head of men reminiscent of Cromwell's Ironsides: austere, evangelical, tough, self-disciplined and sober.[16]

Who better than he to be Britain's agent of divine retribution at Cawnpore and Lucknow? For a public maddened by stories of bestiality and ravenous for vengeance, Henry Havelock was the ideal man to lead the fight-back. Three times he attempted to battle his way through to relieve the civilians and soldiers trapped in Lucknow. On the third attempt he broke through. He died of dysentery on 24 November 1857 just after the siege was lifted. News reached London on 7 January and prompted an outpouring of emotion across the country. Evangelical tract and patriotic street ballad alike mourned him and recounted his daring deeds. Death sealed his mythic status and set in motion a cult of hero-worshipping that was to last well into the twentieth century.

Above all, the heroes of the Rebellion epitomised Victorian masculinity. One hundred and eighty-two men were decorated with the Victoria Cross for conspicuous gallantry under enemy fire, a staggering 13 per cent of the total awarded in the medal's history to date. There were adventure stories galore for the public to feast upon.

These men were among the action adventure heroes of the 1850s. None surpassed them, however, more than John Nicholson, the 'Lion of the Punjab', the 'Saviour of Delhi'. If General Havelock represented diligence and endurance, General Nicholson was a star who burned bright and died young. The Rebellion shone a light on the activities and achievements of the band of brothers who ruled the badlands of the North-West Frontier, the young 'pioneers of Christian civilisation' who raised irregular cavalry units and brought firm government to newly conquered tribal regions. 'Nicholson impressed me more profoundly than any man I had ever met before,' remembered ensign Fred Roberts of when he met him in Peshawar. '. . . His appearance was commanding, with a sense of power about him which to my mind was the result of his having passed so much of his life among the wild and lawless tribesman, with whom his authority was supreme. Intercourse with this man amongst men made me more eager than ever to remain on the frontier.'[17]

During the Rebellion, Nicholson pacified the Punjab at the head of the Moveable Column like the 'incarnation of violence', before racing to the Delhi ridge. In a famous battle, he drove off a larger force that was about to capture the siege train. The presence of Nicholson (who had risen from captain to acting brigadier-general in a few weeks) raised morale from despondency to 'supreme elation'. Leading his men into the streets of Delhi, he was picked out by a sniper and died of his wounds. Present at his funeral were his faithful Pathan, Afghani and Ghurkha followers. 'For him they left their frontier homes, for him they had forsaken their beloved hills to come down to the detested plains,' wrote a British officer; 'they acknowledged none but him, they served none but him.' These stern mountain men broke down in tears as the coffin was lowered into the grave.[18]

Nicholson's death at the age of thirty-four ensured he passed into legend as surely as Henry Havelock. Nicholson was a God-fearing man, an unyielding Protestant Ulsterman. But that was not what spoke so forcefully to the Victorians of 1857 and 1858 or during the years when his legend swelled. Nicholson embodied the virtues of manliness that were becoming so important to the Victorian middle class.

Modern readers are more likely to see General Nicholson not so much as a paragon than as a deeply troubled man.

There was more than a hint of sadism in his brutal treatment of prisoners and subordinates. Nicholson was said to have personally beheaded a thief and kept the miscreant's skull on his desk. He took genuine delight in hunting down and executing rebels, declaring that 'flaying alive, impalement or burning' would have been more to his taste than the merciful way of executing them at the barrel of a gun. With 'a perfectly easy conscience' he was prepared to 'inflict the most excruciating tortures I could think of' on the mutineers. And it was not just his enemies who bore the force of his ire. On his march to Delhi he thrashed a cook boy who got in his way. The boy complained, so he was thrashed again and died. Taciturn, withdrawn and unsmiling, Nicholson surrounded himself with devoted Pathan and Afghani acolytes who obeyed without question, and a handful of British officers who were 'more than brothers'; he had few words for the majority of his fellow countrymen, still less for his countrywomen.[19]

Impressive he undoubtedly was; behind that, however, was a tortured, emotionally stunted man. But the very qualities that make him objectionable now were at the heart of his appeal in the 1850s. His capacity for unflinching brutality was considered to be just what was needed in the moment of the Empire's gravest crisis. And more than that, his brief but glorious military and civil career held up a mirror to the mid-Victorian generation. And they liked what they saw very much.

The legend of John Nicholson helped launch an industry of imperial yarns in which young men govern native peoples with fairness and firmness, with only the fortitude of their character, a band of devoted warriors and a trusty revolver. In the 1850s – in large part thanks to the way the Rebellion was reported – the North-West Frontier became embedded in British popular culture in the same way that the Caucasus was to Russians and the Wild West for Americans. They were places of romance and adventure, where men could be men far from the trammels of civilisation. The virtues attributed to Nicholson – his godliness, unselfishness, generosity and courage – were exactly those needed to rule the outposts of empire. 'He was a man cast in a giant mould,' gushed a soldier present at the siege of Delhi, 'with massive chest and powerful limbs, and an expression ardent and commanding, with a dash of roughness; features of a stern beauty, a long black beard, and a deep sonorous voice.'[20]

Nicholson's beard is worthy of special attention. So was Charles Dickens' for that matter. The beard was de rigueur on the North-West Frontier and in Afghanistan. Officers-cum-officials like Nicholson grew them in order to command the respect of hill tribes, who saw facial hair as primary indicators of masculinity and warlike propensities. When he returned to Britain in 1850, John Nicholson decided to shave his beard in accordance with civilian fashion. The only known photographic portrait of Nicholson shows him uncharacteristically clean-shaven; he looks humdrum, like a nondescript bank clerk.

If he had journeyed home a few years later there would have been no recourse to the razor. Beards suddenly became very popular from the middle part of the decade. In 1853 Dickens' *Household Words* said

John Nicholson, bearded and unbearded

that British males were still 'slaves to razors'. Dickens grew his, one of the most famous nineteenth-century examples, in the middle of 1855. There was a reason for this date. This was when public anger at the aristocratic monopoly of state and military was at its height as the mismanagement of the Crimean War became clear. Dickens was a leading member of the Constitutional Reform Association, a body that argued that the stranglehold of privilege and nepotism that gripped the army and civil service should be broken for ever by opening entry to competitive examination. A beard was decidedly *un*-aristocratic, redolent of the hard life of the frontier. During the Crimean War the military prohibition on beards was lifted and soldiers were encouraged to grow them for the purposes of health – a covering of hair on the face and neck was supposed to protect the wearer against chills. Suddenly it became very fashionable in the mid-50s for middle-class men to shun the razor, in part because it showed solidarity with the rank and file of the military who were suffering because of blunders made by their noble leaders. By 1858 Dickens' small pointy beard had matured into a full door-knocker.[21]

Men of the 72nd Highlanders who served in the Crimean War: the new medium of photography celebrated rank-and-file heroes and helped make the beard the ultimate symbol of manliness

But the primary reason for the revival of the beard was its association with masculinity. The author of a pamphlet titled *The Beard! Why do we cut it off?* said that it was all very well for women and effeminate dandies to give prominence to the 'soft, round beauty of lip and chin'. 'But men,' he continued, 'should represent the sublime rather than the beautiful. His peculiar virtues are of a sterner and more rugged cast, and it is plainly intended that his external appearance should have a corresponding character.'[22]

But could a 'nation of shopkeepers' carry off the warlike beard? During the Crimean War it appeared that middle-class values – the pacific ones on display at the Great Exhibition – had enriched the country but impoverished the British fighting spirit. 'Can we do nothing but make railroads and cotton goods?' spluttered the *Illustrated London News* in 1855. '. . . And are we so demoralised by a long peace – so soaked and sodden in the fat of commercial speculation – that we have lost the robust and manly virtues of our ancestors?'[23]

Emphatically not, the press were saying two years later. Manly characteristics displayed by middle-class heroes like Havelock and Nicholson were exaggerated examples of the virtues of middle-class males in general: self-reliant, self-disciplined and evangelical. For the press, Nicholson was transformed into a bourgeois entrepreneur of empire: 'a specimen of that combination of the man of business and the man of war'. In a discussion of Generals Havelock, Nicholson and Neill, *The Times* said that the 'middle classes of this country' should be proud of such heroes, men 'born and bred in their ranks'. They were 'representatives' and 'reflections' of the 'best and most sterling characteristics' of the middle classes, men not of noble birth or with friends in high places, men 'without a breath of fashion, and without a single drop of Norman blood in their veins'.[24]

The three evangelical generals were stout Anglo-Saxons, 'imbued with the whole tone, temper and ethics of middle class life'. For the press and legions of writers, 'character' was more important than class. Manliness and merit, characteristics that were earned not inherited, trumped breeding and privilege. The middle classes were not after all 'sodden' in selfish commercial pursuits, rendered unmanly and effeminate by hours in the counting house, stock exchange or shop. Where bungling aristocrats besmirched national honour in the Crimea, middle-class officers vindicated it at Delhi, Lucknow and elsewhere on the subcontinent and immediately after in China. 'There is not one single feature of energy, or self-sacrifice, or daring – not one single brilliant trait' that had once been the exclusive preserve of aristocratic military heroes that had not been gloriously demonstrated by Havelock or Nicholson.[25]

And it was not just these famous generals who showed off middle-class martial prowess. Brigadier John Jacob, one of the great soldiers of the frontier, was the seventh child of a Somerset vicar. Major William Hodson and Lieutenant Dighton Probyn VC, two of the most celebrated swashbuckling cavalry swordsmen of the campaign, along with other young officers who received the Victoria Cross, were the offspring of middle-class families. In the 1850s the heroism and daring feats of ordinary Britons were being commemorated and celebrated in the press for an eager public.

'A well-ordered beard has a tendency to give to the face an expression of manly dignity and determination,' said the author of *The Beard!* It was an outward manifestation of the qualities that were prized not just in the soldier, but in every Victorian male: 'courage, daring, energy, firmness, determination'. For did not the businessman, like the soldier, have to exhibit qualities of leadership and every day stand firm against, and subdue, the forces of chaos that beset the free market? Beards remained in fashion for the remainder of the century, getting longer and bushier as the years wore on.[26]

Manliness did not mean machismo. It transcended notions of class and inherent gentility; it was a religious and ethical code of life. Life was a battleground in which character was continually tested; manliness consisted not just in showing resilience amid the vicissitudes, but combining it with generosity of spirit, self-sacrifice, self-restraint and self-discipline. This form of individualism bound together economic notions of laissez-faire, liberal rejection of aristocratic power and evangelical concepts of the self. When *The Times*, in reference to Nicholson, talked of an 'aristocracy of talent' taking over from the aristocracy of privilege, it was pointing directly at these ideas of manliness as opposed to unearned advantage.[27]

Dighton Probyn: as the twenty-four-year-old commander of Sikh irregular cavalry during the Indian Rebellion he was awarded the Victoria Cross; he subsequently served in China in 1860. Such images of swashbuckling cavalrymen glamourised service in British India, cemented the idea of empire, and popularised the beard for Victorian males.

The heroes of the cult of manliness were the soldiers, explorers, missionaries and colonists who were subduing the world and whose deeds were celebrated in the press, the pulpit and in photographs. *The Times* recommended John Nicholson's character as a template for middle-class businessmen. The paper picked out his 'solid simplicity and quiet strength', his 'powerful commanding character, controlled by modesty, tempered by thought and reflection' and his ability to control his emotions and strike at the right moment. In short, he displayed all the characteristics of manliness urged on young Britons. Nowadays business gurus use sportsmen as psychological case studies. In the 1850s young psychopathic generals pointed the way to business success.[28]

According to Charles Kingsley, 'all true manhood consists in the defiance of circumstances'. That might seem clear-cut in moments of peril for manly men on the frontier. But it was less so for sedentary chaps engaged in office work on the home front who had few chances to emulate such athleticism and triumph over adversity. In other words, the bar of manliness was set very high.[29]

But masculinity was bolstered most obviously by relations between the sexes. If men were enjoined to be stern, fierce and disciplined, women were prized for the opposite – for their tenderness and love. According to Sarah Ellis in her *Women of England* (a bestseller in Britain and America), manliness was created and sustained by womanliness. By this she did not mean women who were 'gentle, inoffensive, delicate and passively amiable'. The ideal of femininity that she espoused was, like manliness, based on 'individual exertion', self-scrutiny and the 'formation of character'. Women should, she said, exhibit 'the energy of a sound understanding and the grace of the accomplished mind . . . the disinterested kindness of a generous heart'. Like men, they should show 'promptitude of action, energy of thought and benevolence of feeling'.[30]

Ellis' conception of womanliness formed itself around notions of manliness. In Britain the new laissez-faire economy encouraged aggressive competitive individualism, whether 'in the mart, the exchange, or the public assembly'. 'Greater facilities of communication' were rousing men 'to tenfold exertion in the field of competition'; in the modern age demands of commerce and trade were so intense that 'to slacken in exertion is to fail altogether'. At the same time Britain's 'adventurous sons' were sent to subdue the most dangerous parts of the globe. Whether they contended with market forces in Britain or exhibited 'noble daring' on the frontiers, men in the mid-nineteenth century had to engage in 'warfare' and 'strife'.[31]

The demands thus placed on men meant that 'their whole being is becoming swallowed up'. And if men had to step up to the challenges of the modern world and face them with determination and stoicism, women had a corresponding role to play. As Sarah Ellis argued, the 'best energies of the female character' were developed and honed in a particularly feminine 'sphere of action' where she could have the most influence on the affairs of society.[32]

'The sphere of a woman's happiest and most beneficial influence is a domestic one,' wrote Mrs Ellis. It was mothers and wives who had the 'holy duty of cherishing and protecting the minor morals of life, from whence springs all that is elevated in purpose and glorious in action'. Like manliness, Victorian womanliness was based on middle-class values. Ellis did not want her sex to emulate upper-class women, with their armies of servants and obsession with public philanthropy. A woman's character was formed and elevated by self-sacrifice – by 'acts of duty faithfully performed' in creating the perfect home. Her attention to her home should be 'microscopic' and never-ending. She should not simply strive for clinical 'household management', but rather create a loving and moral environment. Women were the guardians of national morality and they laid the foundations for Britain's greatness in thousands of small tasks performed in the domestic sphere.[33]

Yet even as the Rebellion boiled and raged, this happy idyll was under threat. In 1857 Parliament debated the Matrimonial Causes Bill; on 1 January 1858 the controversial and bitterly opposed measure passed into law. From that date on, men and women could terminate their marriages not by prohibitively expensive Acts of Parliament, as was the case hitherto, but by civil action in a new divorce court. Now relatively straightforward and more affordable (about £100), divorce was open to every class. A man could secure the end of his marriage simply by providing evidence of his wife's adultery; a woman had to prove that her husband was guilty of incest, bigamy, bestiality, cruelty or desertion as well. In the first year of the new law 300 or so divorces were granted, compared to just three the year before. Opponents of the Act were braced for a tidal wave of divorce and sordid tales of women's adultery that would tear asunder the happy home.

So, the Rebellion coincided with anxieties about femininity and marriage. The response was to exaggerate existing notions of womanhood and domesticity. Women's 'tenderness' and their creation of a happy home meant that husbands who engaged in the 'warfare' of daily life – in a real and metaphorical sense – could find a haven from the relentless and morally grinding pursuits of money, success and glory. The mid-Victorian cult of domesticity was even more important than the beard in affirming manliness. The sanctified home was rigorously defended from the sordidness of the marketplace, the army and politics: the moral conscience that women provided in the home allowed men to unleash their masculine instincts in their own sphere without becoming morally degenerate.

That was one of the reasons why the Indian Rebellion was such a profound shock. It represented the fatal intrusion of the brutal world into the haven of the home. In a famous cartoon from *Punch* titled 'English Homes in India', a lady with small children is alarmed as two bloodthirsty and lecherous mutineers burst into the scene of domesticity. This should be read in its literal sense: this is a parlour in India where enormities are about to be inflicted on innocents. But the Englishness of the setting also represents the way in which news of atrocities burst into and contaminated the hallowed middle-class home in Britain. Anarchy had breached the gates and crossed the threshold, and it touched directly on ideals of manliness and womanliness.

The idealisation of women as 'angels in the home' and men as their protectors and deliverers only helped to intensify the violence of retribution. British soldiers in India, egged on by the hysteria back home, meted out countless acts of reprisal to avenge wronged British womanhood.

In the painting *In Memoriam* by Sir Joseph Paton, a saintly British mother, surrounded by her daughters and the accoutrements of a

In Memoriam *by Joseph Noel Paton,*
engraving by William Henry Simmons

Victorian home, clasps a Bible and looks heavenward. In the original canvas the men bursting through the door were Indian sepoys bent on rape and slaughter. When *In Memoriam* was unveiled it provoked such a visceral response that Paton turned the sepoys into Scottish soldiers coming to rescue the women and children. Like the *Punch* cartoon, domesticity prevails to the bitter end. Here was the defining image of the Rebellion for the British: the wife/mother as a martyred saint – pious, stoic and loving to the last. The virginal Ulrica Wheeler showed another side of womanhood: the avenging angel who killed her assailants and finally herself rather than compromise her virtue.

One night in 1907 Florence Leach, a missionary doctor, was called to the bazaar in Cawnpore to tend an elderly Muslim lady who was dying. In perfect English she told Florence that she was the daughter of Sir Hugh Wheeler. Ulrica, patron saint of Victorian chastity and stoicism, had not in fact chosen to commit suicide at the thought of sexual relations with an Indian, but had married Nizam Ali Khan, mutineer of the 2nd Cavalry. He was not her assailant, but her rescuer from the Satichaura Ghat.

It was little wonder that Ulrica lived out her days anonymously in the Cawnpore bazaar. Elevated so high in the pantheon of British womanhood, the truth would provoke a furious response. Recurrent rumours that she had survived and chosen happiness with an Indian over noble self-sacrifice had emerged since the 1870s. In her days of sainthood she was held up as an icon of pure British womanhood; in her disgrace it was remembered that she was not British at all: she had an Indian mother.[34]

The image of women as passive and tender angels who were only roused to glorious action at the thought of sexual violation by an Asian man joined the long list of myths associated with the Rebellion. Women did what was necessary to survive. The idealised notion of womanliness took very little account of women's tales from the frontier. On the road to California, in farms in New Zealand and elsewhere in the new world of the 1850s women took on supposedly male roles. Women even of high birth and refinement discovered not only that they could cook and wash, but that they could handle firearms, run farms and set up profitable businesses. Most of this was brushed over in the promotion of women as saints and angels, a process encouraged by the way the story of the Rebellion was told. Many found this simplistic narrative objectionable. Ruth Coopland, who watched her husband being killed by rebels at Gwalior and then trekked, eight months pregnant, through hostile territory to Agra wrote: 'Some men may think that women are weak and only fitted to do trivial things, and endure petty troubles . . . but there are many who can endure with fortitude and patience what even soldiers shrink from. Men are fitted by education and constitution to dare and to do; yet they have been surpassed, in presence of mind and in the power of endurance, by weak women.'[35]

The responses to the Rebellion changed Britain and its relationship to the world. For one of the uprising's earliest historians, the war provided 'perhaps the most signal illustration of our great national character ever yet recorded in the annals of our history'. For another writer it exhibited 'the indomitable spirit and enduring energies of the British character, when called into action amidst scenes of unparalleled horror and acute personal suffering'.[36]

The use of the word 'character' is worthy of special attention. For the British the events of 1857–8 resolved any doubts they may have had about their right to rule large portions of the globe. The celebrated national character equipped them, or so it seemed to them, with the virtues and talents to construct and administer an empire. 'The real aristocracy of a country – that which raises a country – that which strengthens a country – that which dignifies a country – that which spreads her power, creates her moral influence, and makes her respected and submitted to, bends the hearts of millions, and bows down the pride of nations to her', *The Times* waxed, '. . . – this aristocracy . . . is an aristocracy of *character*.' By emphasising character, commentators took the argument beyond notions of racial superiority.[37]

Race was important in this; but national character was the product of many other things. Education and religion, liberties and laws, political institutions and forums for public debate, entrepreneurialism and industry, history and tradition all converged to shape a people's character. Britain prospered in the world, it was believed, because of its domestic freedoms and political arrangements. Never did complacency run so high in Britain as at this moment in history: free trade and the non-interventionist state had awoken the vitality of the people, and this in turn allowed them to spread their values across the globe.[38]

If the reappearance of the beard betokened a crisis of masculinity in Victorian Britain, it also provided follicle evidence that society had been militarised to a degree not seen for generations. 'If anything were needed to show that we were at heart a nation of soldiers,' commented the *Illustrated London News* musing on the reawakening of the martial spirit, 'this mutiny has effected it.' Until recently India and the profusion of 'small wars' had attracted attention, to be sure; but they were marginal, as was the Empire in general. Thanks to the effect of the telegraph on news, world affairs were at the centre of public attention for the first time. The euphoria of regaining India in 1858 marked a heyday of national self-confidence.[39]

How different from the heyday celebrated just seven years before. Back at the time of the Great Exhibition the idea of war and empire in its traditional form seemed outmoded. Free trade and electric communication would render these ancient things redundant; empires would wither away, war would become an irrelevance and peoples all around the world would join the march of progress. The Rebellion dispelled these airy utopian dreams. By rebelling, the Indians had shown that non-European peoples would not be transformed automatically by the power of trade and communication. They needed to be shepherded or coerced towards the future by more enlightened races – by people like the British who possessed the 'character' to remake the world.

The mood was markedly bellicose in the years following the Crimean War. That war begat the spat with America over Nicaragua and the conflict in Persia; the 'Chinese Election' of 1857 was further proof of the public's appetite for foreign adventures. But it was the Indian Rebellion that was the turning point in the British public's attitude to empire.

The way it was reported provided the essential ingredient in this transformation. It was a drama of rescue and revenge by instalments, played out in a way that guaranteed breathless suspense. It was a news event without equal: immediate enough to provoke a powerful

JUSTICE.

Britannia roused to vengeance: Punch *reflects the national mood – and the sense of national mission – at the height of the Indian Rebellion*

emotional response, but distant enough to become a nineteenth-century soap opera. The country was roused to a pitch of horror and revulsion, and then provided with saviours in the form of Havelock, Nicholson and a cast of colourful, dashing officers and men. The vile and blood-soaked reprisals meted out on Indians were celebrated with extraordinary vigour back in Britain. For here was an empire of moral, Christian purpose at war with bestial savagery. Before the Rebellion the British were lukewarm about empire; afterwards the mass of the public was fully signed up to the country's providential mission of spreading civilisation and religion by force of arms. Henceforth the British were prepared to take a tough line with Indians,

Chinese, Maori, Africans and anyone else who threatened to stand in their way. The battle for empire became the battle against evil in the world.

Far from making people more cosmopolitan and internationalist, the telegraph and the new press seemed to do the opposite. They awoke the imperialist instinct. 'An English minister must please the newspapers,' sighed Lord Aberdeen, 'and the newspapers are always balling for interference [with foreign countries]. *They* are bullies, and they make the government bully.'[40]

Thousands of stories, prints and photographs recast the Empire in the public imagination. Above all, after the (mainly false) tales of rape and torture committed by Indian insurgents, people became convinced of the righteousness of the British Empire and its urgent historic role in furthering civilisation. Even the army, so long regarded as a sink of licentiousness and atheism, was recast as a Christian fighting force thanks to the extraordinary impact of Havelock. Public opinion shifted from considering that evangelicals had no place in the rough, tough army to believing that manly Puritans in fact provided the ideal material for modern soldiers.

Pride in empire, in the army and the navy, in the historic destiny of Britain as an agent of change in the world became deeply ingrained in a way that never existed before. These feelings engulfed sections of society that had held aloof from, or that had been downright hostile to, the idea of foreign intervention and empire-building: middle-class evangelicals, working-class Methodists and Baptists, free-trade liberals and radicals. The jingoistic moral fervour that fuelled empire-building and defined Britain and the world in the last four decades of the nineteenth century and the first four of the twentieth began with the Rebellion. 'The Indian empire,' declared *The Times*, 'is the triumph of the middle classes of England.'[41]

British self-confidence was at its zenith in 1858. 'The ubiquity and the universal influence of this country were never more signally proved than during the present year.' Future historians would, according to *The Times*, revel in the accomplishments of that year.

The British had proved their valour and military prowess by

A LESSON TO JOHN CHINAMAN.

MR. PUNCH. "GIVE IT HIM WELL, PAM, WHILE YOU ARE ABOUT IT!"

Palmerston chastises Imperial Commissioner Yeh in a Punch *cartoon, 1857*

crushing 'the most terrible military rebellion of modern times'. They had sensationally 'opened the vast empire of China to the world'. At the same time, another new middle-class hero, David Livingstone, had reaffirmed British pluck. His *Missionary Travels* was published at the end of 1857 and by the autumn of the following year he was back in Africa, endeavouring to open the Zambezi to 'Christianity, Commerce and Civilization'.[42]

Britain was at the pinnacle of its power in 1858. But while attention was riveted on the corners of the globe it became impossible to ignore problems closer to home any longer. In the summer temperatures climbed to the mid-thirties centigrade. As they did the stench from the River Thames became unbearable. Even MPs were revolted by the stink of effluent rising up from the river that ran alongside their brand-new and imposing Houses of Parliament.

Population growth, industrial progress and the introduction of the flushing lavatory meant that thousands of tons of sewage were being discharged directly into the Thames. No one would ever forget the smell of rotting shit coming up from the river during the burning-hot summer of that year, or the way it turned the stomach. Worse still, waterborne diseases became deadly killers. One MP asked the Commissioner of Works, Lord John Manners, what the government intended to do about the fact that the country's noblest river had been turned into a cesspool. 'Her Majesty's government,' Manners replied to benches of MPs holding handkerchiefs to their noses, 'have nothing whatever to do with the state of the Thames.'[43]

Nothing symbolised so pertinently the Britain of the 1850s than the Great Stink of 1858. These years were the heyday of liberty – liberty in the sense of non-intervention on the part of the government in the realms of public health, labour relations, conditions in factories, education, parliamentary reform and a slew of other domestic issues. Attention was fixed on the wider world and Britain's place in it. As Britain prospered abroad, the fabric of the country was neglected and outbreaks of cholera carried off thousands. The *Illustrated London News* put this most memorably: 'We can colonise the remotest ends of the earth; we can conquer India; we can pay the interest of the most enormous debt ever contracted; we can spread our name, and our fame, and our fructifying wealth to every part of the world; but we cannot clean the River Thames.'[44]

The paper made a very revealing point, one that speaks volumes about the Britain of the 1850s. Henry Adams, recently graduated from Harvard, described the financial epicentre of the world in the year of its zenith, 1858: 'London was still London. A certain style dignified its grime; heavy, clumsy, arrogant, purse-proud, but not cheap; insular but large; barely tolerant of an outside world, and absolutely self-confident . . . Every one seemed insolent, and the most insolent structures in the world were the Royal Exchange and the Bank of England.' The conqueror of India, the subduer of China, the unrivalled metropolis of the world, Britain would not let the Great Stink ruin the party. Even as its stench rose out of the river another momentous triumph arrived to stoke the mood of euphoria.[45]

13

Master of Time

NEW YORK–LONDON

The Telegraph has more than a mechanical meaning; it has an ideal, a religious, and a prospective significance, far-reaching and incalculable in its influences.

Charles Briggs and Augustus Maverick[1]

On 5 August 1858 the British battleship HMS *Agamemnon* limped into Valentia Bay in County Kerry, Ireland. She had long used up her coal and most of the decks had been stripped and burnt to provide enough fuel for her to fight against the winds of the wild Atlantic and inch towards the haven of Lough Kay. There she bobbed in the shadow of the wild and rocky mountains on the morning of 5 August as light and helpless as a cork, fortunate to have survived the high seas that had tossed her mercilessly and the ferocious storms that had battered her almost to pieces.

Agamemnon was a pre-eminent symbol of British power in the 1850s: when she was launched in 1852 she was the most heavily armed steam-powered warship in the world with 91 guns, and the Royal Navy's first battleship to be constructed with steam engines as part of its design. She had served as flagship in the Black Sea theatre of the Crimean War. In August 1858, however, she was the symbol of something else. She sailed under a banner that combined the crosses of the Union Jack with the stars of Old Glory.

A length of cable was paid out of the leviathan battleship and taken

The Agamemnon *caught in a storm on 20–21 June 1858*

to a telegraph office at Knightstown on Valentia Island. There the end was attached to a galvanometer and an electrical pulse travelled down 2,000 miles of cable insulated with gutta-percha. The signal was received on the far side of the Atlantic, at a telegraph station at Trinity Bay, Newfoundland, a few seconds later. News of the success was telegraphed from Newfoundland to New York: THE ATLANTIC CABLE HAS BEEN LAID!

'This brief announcement was discredited by the majority,' commented a New York paper. People could be forgiven for thinking all this was a hoax. The previous year, *Agamemnon* and the USS *Niagara* had met in mid-ocean and begun paying out the cable, the American ship bound for Newfoundland and the British one for Ireland. On the first attempt the cable snapped and could not be recovered. In July 1858 a second attempt was made, but *Agamemnon* got caught in the worst Atlantic storm in recent memory. On 29 July the two warships rendezvoused again in mid-ocean, spliced the cable, and headed off in their different locations. This time the cable was laid successfully; *Niagara* brought its cable end to Trinity Bay, and *Agamemnon* fought the wind to Ireland. After so many disasters it was no wonder the

New technological industries: machines covering the Atlantic Cable with gutta-percha at the Gutta Percha Company, Wharf Road, Islington

Coiling the Atlantic Cable at the Glass, Elliot & Co. Submarine Telegraph Company works, Modern Wharf, Greenwich

public thought this pie in the sky. But then on 16 August the first electric message sent between London and New York was transmitted: 'Europe and America are united by telegraphic communication. Glory to God in the highest, on earth peace, goodwill towards men.'[2]

Church bells pealed across New York as soon as the telegram arrived; the cacophony increased when factories and shipyards opened their whistles and let out shrill bursts of steam and the streets reverberated with the booms of a 100-gun salute fired in Central Park. By the evening a fireworks display had been organised in the park, attended by 100,000 people (out of a population of about 600,000); public buildings, theatres, hotels, businesses and houses on Broadway and elsewhere were decorated and illuminated. Multicoloured lights on the rigging of ships of all nations speckled New York harbour; the men on the vessels joined the New Yorkers in setting off rockets all that night and firing salutes.

But that was only the prelude to the official celebrations. An estimated 500,000 people lined Broadway and 5th Avenue, or crowded at windows or perched on roofs, on 1 September to watch ten regiments of infantry and cavalry, troops of artillery and military bands lead a gigantic procession from Battery Park to the New York Crystal Palace (situated in what would become Bryant Park). Following them came the stars of the show: first, an American Express carriage pulled by twelve plumed horses on which sat an enormous section of the submarine cable coiled in a pyramid and a working telegraph machine. It was followed by the hero of the day, Cyrus Field, and sailors from the *Niagara* bearing aloft a model of their ship. Behind them marched about 15,000 people under a forest of banners: veterans and representatives of New York's associations, trades and charities. After the parade the dignitaries attended banquets, heard speeches and applauded as gold medals struck by Tiffany & Co. were presented to those involved in the great work.

The show-stopper came at ten o'clock in the evening when the city's volunteer firefighters led a torchlit pageant through the illuminated streets, as bright as at noon, dragging their highly polished

Broadway, 1 September 1858

fire engines decorated with flowers, banners and lights of every colour. The firefighters held flaming torches above their heads or let off Roman candles and rockets as they marched. Proceedings were brought to a thunderous climax with another extravagant fireworks display.[3]

One of the reasons why the amazing success of the Atlantic Cable was so celebrated was that it came hard on the heels of a miserable twelve months. The world had been rocked in the last half of 1857 and the beginning of 1858 by the Indian Rebellion and, simultaneously, a global financial crash.

The Panic of 1857 seemingly brought the great boom of the 1850s to an abrupt end. It began in the region that had seen the most remarkable growth, the American West, and spread like contagion back to the financial markets on the east coast and in Europe. There were a number of intertwined causes. One was political: with violence affecting Kansas and uncertainty about the spread of slavery into the west, migration began to slow. Then the newspaper columns that reported financial data telegraphed from the Liverpool markets started to show a daily downward movement in the prices of agricultural produce. During the Crimean War, Russian grain exports had been embargoed. Now, with peace, the Western European markets were being flooded. That, combined with bumper European harvests, meant demand for prairie grain dried up. Farmers and financiers had gambled a lot on orders from Europe rising year on year; the realisation that the end of the Crimean War could dash all these hopes took hold very slowly. 'This year the Europeans do not want our breadstuffs,' lamented *Harper's Weekly* on 22 August.[4]

Railroad companies had been too easy with their credit, allowing farmers to buy more land from them than they could farm when grain prices were $2.19 a bushel. When the good times came to a sudden end and the price slumped to eighty cents a bushel, these farmers found themselves terribly overextended. Fewer people coming west, less agricultural goods being exported east and declining property values meant that railroad income started to plummet. The international bond market began to react.

On 24 August the Ohio Life Insurance and Trust Company – a New York-based bank that had borrowed heavily to gamble on risky railroad and land ventures – failed. This precipitated a nationwide panic, exacerbated by the instantaneous transmission of news by telegraph. Investors, fearful that more banks would follow Ohio Life's demise, began a run to convert their deposits and paper money into gold and silver. Mortgages across the west foreclosed, businesses collapsed, banks shut their doors. Then on 11 September the SS *Central America*, bound for New York from the terminus of the Panama Railroad, was caught in a Category 2 hurricane off the Carolinas. The ship went down with the loss of over 400 lives and fifteen short tons of Californian gold coins and ingots. The treasure included an astonishing eighty-pound gold brick, the Eureka ingot.*

The loss of this gold, combined with the collapse of railroad securities, foreclosures in the west and a panicky demand for hard cash put American banks under immense strain. Added to that, imports from Europe were now being paid in specie rather than in agricultural exports, resulting in a further drain of reserves. In October bankers suspended payment in coin, forcing the public to rely on banknotes. Western banks no longer inspired much confidence; their paper money began to lose value. People who held these banknotes found they could not be exchanged for anything like the amount printed on them, or that they were entirely worthless. At the same time, the Indian Rebellion and the war against China drained British bullion reserves. The Bank of England's lending rate went up and the credit markets tightened.

It was no secret that the outrageous economic growth of the western states was dependent upon debt and faith in the future. The bubble was pricked in August 1857. Confidence gurgled down the drain. Lots in western towns that were changing hands for over $1,000 dollars in the first few months of the year were lucky to fetch $10 by the fall; most were unsellable. Railroad securities were tied to the price of land. Several companies, including the mighty Illinois Central, defaulted; some went under.

The government of Minnesota, which had recently raised enormous loans for the construction of railroads, was bankrupted. No one wanted to pay for iron roads now. Some 700 paper towns in the territory, in which land had been sold to speculators in expectation of imminent railroad construction, disappeared from the map. Actual budding cities were abandoned. In some cases these ghost towns survive to this day. Laurence Oliphant's gamble on the future met a sorry end. He had purchased a log cabin in Superior City, believing that its growth was as certain as night follows day; in his absence the

* It was recovered from the wreck of the *Central America* in 1988 and sold in 2001 for $8,000,000.

wannabe Chicago grew to 2,500 inhabitants. In the aftermath of the Panic, Oliphant believed that Superior was deserted (he was a long way away, en route to China); in fact the population fell to 500 die-hards subsisting amid the ruins of their dreams.* His cottage, which had surged in value, became worthless overnight and he had to sell the doors and windows to pay the tax. It was not until the mid-1870s that Oliphant was able to sell and realise the amount he had paid for his land over twenty years before.[5]

Panic in the west flared back to New York, where nineteen banks failed and unemployment swelled. In London that November it was reported that 'feverish anxiety pervades the mercantile class. Every fresh steamer brings gloomier accounts from the United States. All the merchants and manufacturers connected with American trade are startled and trembling.' Many of the bill brokers of Lombardy Street – which supplied the world with short-term credit – were revealed when the tide went out to be swimming naked. Over thirty houses crashed, owing £13 million and possessing £7 million worth of toxic assets; only two of them had operating capital in excess of £100,000. Many more just about rode out the storm, propped up by the Bank of England. Lending on the international money markets was at a standstill.[6]

With a shortage of money in circulation and a severe credit crunch, trade in the American West was paralysed. St Paul – and countless other cities on the frontier – relied on imports of food and basic necessities. The delusions of the decade became apparent: Minnesota simply had nothing to export, save fur, fish and busted dreams. The lumber trade was dependent on the property boom, now over. The wheat fields were stripped bare by locusts. About two-thirds of the boom-time population deserted that once-bustling, madly self-confident city, and four out of five businesses sank. Those remaining

* Superior was once again touted as another Chicago in the 1880s when the Northern Pacific Railroad arrived. Property prices boomed and then collapsed in the Panic of 1893. It was Superior's Minnesotan neighbour, Duluth, that became the main export hub for the region. Even so, it did not quite become another Chicago. Today Duluth has a population of 86,000 and Superior 27,000.

settled in for a winter of food shortages, unemployment and misery. In 1856–7 St Paul had boasted a spectacular number of self-proclaimed paper millionaires. A year later these swells 'could not now get credit for a barrel of flour, or even a grocery bill of three dollars'. A traveller who passed through Illinois found large populations crowded into urban settlements, which stood as monuments to the 'folly of extravagant speculation'.[7]

Little wonder, then, that St Paul, even as it was coming to terms with its fall from giddy progress to poverty, joined many other cities in America in holding a street party, carnival and fireworks display to mark the success of the cable. This was the nineteenth-century equivalent of putting a man on the moon, the moment when the world changed for ever. In 1858 as well as in 1969, the dramatic triumph of technologies still in their infancy showed that nothing was impossible for humankind and promised an age of accelerated change beyond all comprehension. Both breakthroughs seemed far in advance of their time. 'Instead of proceeding by slow degrees,' commented *The Times*, 'the projectors have leapt at once to a gigantic success.' The seabed in the mid-nineteenth century was as unknown and as unknowable as space was until the late-twentieth. The minds of people boggled at the very idea of thoughts being flashed in seconds 2,500 miles along a thin thread of cable intruding on the alien world of the Atlantic floor among creatures unseen by mankind. The ocean was the frontier beyond frontiers.[8]

People wanted to *own* a piece of history. Tiffany got in on the act, buying up surplus sections of the cable that were cut into four-inch-long pieces and sold cheaply so that they could be purchased by as many Americans as possible. Tiffany also set these bits of cable in silver or fashioned them into such things as earrings, bracelets, charms, stamps for sealing wax, walking-cane handles, candlesticks and centrepieces for dining tables.[9]

The men, women and children who partied in cities throughout the United States in the fall of 1858 were conscious that they were living on a precipice in time. The headline in *The New York Times* read:

FIRST NEWS DISPATCH BY THE ATLANTIC CABLE

Highly Important Intelligence

PEACE WITH CHINA
LATER FROM INDIA

The very first news item sent from London announced the opening of China to the world with the Treaty of Tientsin and the final crushing of the Indian Rebellion ('Later from India' meant 'latest'). Clearly the convergence of these events heralded a new dawn for humanity. And, most importantly, America was plugged into the rest of the world, now able to participate in such momentous developments rather than distantly spectate. The Panic of 1857 had shown that, like it or not, the health of the American economy was dependent upon 'such fantastic contingencies as plenty or famine, peace or war, in the other quarters of the globe'. Rebellions in India, wars in China, crop failures in Europe, revolutions and elections in other countries and a host of other contingencies: all these affected the prices of wheat and cotton. Now Americans would learn of them at the same time as everyone else.[10]

But more than that, electric communication would usher in an age of universal peace. Now that the Atlantic Cable, the hardest challenge for telegraphy, had succeeded, there was no place in the world, however remote, that could not be plugged into the network. 'Now the great work is complete,' wrote the American journalists Charles Briggs and Augustus Maverick, '. . . It is impossible that old prejudices and hostilities should longer exist, while such an instrument has been created for an exchange of thought between all the nations of the earth.'[11]

Already the British were putting in motion a cable to India. 'We must go on, on, on,' urged *The Times*. 'We have proved that we can annihilate space and time.' The paper wanted the tentacles of electric communication to delve under the sea and embrace South-East Asia, Hong Kong, China, Japan, Australia and New Zealand now that Cyrus Field had proved 'that the work is so easy'. Briggs and Maverick looked forward to a world girdled with cables, which would 'inaugurate new realisations of human powers and possibilities . . . [The Atlantic Cable] marks an era in the unfolding of the human mind. The Telegraph has more than a mechanical meaning; it has an ideal, a religious, and a prospective significance, far-reaching and incalculable in its influences.'[12]

If the great crash of 1857 threatened to bring an end to the utopianism of the decade, the great cable of 1858 electrified it.

But just as there had been a financial crash, so too did the technology bubble burst. Even as elation gripped the United States, problems were being reported with the cable. The speed with which messages were transmitted was sometimes good, often agonisingly slow. Sometimes they never got through or they were garbled. According to Cyrus' brother, Henry Field, 'these flashes of light proved only the flickering of the flame, that was soon to be extinguished in the eternal darkness of the waters'. The last complete message was received on 1 September, the day New York erupted in festivities. At the Valentia end the chief engineer, Edward Whitehouse, sent shocks of 2,000 volts through the cable in a doomed attempt to revive it. He was blamed when the cable died on 28 October.[13]

More likely the fault lay with the cable itself. During the attempts to lay it in 1857 and 1858 the great wire had received rough treatment in the storms that buffeted the ships; it had been coiled and re-coiled in the subsequent voyages. The integrity of the gutta-percha insulation was compromised when it was left in the sun or exposed to high temperatures in the holds of the ships. No one ever discovered the exact causes of the failure, but, said Henry Field, given the handling of the cable the 'wonder is, not that the cable failed after a month, but that it ever worked at all'.[14]

After the euphoria came bitterness and recrimination. Many Americans believed there had never been a cable at all; the whole thing had been a con, a way of palming off costly commemorative merchandise on a gullible public. Investors and engineers squabbled among themselves, levelling blame wildly. The prospect of worldwide electric communication took a further blow the following year when a British attempt to link London and Calcutta ended in unmitigated disaster. The 3,500-mile submarine cable between Suez and Karachi was, like the Atlantic Cable, inadequately armoured. The iron rusted, allowing microbes and worms to bore through the gutta-percha. In both cases, the Atlantic and Red Sea cables, there was no means of raising the cable to test for failures once it was lost on the seabed. That fact alone made the whole project unfeasible in the long term.

Huge amounts of money had been speculated for zero return; public expectations had been raised to fever pitch and then dashed. The mood was darker in New York than in London (where the response to the initial success of the cable had been noticeably subdued). *The Times* and Cyrus Field both said that the endeavour should be regarded as a gigantic experiment. And so it was; the whole thing was fuelled by utopianism more than scientific certainty; hope galloped far ahead of available technology. Science advances by failure, and a generation of young scientists and engineers took advantage of the gargantuan experiments to refine their knowledge and work towards a solution. But try telling that to investors, who had tipped millions of dollars into the waves and did not fancy a repeat 'experiment'. Intercontinental communications were on hold.

Crushing anti-climax it may have been but, once savoured, the experience of instantaneous long-distance communication could not be forgotten. The world was too slow. 'In these days,' commented *The Times*, 'when intelligence is money, power, dominion, life, a month is a thousand years.' Alongside the gold rush, there was a news rush. Entrepreneurs understood that fortunes could be made remaking the way that the world's information was organised and shaving time off its exchange. Investors had always put a premium on red-hot news; now the public wanted piping-hot world news before it became tepid. The Associated Press news yacht lay ready to intercept mail steamers that were scheduled to drop off telegraphic dispatches at Cape Race in time to be transmitted to meet the evening papers' deadlines across the US. The ten-day crossing from Britain to Newfoundland was cut to seven soon after the cable fizzled out.[15]

The way that news was gathered, organised and sold underwent transformation. The telegraph broke the old monopolies that controlled the flow of information. By pooling the costs of data collection and transmission, organisations such as AP in the United States, Havas in France and the German Wolffs Telegraphisches Bureau made international news affordable for small provincial papers.

But because it stood at the centre of a web of telegraph, mail and shipping lines that encompassed the entire planet, Britain was best placed to dominate this emerging business. One of the reasons that the British were less wild in the celebration of the Atlantic Cable was because they stood firmly in the main current of international news. America, awakening to its place in the world in the 1850s, was outside it and had to wait between seven and ten days to hear epochal news events. London became the clearing house of world news, where it was repackaged for markets in Europe and North America.

That was what drew Julius Reuter to London back in 1851. His first ventures had been in supplying news of the European money markets to private subscribers from Aachen and then London. In 1857 the Indian Rebellion represented his opportunity to expand his field of operations. Events in India were potentially epoch-defining – the fall of British power on the subcontinent would reshape the politics of the entire globe – and readers in all countries were eager for the latest developments. Reuter negotiated access to the official government dispatches as soon as they came down the wires from Trieste. He telegraphed the headlines hot from India to newspapers across Europe and to Havas and Wolffs Telegraphisches Bureau.[16]

Just over a year later Julius Reuter went from being an exporter of world news to an importer, turning his network of agents and connections at foreign newspapers into a news-gathering service. Reuter offered his European telegraphic news service to *The Times*, but he was rebuffed, told that 'they could do their own business better than anyone else'. Reuter then went to the *Morning Advertiser* and offered them a free trial of his news service. The experiment was a success. For £30 a month the *Morning Advertiser*, other national dailies and much of the provincial press received the latest telegraphed news from Europe. This way of pooling the costs of foreign news helped break *The Times*' monopoly on international affairs. Newspapers that could not employ foreign correspondents or receive a stream of expensive telegraphs now had an affordable way of printing the news of the world without driving up the cover price of their papers. In many instances, Reuters provided news quicker and of better quality than *The Times*. A few months after rebuffing Julius Reuter, *The Times* was forced to purchase some of his news. Because the Paper of Record did not want to credit an outside agency, Reuter charged them extra. When the paper gave Reuters its first credit in December 1858 it paid half price for the story. By then readers of papers around the world were becoming used to seeing the name REUTER attached to breaking news.[17]

Within a few months of supplying the world's press with stories, Reuter was selling news provided by his correspondents in over 100 datelines around the world – in Europe, North and South America, Australia, New Zealand, China, Japan, India, Cape Town and points between. Most importantly, he was regularly scooping *The Times*.[18]

His first major breakthrough came on 10 January 1859. That morning King Vittorio Emanuele II of Sardinia-Piedmont opened his parliament at Turin. The short abstract of the speech, little more than 150 words long, reached Reuter's office at 1.30 p.m. and it was immediately inserted into the third edition of that day's *Times*. 'This is a moment of suspense,' declared the paper's editorial the next morning, 'when the words of sovereigns are substantive events.' Reuter's telegram was the scoop of the decade: as soon as it became public it was clear that Europe stood on the brink of war. The first half of the telegram concerned municipal reform in Sardinia and the failure of the silk crop. But it was the second half that was sensational. The king talked vaguely about clouds on the political horizon, but then declared his country's love of liberty: 'it respects treaties but is not insensible to Italy's cry of anguish . . . Let us await the decrees of Providence.'[19]

Behind these enigmatic pronouncements the meaning was all too clear. At a New Year's Day reception Emperor Napoleon III had used aggressive language aimed at Austria. Join the dots and it was certain that Sardinia-Piedmont was about to ally with France in a war against the Austro-Hungarian Empire with the intention of throwing the Austrians out of Lombardy and Venetia and unifying the whole of Italy. *The Times* decoded the 150-word summary of Vittorio

Emanuele's speech and extracted from it a long article forecasting the end of the balance of power in Europe and the possibility of a general war involving all the great powers – France, Austria, Prussia and Russia. It might also mean that Britain would be dragged into war against an aggressively expansionist France. A few days after his scoop Reuter reported that 1,000 veterans of Napoleon Bonaparte's Grande Armée greeted Napoleon III's son at Turin with the cry *Viva Italia!* In Austrian-controlled Milan graffiti read *Viva Verdi!* – the composer's name being a codeword for national independence. The letters stood for *Vittorio Emanuele R*e *D'Italia*.[20]

News of these events reached America by fast steamship and telegraph a few days later, bringing details of turmoil on the European money markets that saw £60 million wiped off the value of shares in a single day in anticipation of a Continental war. The spectre of a multi-power conflict moved markets in the States as well. For one thing, another European war would mean revived demand for prairie grain. 'This is a sad thought . . . that we are to gain by the sufferings of our brethren in Europe,' sighed the *American Agriculturist*, 'but we are only recording financial facts.'[21]

Two weeks later Reuter got an even bigger scoop. The text of the speech that Napoleon III was about to give to the Chamber of Deputies was handed to Reuter's Paris correspondent in a sealed envelope. At midday the emperor got to his feet and began to speak. Simultaneously the Reuter's correspondent opened the envelope and began tapping out the text of the speech to London. The message got straight through as Reuter had paid for exclusive use of the submarine Channel cable for an hour starting at noon on the dot. This was a debut in the history of journalism: never before had the text of a speech by a world leader been pre-released under embargo or reported in real-time in a foreign country. By 2 a.m. special editions of the London papers were on sale with the full text of Napoleon's speech, the words barely out of his mouth. The news was worth paying for: in his speech the French Emperor reiterated his friendship with Britain and told Europe that 'Peace, I hope, will not be disturbed.' 'The world will breathe again,' said *The Times*.[22]

Julius Reuter chose the ideal moment to enter the news business.

The world was in ferment in 1859. Reuter, with his excellent contacts in Vienna, Paris and Germany, had a ringside seat as Europe edged closer to war in March and April. Few newspaper readers failed to notice his name heading prominent columns of foreign intelligence every day. Thanks to Reuter and other entrepreneurs who harnessed old and new technologies to annihilate time, people in Europe and America experienced for the first time something we take for granted – instantaneous news. Readers of newspapers now read dispatches from the far corners of the globe at the same time, or sooner, than government ministers.

The ubiquity of information – and its fleeting shelf life – is the hallmark of modernity. In the late 1850s a popular press with up-to-the-minute international news became a reality. Better-informed publics reacted instantly to international events, and in doing so began to shape those events. The ways that governments conducted business between each other and the ways that wars were fought would never be the same again. Perhaps one of the popular perceptions of the Victorian period is its illusion of stability and its unchanging nature. But in reality it was a time of unceasing transformation. Within a few months of bursting upon the scene from complete obscurity, Julius Reuter was presented to Queen Victoria by Palmerston and his name was ubiquitous. With his pince-nez spectacles, shrewd business sense and extravagant side-whiskers, he was a true representative of the age, not to say a re-maker of it.

People were intrigued by this new force in the world. 'WHO IS MR REUTER?' demanded a newspaper headline of the man who miraculously got the news before anyone else; 'is this Mr Reuter an institution or a myth?'. The *Birmingham Journal* summed up his significance: 'Reuter is not only the man of the time, but the master of Time.'[23]

14

Best of Times, Worst of Times

BEIJING, TURIN, MONTGOMERY

O mia patria, si bella e perduta
[O, my country, so lovely and so lost]
Verdi, Chorus of the Hebrew Slaves, *Nabucco*

On a day in August 1859 telegraph operators in Europe and North America were amazed when messages of unstoppable gibberish started bombarding their receivers. Frederick Royce, a telegraph operator in Washington, DC, was in the middle of sending a message when his apparatus gave him a severe electric shock. A fellow operator saw a spark of fire flash from his head. He was not alone in suffering electrocution over the course of a strange few summer days.

Across the world telegraph poles and machines started spewing impromptu fireworks displays as the wires charged to thousands of volts. Operators hastily disconnected the machines from their batteries, but in a few cases messages were exchanged between stations faster than usual. In many places 'fantastical and unreadable messages came through the instruments', according to the *Philadelphia Evening Bulletin*, despite having no power source. Electric communication networks went down everywhere in the world.

Meanwhile gold miners camped in the Rockies were awoken by a strange light. They began to prepare breakfast thinking dawn had come. Frost settled in Illinois and Michigan despite it being the dog days of summer. A physician in Newburyport, Massachusetts, described the night sky as 'a perfect dome of alternate red and green streamers'. New Yorkers and San Franciscans crowded onto sidewalks and roofs to take in the 'varied and gorgeous' colours that streaked across the sky and provided enough light to read newsprint. Havana was illuminated in a supernatural red glow; in San Salvador it looked as if the roofs of houses and leaves on the trees were soaked in blood. The seas off South Carolina seemed to be on fire. Overhead conditions in Britain were cloudy, but 'luminous waves' were observable in many places and the stars appeared bigger and brighter. On the other side of the world, in Australia and New Zealand, people as far north as Brisbane were treated to a kaleidoscope of colour, from silver and gold to bright green to a livid red that resembled 'the reflection of an immense bushfire at a distance'.[1]

The amazing columns of light and the anarchy on the telegraph lines occurred on two dates in 1859 very close to each other. On 28 August the self-recording magnetometer at Kew Observatory registered a severe shock to the planet's magnetic field. Four days later the first ever observations were made of a super flare on the surface of the sun. A few hours after, on 2 September, Kew's magnetometer recorded a second and even more violent geomagnetic disturbance.

These phenomena were caused by two massive solar coronal mass ejections – sudden belches of ultraviolet radiation from the sun – that raced towards the earth at 1,500 miles per second. The solar wind brought with it the magnetic field of the sun and injected immense magnetic and particle energy into the earth's magnetosphere. With our complete dependence on electricity, satellites and radio waves, a solar flare of such a magnitude directed at the earth today would be catastrophic, causing trillions of dollars' worth of damage and a technological apocalypse. The Great Geomagnetic Storm of 1859 gave millions of people a uniquely magnificent and extensive display of the *aurora borealis* (the northern polar lights) and the *aurora australis* (the southern lights) and brought mayhem to the world's telegraphic networks. Those telegraphs that mysteriously continued in operation without their batteries did so because they ran on geomagnetically induced current.

Because electrical communications were in their infancy, the solar storm turned out to be a spectacular event without serious repercussions. But its dramatic effect on telegraph systems remains a powerful warning of just how vulnerable the modern world is to uncontrollable extra-terrestrial events.

A space storm and great plumes of light might be taken as a portent of mighty change. But people in the summer of 1859 did not need to be told by the night sky that they were living through a moment of upheaval. That year was crowded with momentous events all over the world.

That is certainly true when it comes to our concept of the origins of life on this planet and our place in the universe. On 24 November, the day it was published, the entire print run of Charles Darwin's *On the Origin of Species* sold out. The way that Darwin's theory of natural selection revolutionised our understanding of the world and threw down the gauntlet to religious and scientific assumptions needs hardly be stated: aside from Karl Marx, no other thinker of the nineteenth century has such a continuing hold on the twenty-first. *Origin* did not change the world overnight, however. The first edition was modest – 1,250 copies – and although it was widely reviewed in the serious periodicals many of the mass-market dailies and weeklies, such as the *Daily Telegraph*, *Lloyd's* and *Punch*, ignored the book entirely for some time. But by 1867 the *Saturday Review* was able to say of the *Origin* that 'so rapid has been the hold that it has taken on the public mind, that the language incident to the explanation of the "struggle for life" . . . has passed into the phraseology of everyday conversation'.[2]

Darwin's theory was undeniably an intellectual grenade that exploded in scientific and theological circles and in universities around the world. But once the theory left Darwin's study and entered the public arena it gained a political significance beyond anything Darwin intended. As one writer said: 'The hypothesis is clamorously rejected by the conservative minds, because it is thought to be revolutionary, and not less eagerly accepted by insurgent minds, because it is thought to be destructive of the old doctrines.'[3]

One of the major reasons why the *Origin*, a scientific book, spoke so powerfully to the age was because its language and arguments chimed so agreeably with prevailing notions. It was the philosopher and sociologist Herbert Spencer who came up with the phrase 'the survival of the fittest', the words most commonly associated with Darwin and which Darwin borrowed to use in later editions of the *Origin*. In the modern world, according to Spencer, unrestrained market forces took the role of nature in sifting the weak from the strong, rewarding thrift and punishing sloth. In the free-market economy, people adapted themselves to prosper in the way that species adapted to their environment over time; any attempt to artificially support the impoverished by paternalistic welfare legislation was diametrically opposed to the brute laws of nature. Darwin's 'bulldog', the biologist T. H. Huxley, wrote that the theory of evolution was hailed as 'a veritable Whitworth gun in the armoury of liberalism'. The struggle for survival writ large in nature justified and reinforced notions of laissez-faire economics, free trade, self-improvement, competitive individualism and manliness.[4]

The popularisation of the theory of evolution and its impact on mainstream society reshaped world views. A nation was like an organism, evolving over time or perishing in the brute competition with rivals. As the British diplomat Rutherford Alcock put it in relation to China, there was a 'natural and moral law' that governed the growth and decay of nations. When two civilisations come into contact, he said, it has 'had but one result: the weaker has gone down before the stronger'.[5]

This put a new, frightening complexion on imperialism based on racial superiority. If helping the poor at home was a violation of the laws of evolution, expecting less civilised peoples to raise themselves up to Western ways by persuasion alone was also doomed to failure. One race had to dominate another. That seemed pertinent in light of the recent Rebellion in India.

Darwin borrowed language from the field of political economy, and in turn his language and arguments fed back into contemporary political debates. That's what gave his theories such potency. The impact of *On the Origin of Species* is ambiguous. It caused people to rethink the world and throw off superstition. But it gave rise to darker ruminations. Suddenly it seemed that highly civilised societies, like Britain, were vulnerable to rapid terminal decline because they could afford to weaken their own gene pool by artificially supporting the slothful, imbecilic and morally unfit. The language of evolution also encouraged a way of looking at the world based on ideas of racial superiority and war. And it did appear that the years after 1859 were marked by a struggle for 'the survival of the fittest' among the nations of the world.

The year 1859 was a bumper one for the publishing industry. John Stuart Mill's *On Liberty*, one of the key texts of modern politics, came out in that year. The great celebration of laissez-faire, individualism and manliness, Samuel Smiles' *Self-Help* (in which heroes such as John Nicholson were once again held up as models of self-improvement) became the international bestselling self-help book of the nineteenth century. The great assault on liberal capitalism, Marx's *A Contribution to the Critique of Political Economy* – the basis for his later *Das Kapital* – appeared in German.

A fourth book that appeared in 1859 remains one of the most-read novels of all times. Charles Dickens' story of the French Revolution, *A Tale of Two Cities*, captured, in its famous opening sentence, the ambiguities and tensions of the late 1850s: 'It was the best of times, it was the worst of times, it was the age of wisdom, it was the age of foolishness, it was the epoch of belief, it was the epoch of incredulity, it was the season of Light, it was the season of Darkness, it was the spring of hope, it was the winter of despair, we had everything before us, we had nothing before us, we were all going direct to Heaven, we were all going direct the other way – in short, the period was so far like the present period, that some of its noisiest authorities insisted on its being received, for good or for evil, in the superlative degree of comparison only.'

In his review of *On the Origin of Species* Huxley said that Darwin's 'species question' chipped away attention even at the greatest issue of the day: Italy. When the public read the lines 'It was the best of times, it was the worst of times', in the first instalment of *A Tale of Two Cities* on 30 April 1859, Austrian troops crossed the Ticino River and invaded Piedmont after weeks of deliberate provocation by Vittorio Emanuele's troops. Since 1 January Europe had teetered on the edge

Italy in January 1859
0 25 50 75 100 miles
0 50 100 150 km
Kingdom of Sardinia
Part of Austro-Hungarian Empire
SWITZERLAND
AUSTRO-HUNGARIAN EMPIRE
Milan
LOMBARDY–VENETIA
Solforino
Venice
Turin
PIEDMONT
FRANCE
PARMA
MODENA
Genoa
ROMAGNA
OTTOMAN EMPIRE
Nice
Florence
Ligurian Sea
TUSCANY
PAPAL STATES
Adriatic Sea
Rome
CORSICA
Naples
SARDINIA
KINGDOM OF THE TWO SICILIES
Tyrrhenian Sea
Ionian Sea
Mediterranean Sea
SICILY
N

of crisis. Now France joined Piedmont in a war to oust Austria from Italy.

The crisis had been cooked up by the master political strategist Camillo Benso, Count of Cavour, Prime Minister of Piedmont, with the aim of enlarging his country and making it a great European power. First, Austria had to be defeated and thrown out of Italy. When they declared war the Austrians expected to make light work of little Piedmont before Cavour's French allies could arrive. But their advance was hesitant and the French made excellent use of the telegraph and rail networks. *The Times* correspondent wrote that the French defeated the Austrians at Montebello because 'train after train arrived by railway from Voghera, each train disgorging its hundreds of armed men and immediately hastening back for more'.[6]

With extraordinary speed the French got 600,000 soldiers and 125,000 horses to the plains of Lombardy. It was the first major use of the railway as an instrument of war and it proved decisive. In early June the French used the rails to outflank the Austrians and force them to abandon Vercelli. As one member of Britain's House of Lords put it, with electric communication and railways, modern war consisted of 'a word and then a blow'.[7]

Modern communications meant that every twist and turn of the conflict was followed almost in real-time in Britain. Reuters news agency's flood of telegraphs from the front arrived several times a day. These dispatches were shipped across the Atlantic and filled the pages of American newspapers as soon as they were telegraphed by AP from Newfoundland. The blanket coverage provided by electrical means made the war seem involving and immediate. This feeling of visceral proximity was heightened by the widespread use of photography; the resulting images were used as the basis for hurriedly executed engravings of recent events. Papers such as the *Illustrated London News* and the *New York Times* published maps of the battles. Not far behind the telegram bulletins, carried by rail and steamer, came graphic reports filed by a legion of special correspondents embedded by British, French and American newspapers in the contending armies. From the safety of their offices, journalists picked over the mountain of official reports, eyewitness accounts, telegrams and official bulletins and kept up a daily running commentary.[8]

This was war as spectator sport with newspapers revelling in the first conflict of the media age. But interest went far beyond voyeurism. For a start, the cause of independent Italy had long been embraced by the British and American publics. A unified Italy free from foreign domination would be a victory for liberalism and constitutionalism over the malign forces of autocracy that loured over Europe. Italian patriots and nationalists, the heroes of the abortive revolutions of 1848 such as Mazzini, were lionised in Britain and America and imbued with romanticism. Many exiled Italian freedom fighters lived in London, where they were courted by high society and politicians; significant numbers had made their way across the Atlantic. Early in the war the New Orleans *Daily Picayune* declared that 'American feeling is decidedly anti-Austrian . . . because hatred of tyranny is an American instinct.' President James Buchanan reiterated sentiments widely expressed in the press: 'The sympathy for poor down-trodden Italy is very strong in this country'.[9]

The urban landscape of 1850s Britain was testament to the passion for all things Italian. Burnley Mechanics' Institute (1854), Blackburn Town Hall (1856) and Manchester's Free Trade Hall (1853–6) were built in the palazzo style; Bradford Town Hall has a stunning Tuscan *campanile* modelled on the Palazzo Vecchio in Florence. The public-spirited burghers of prosperous northern manufacturing towns saw themselves as standard bearers of the civic humanism of the Renaissance, heirs of the merchant princes of Venice or the Florentine banking dynasties. The liberation of Italy, the cradle of modern politics and capitalism, from domination by foreign empires was essential in the building of a new, rejuvenated Europe.[10]

Many Britons and Americans were emotionally invested in the struggle for Italian independence. But the war achieved such publicity because the fate of Europe, and in large measure the world, hung in the balance.

Napoleon III's adventure in northern Italy was regarded as a revival of Napoleonic ambitions. The very real possibility that the Franco-Austrian War could spark a general war between the great powers explains the avidity with which the war was followed round the world. British and American commentators foresaw a situation in which Russia and the German states intervened to protect Austria. And if the worst-case scenario – a full-blown Continental war – did not happen then the very least that could be expected was a France that dominated Italy, and therefore Western Europe. The name 'Waterloo' was scattered across the press, a reminder of the last time Britain had to contain an expansionist French Continental empire. 'The Empire [of Charlemagne and Napoleon Bonaparte] is indeed revived,' declared the *Illustrated London News* portentously. 'France is herself again.'[11]

How should Britain respond? If Europe broke down into war it would have to intervene in some form. It would in all likelihood face invasion from France. This put the country in a difficult position. Its army was small and scattered round the globe. European rulers such as Napoleon III knew that the mere threat of invasion was enough to neutralise Britain while they reordered Europe to their liking. Conscious of Britain's vulnerability to invasion but determined to assert its power in Europe, the people took matters into their own hands. Thousands of men of all classes in towns, cities and villages began arming themselves, formed rifle clubs and drilled for war.

As it turned out the Italian war was short. The Piedmontese and French defeated Austria at the battles of Magenta on 14 June and, twenty days later, Solforino. On 11 July the bells of Big Ben pealed for the first time over London. Coincidentally, it was the day that Napoleon III signed an armistice with Austria at Villafranca. Fearful that Prussia and the German states would intervene by crossing the Rhine and alarmed that the public would turn against the war after the bloodbath at Solforino, Napoleon sought an immediate end to the war before his aims were fulfilled. Austria ceded Lombardy to France, which then ceded it to Sardinia.

Napoleon's early exit from the war sparked outrage in Italy. Austria remained in possession of Venice. During the war the people of the central Italian duchies of Tuscany, Modena, Parma and Romagna had revolted against their pro-Austrian rulers and were seeking to join Piedmont. At Villafranca Napoleon agreed to restore the Habsburg dukes. Cavour resigned in protest; Verdi raged 'Where then is the independence of Italy, so long hoped for and promised? . . . Venice is not Italian? After so many victories, what an outcome . . . It is enough to drive one mad.'[12]

Dominated as ever by foreign powers, Italy bubbled and frothed. The peoples of the duchies took matters into their own hands and voted to form the United Provinces of Central Italy in defiance of Austria and France. The power vacuum in the peninsula threatened to suck in the whole of Europe. In the late summer and autumn of 1859 there were countless possible scenarios. Would France and Austria go to war again? Or might they unite to dismember Italy? What new combinations would be formed among the great powers as the equilibrium of Europe underwent violent convulsions?

It was not just the European situation that was so alarming. France seemed to be mounting a bid for power on an even wider scale. When Ferdinand de Lesseps, a French career diplomat, swung a pickaxe in Egypt in April that year he signalled a new challenge to Britain's global supremacy. His symbolic blow cut the first soil in the construction of the Suez Canal. The canal would, it was feared, give France a short-cut to Asia, creating a military highway that would allow Napoleon's empire to eclipse Britain militarily and commercially in India, South-East Asia, China, Japan and the Pacific.

Britain had an answer to the Suez Canal, and it was a triumph of her industrial might. A few days after the *aurora borealis* Isambard Kingdom Brunel's SS *Great Eastern* made her maiden voyage. Brunel's vast ship would revolutionise the global economy, it was boasted. Six times larger than any ship afloat, she was capacious enough to carry coal to power her to the ends of the earth and profit from economies of scale. With a speed of 14 knots she could beat the clipper trade out of existence.

What price, then, the Suez Canal? The *Great Eastern* could convey British trade and emigrants to India, Singapore, Shanghai and Melbourne in bulk and at speed. And not only that. The ship could project British military power around the planet. It was supposed that an army of 10,000 men could be dispatched on such a leviathan in super-quick time wherever it was needed. Build a few more *Great Eastern*s and Britain would dominate the world. The great ship would mean a 'closely consolidated empire'. No more need to disperse British military muscle to gloomy garrisons in distant parts – regiments could race to where they were needed. And the flow was not one-directional: troops from Australia, New Zealand, Canada and India could be thrown into a European conflict. 'M. Lesseps might sink all the money in France in his Suez Canal and Mr Buchanan assert American morality and manifest destiny' (as he was doing in Central America and Cuba), but Britain's industrial pre-eminence meant that her supremacy as a military and commercial superpower was unassailable.[13]

The press raised the concept of *Great Eastern* as a weapon of war as a way of putting fear into the hearts of those who dared challenge Britain's global hegemony in 1859 – Napoleon III and President Buchanan chief among them. It was so fanciful as to be laughable. Even without a Suez Canal the old enemy was competing aggressively for the spoils of Asia. In 1859 Napoleon III's troops were busy carving out an empire in Vietnam and the Pacific islands. The invasion panic grew more intense in November, when the French navy launched the most dangerous ship in the world, one that could make mincemeat of the *Great Eastern* and any other ship besides. The first ironclad warship in history, *Gloire* was a direct threat to Britain's wooden wall – her trusty ships of the line, the hulls of which were un-armoured.

Britain had built its fortunes and world dominance on the balance of power in Europe that had existed since 1815 and on its unchallengeable Royal Navy that gave it absolute security from invasion. The launch of *Gloire* at the height of the European crisis of 1859 ended that happy situation at a stroke. In an age of steam-powered, ironclad warships and exploding shells, Britain's pride, her wooden ships of the line, suddenly became antiquated. Not since 20 October 1805, the day before the Battle of Trafalgar, had Britain seemed so vulnerable. No one expected the French tricolour to fly from Buckingham Palace, to be sure; but the image of Britain's ports and warehouses, containing the wealth of the world, consumed by fire sprang readily to the imagination.

Trample too much on Napoleon III's dreams and that might happen. The appearance of *Gloire* on the waters coincided with the British government's attempt to intervene diplomatically to help create an independent central Italian state. The French press and military responded to what they regarded as unacceptable meddling by ramping up their aggressive language and talking openly of avenging Trafalgar and Waterloo by invading their cross-Channel rival. Thousands more ordinary Britons trooped off, weapons in hand, to report for duty at their local rifle corps. War between Britain and France, the *New York Daily Tribune* judged, 'appears a question of time only'.[14]

Napoleon III's Italian adventure threatened not only the balance of power in Europe, but throughout the world. A war between Britain and France would have fundamentally altered global trade, commerce and finance, with consequences for most of the human race.

The convulsions besetting Europe in the summer of 1859 were not the only upheavals in the world. Britain and America came close to war, this time involving a dispute over the boundary between the far north-west of the US and British Columbia. It arose when an American settler on remote San Juan Island shot a trespassing pig belonging to his neighbour, a British subject. Both Britain and the US claimed the island. By the end of the summer the British Pacific Fleet was in an uneasy stand-off with US military forces. For a few weeks the possibility of the spat turning into a war was all too real. *The Times* declared: 'To say we shall not and will not go to war about this trifle is too much, because we must assert our rights if they are plain.'[15]

In the space of a few months Britain had collided with two of the world's greatest powers, the United States and France. And at exactly the same time news came in of a much more serious collision with a third great world power.

As Napoleon III stormed across the plains of Lombardy, James Bruce, Britain's Ambassador-designate to the imperial Chinese court, was preparing to travel to Beijing to ratify the Treaty of Tientsin. When his escorting gunboats approached the Taku Forts at the mouth of the Peiho River en route to the Chinese capital they were warned that any attempt to bring armed force to the capital would be resisted. The Manchu imperial government was prepared, with the greatest reluctance, to ratify the Treaty if the ambassador came with no more than ten unarmed companions and obeyed the elaborate rules of etiquette laid down by court protocol. When the representative of the United States went to Beijing earlier that year to ratify the Treaty, his party was escorted all the way to the city and they were not permitted to use their sedan chairs. Instead they travelled in closed carts, unable to gaze or be gazed upon. They were kept under virtual house arrest in the capital and could not present a letter from President Buchanan to the emperor because they refused to kowtow. Exchange of the ratified treaties took place not in Beijing but at the mouth of the Peiho. News of the failure, when it came via London and the Newfoundland telegraph, jostled for space on the pages of America's newspapers with updates on the crisis in Europe.[16]

Bruce had no intention of being treated in such a subordinate manner. The humiliation of the Americans showed that the only way to open up relations with the Chinese court on a footing of equality was by force.

A year before, Manchu troops had deserted the flimsy and poorly armed Taku Forts on the mouth of the river when Elgin proceeded to Tientsin to negotiate with the imperial commissioners. The British, with contempt for Chinese soldiery, expected to bypass the forts with the same ease on 25 June 1859. They attacked with fatal complacency. By the end of the day they had lost six gunboats and ninety-three sailors to a deadly and accurate artillery bombardment; a further 111 were wounded. Soldiers and marines sent on a lunatic frontal assault on the forts got stuck in a quagmire and were mown down; 426 men were lost on land and 345 wounded.

The British press howled in fury when the news broke in September. Here was another manifestation of Oriental treachery and faithlessness. But much worse was the 'very painful and very humiliating fact that the Chinese defeated us – we can scarcely credit it as we write the words – that they defeated us with heavy loss'.[17]

Once again Britain's prestige in Asia had been tarnished. Retribution was on everyone's lips – retribution for the insolence of the Chinese and the humiliation suffered by the Royal Navy. Britain had to restore its prestige in the East and, in the words of *The Times*, 'teach such a lesson to these perfidious hordes that the name of "European" will thereafter be a passport of fear . . . throughout their land'.[18]

The time had come to give the emperor an electric shock so that he would at last see reason. That meant one thing: capturing Beijing itself. 'The occupation by a barbarian army of a capital into which even a barbarian diplomatist is not to be admitted,' said Palmerston, 'would go further to proclaim our power, and therefore to accomplish our ends, than any other military success.'[19]

Less than a year after his triumphant return from China Lord Elgin was making his reluctant and weary way back to the Far East to finish what he had started. The earl travelled east in a foul mood, missing the comforts of home life even before he crossed Suez. This time he had a serious force at his disposal: 13,000 British and Indian troops, 7,000 French soldiers (France and Britain were friends in China, if not Europe), seventy warships and 200 transports. The defeat of the Indian Rebellion and the revamped Indian army meant that Britain could use its crack Punjabi, Pathan and Sikh soldiers to project its power into Asia. The army's medical services and logistical support had advanced light years since the Crimean debacle. With Enfield rifles and brand-new breech-loading Armstrong guns capable of firing exploding shells over a range of five miles, this was the best-equipped British force in recent history. It arrived at the Peiho River on 3 August 1860.

The invaders made light work of the Taku Forts this time and marched on Tientsin. As usual the Manchu commissioners began to negotiate and offer concessions. But Elgin's attitude had hardened and he ordered his force on to Beijing. In desperation, the Chinese took hostages. Among the thirty-nine British, French and Indian prisoners were Harry Parkes, the man who had provoked hostilities years before; Elgin's private secretary; and *The Times*' special correspondent. Taken to the Board of Correction in Beijing, they were held in appalling conditions, tortured and interrogated.

The time for talking was over. That same day 3,500 French, British and Indians went into action and defeated the 30,000-strong Manchu army near the village of Zhangjiawan. The allies engaged the huge Chinese army again on 21 September. A fearsome horde of elite Mongolian cavalry that filled the land 'as far as we could see' gave a bloodcurdling war cry and charged en masse, attempting to turn the allied flank. Sikh, Punjabi and Pathan irregular horsemen counter-charged into the clouds of dust alongside British Dragoon Guards. 'The result was most satisfactory,' wrote Colonel Garnet Wolseley. 'Riderless Tartar horses seemed to be galloping about in all directions, and the ground passed over in the charge was well strewn with enemy.' The artillery then opened fire on the retreating Mongols, 'every Armstrong shell bursting amongst them and bringing down the enemy in clumps'.[20]

The barbarians were at the gates of Beijing and the Xianfeng Emperor fled beyond the Great Wall with his concubines as soon as the battle was lost, leaving power in the hands of his twenty-seven-year-old half-brother, Aisin Gioro Yixin, known as Prince Gong.

Beijing was encircled. The French troops encamped in the grounds of the Yuan Ming Yuan, the imperial Summer Palace, an idealised vision of China such as one might see in Chinese paintings; it was among the chief wonders of Asia. In this oasis of tranquillity, which covered an area of 860 perfectly landscaped acres, were lakes, canals, enclosed gardens, reception halls, temples, terraces, palaces, statues and pavilions. 'Pebbled paths led you through groves of magnificent trees, round lakes, into picturesque summer-houses, over fantastic bridges,' rhapsodised the translator Robert Swinhoe, who arrived at the palace soon after it was secured. 'As you wandered along herds of deer would amble away from before you, tossing their antlered heads. Here a solitary building would rise fairy-like from the centre of a lake, reflecting its image on the limpid blue liquid in which it seemed to float, and then a sloping path would carry you into the heart of a mysterious cavern artificially formed of rockery, and leading out on to a grotto in the bosom of another lake.'[21]

'No one just then cared for gazing tranquilly at the works of art.' The serenity of the palace was turned into a tumultuous orgy of spoliation by French troops, and then their British counterparts. Many of the men 'were armed with large clubs, and what they could not carry away, they smashed to atoms'. The soldiers lit their pipes with irreplaceable Chinese books and manuscripts; they took pot shots at the chandeliers and bayoneted pictures. Many of the looters did not really appreciate the value of what they were destroying, either in a

The walls of Beijing

cultural or a material sense: 'quantities of gold ornaments were burnt, considered as brass'. The vandalism was shameful and the palace was 'littered with the debris of all that was highly prized in China'.[22]

The sacking of the imperial palace was the least of Elgin's concerns. The war that had started over a pirate boat in the Pearl River had transmogrified into an epochal struggle to drag China into the world system; now, in early October, it was all about the fate of the thirty-nine hostages.

With the allied army camped outside Beijing, Prince Gong had very little room for manoeuvre. The immense capital, one of the most remote and isolated in the world, only ever visited by a smattering of Westerners, sat behind walls that were forty feet high and sixty thick and had a perimeter of thirty-eight miles. Stand at any point and they vanished into the distance. These formidable walls were punctuated with colossal guard towers, watchtowers and tiered gatehouses; beyond them the besieging army could see palaces, pagodas, copulas and minarets pointing up to the sky. Elgin informed Gong that unless the prisoners were released by 13 October Beijing would be stormed.

The morning of the deadline was tense. No one expected Gong to capitulate. But, as the artillerymen stood ready to bombard Beijing as the seconds ticked down to the midday deadline, the Andingmen gates swung open. Elgin, escorted by 500 troops, took up residence in the Temple of Earth where emperors made offerings to heaven at the summer solstice. A few minutes later the Union Flag was flying from the wall of Beijing, 'the far-famed celestial capital, the pride of China, and hitherto esteemed impregnable by every soul in that empire'. Then the guns were hauled up onto the high walls and trained on the streets and myriad tiled roofs and towers of the city that crowded for miles into the distance. Even from this vantage point the southern walls were too far away to be seen.[23]

Summer was over and cutting winds were gusting into Beijing. On 18 October the city was plunged into near-darkness by a great pall of smoke billowing in on the north-west wind. The Summer Palace was being burnt on Lord Elgin's orders as 'a solemn act of retribution' for the torture and murders of the hostages. Robert Swinhoe rode up to the palace: 'As we approached . . . the crackling and rushing noise of fire was appalling . . . the red flame gleaming on the faces of the troops engaged made them appear like demons glorying in the destruction of what they could not replace.'[24] 'When we first entered the gardens,' wrote Colonel Garnet Wolseley, the expedition's quartermaster-general, 'they reminded one of those magic grounds described in fairy tales; we marched from them upon the 19th October, leaving them a dreary waste of ruined nothings.'[25]

Elgin acted as he did because he wanted to inflict a grievous punishment on the Qing dynasty without harming the Chinese people. Burning the sacred palace was considered the most crushing of all blows, for it would rob the people of their belief in the empire's universal sovereignty and convince them of the invincibility of Europeans. Elgin also wanted to impress upon Gong and the Manchu officials that the allies would no longer wait, no longer talk or compromise. According to Wolseley the burning of the Summer Palace 'was the stamp which gave an unmistakeable reality to our work of vengeance'. As the acrid smoke hung over Beijing on 19 October Prince Gong capitulated to all Elgin's demands.[26]

Emperors and emperors alone enjoyed the right of being carried in their sedan chairs by eight men; members of the imperial family were allowed no more than four. On 24 October 1860 sixteen Chinese porters in scarlet uniforms bore the richly decorated sedan chair in which the 8th Earl of Elgin sat. His guard of honour consisted of 100 cavalry and 400 infantry and two bands playing 'God Save the

Queen'. They entered through the Andingmen gates and proceeded through the main street, lined on either side by British troops. In all there were 8,000 barbarian soldiers in Beijing that day, marching to the music of their bands under regimental and national banners. So great was the press of crowds from the million and a half residents of Beijing eager to see the conquering forces that clouds of dust swirled up and over the streets. Elgin was borne along the streets and into the Hall of Ceremonies in the Board of Rites, which stood on present-day Tiananmen Square.

Throughout the ceremonies Elgin was icy and irritable. He had arrived three-quarters of an hour late, a discourtesy that was taken as a calculated snub. Prince Gong glowered with anger and shame as he ratified the Treaty of Tientsin. The unthinkable had happened: Beijing was under military occupation and a barbarian sat in the Hall of Ceremonies as an equal. China paid 4 million Mexican dollars in reparations and ceded Kowloon to the British. In March the following year China's humiliation was complete when Britain, France, the United States and Russia established embassies in the capital.

Back at the beginning of the decade it was widely held that the world would be transformed organically. The progressive forces of the age – multiplying trade, spreading knowledge, new technologies and long-distance migration – would rejuvenate the entire planet. By the summer of the great solar storm few could hold on to that faith. Technologies that had been seen as utopian quickly became the tools of the modern power state. Telegraphs, railways, mass media and other developments had empowered governments to mobilise their publics and military swiftly and efficiently for national ends. In Europe and Asia in 1859 and 1860 change was being realised by the exercise of brute power.

And it was not just modern states that were coming to see violence as the agent of radical transformation. The summer of 1859 is remarkable for the sudden prominence of individuals who were prepared to hasten the unfolding of history by acts of direct, incendiary personal action. The modern age of mass media and instantaneous communication gave them the celebrity to electrify an edgy, disordered world.

Three men in that year, in three continents and very different contexts, arose to take history by the scruff of its neck: Giuseppe Garibaldi in Italy, John Brown the radical abolitionist in the US, and in Japan the samurai scholar Yoshida Shoin. All three were intolerant of discussion and compromise and believed that symbolic demonstrations of self-sacrifice could transform the world. In the face of Western intrusion and the fatal complacency of the ruling class, Yoshida taught his students that the only way Japan could be shaken out of its lethargy was for 'men of high purpose' – *ishin-shishi* – to commit acts of terrorism and glorious martyrdom. 'Life and death, union and separation, follow hard upon one another,' Yoshida told his students. 'Nothing is steadfast but the will, nothing endures but one's achievements. These alone count in life.' On the other side of the world, in New England, John Brown frequently quoted Hebrews: 'Without the shedding of blood there is no remission of sin.' A few years before, in Kansas, he had killed enslavers; in recent years he raided Missouri and freed slaves. Similar sentiments were true for the great Italian patriot. When Laurence Oliphant encountered Garibaldi in a secret smoke-filled room in Turin, he advised the guerrilla general to settle an issue in parliament by constitutional means. 'Oh,' snorted Garibaldi, '*interpellatione, sempre interpellatione!** I suppose a question in the Chamber is what you propose: what is the use of questions? What do they ever come to?'[27]

Yoshida held that the *bakufu* was supinely letting in Westerners and suppressing the vitality of the country; the country was sleepwalking towards disaster. In the United States, government was deadlocked as North and South became resolutely and implacably opposed. Both Yoshida and Brown wanted to awaken dormant forces by igniting revolution, Yoshida by a spate of assassinations of high-ranking officials and Brown by instigating a slave rebellion in Virginia. In Italy, the cause of national unity and independence was the victim of the machinations of great power politics between France and Austria. Only by radical, revolutionary action on the part of the people,

* Literally, 'Interpellations, always interpellations!', or, in other words, 'Parliamentary procedure, always parliamentary procedure!'.

Garibaldi believed, would the diplomatic stalemate be broken and Italy 'redeemed'.

In contrast to the chicanery and duplicity of professional politicians in 1859, Giuseppe Garibaldi shone out as a patriot and man of action. Garibaldi's life story was already legendary by then. His military reputation was secured when he led forces during the Uruguayan Civil War (1842–8); during the Italian revolutions of 1848–9 he commanded rebel forces in the short-lived Republic of Rome. In the years that followed, Garibaldi travelled the world as captain of a merchant ship, spreading his fame to the US, Asia and Britain. When the Franco-Austrian War began in April 1859, Garibaldi was a living symbol of Italian independence. Too famous to ignore, he was made a major-general in Piedmont's army. But Cavour and Napoleon did not like the nationalistic enthusiasm evinced by Garibaldi and his followers, fearing that it would upset their finely calibrated realpolitik. He had to be managed carefully.

Palmed off by Cavour with 3,000 poorly armed and badly trained men and sent out of harm's way to harry the Austrians in the Alps, the veteran guerrilla commander astonished the world. His raw recruits – known as the *Cacciatori delle Alpi*, the Hunters of the Alps – streamed out of the mountains and won the first battle of the war against superior Austrian troops at Varese near Lake Maggiore and then at San Fermo near Lake Como.

Garibaldi's campaign transformed a war between two great powers into a crusade of liberty. The residents of the towns he liberated welcomed him as a conquering hero, almost a messiah. Even more importantly, he issued proclamations to the people of Italy rallying them to a patriotic movement: 'You are called to a new life and you must respond to the call, as our fathers did'. Within a short space of time the guerrilla army swelled to 10,000 as men from all over Italy heeded his call. According to *The Times*' special correspondent, 'the hero of Montevideo' bewitched the Italian people and 'has enlisted young men of the highest classes, artists, literary men, professors and scholars, in his ranks as mere privates'. Volunteers from Switzerland, France, the United States, Britain and even a man from China rallied to this romantic cause of national liberation.[28]

The material of the age and the hero of the age combined:
a gutta-percha medal showing Giuseppe Garibaldi

With regular telegraphic dispatches from the Alpine front, readers in many countries were with Garibaldi's advance almost every step of the way. 'It is something to look upon a man like this,' stated a breathless editorial in the *Weekly Arizonian*, '– to follow in the smoke of battle, and through Alpine and Apennine fastnesses and Como valleys, an unconquerable spirit.' The world went Garibaldi-mad in 1859. Journalists went to interview him, wildly embellished biographies were published in book and newspaper form in numerous languages, plays hastily written and put on the stage, portraits and memorabilia sold by the cartload. He was a hero fit for the media age. 'The life of Garibaldi reads like a romance,' wrote his admirer in faraway Arizona. 'It is full of melodramatic and tragic interest, of hair-breadth escapes, of battle-fire, of trials and triumphs.'[29]

The shabby armistice signed at Villafranca did not halt the patriotic movement. Rather, it spurred it on. If Napoleon thought he could divide up Italy as he wanted with the Austrians, he did not count on the passions stirred up by Garibaldi. And if Piedmont was set only on enlargement, the people could take matters to the ultimate conclusion: national unity. Garibaldi urged Italians to join the people in the central duchies and fight for freedom against France and Austria; he campaigned for funds for a 'Million Rifles'.

Garibaldi's political power in 1859 derived from his apparent un-political intentions. To many around the world, and in Italy, he was, in the words of the *Weekly Arizonian*, the 'Messiah of Liberty': 'All men who feel the liberty-throb cannot but admire an earnest, gallant, honest republican.' With his exotic life story, manliness, well-broadcast sexual conquests and his seemingly apolitical dreams for a united Italy, he had extremely broad appeal, and correspondingly broad support. *The Times* emphasised Garibaldi's 'strength of simplicity of character' and the steadfastness 'of purpose, of fortitude, and of courage'. 'The simple, frank, open-hearted soldier – the man of midday – has nothing in common with the gloomy, dreaming, burrowing conspirator – the man of midnight.'[30]

Here was a man for the times. And this celebrity gave him real power. Hero-worshipped throughout Europe and America, not to say Italy, he was a weight to place in the scale against the superpowers. The fate of Italy, and perhaps Europe, rested on his next moves. With Italian nationalism awoken and providing an inspiration across the globe, which country would dare stamp out 'an independence so boldly asserted, so manfully defended'?[31]

According to the New York magazine *Harper's Weekly*, 'There are conjunctions in every age when the triumph or defeat of great principles hinges on the fortunes of individuals.' It was referring to Garibaldi, but the same was true at exactly the same time for Brown and Yoshida. Both doubted the capacity of conventional politics radically to transform their societies. 'Talk! talk! talk! That will never free the slaves,' John Brown shouted at yet another abolitionist meeting. 'What we need is action – action.' At his *Shoka Sonjuku* – the 'school under the pines' – Yoshida prepared his samurai students for acts of noble self-sacrifice. 'Neither the lords nor the shogun can be depended upon, and so our only hope lies in grassroots heroes,' Yoshida told his disciples. 'Once the will is resolved, one's spirit is strengthened. Even a peasant's will is hard to deny, but a samurai of resolute will can sway ten thousand men.'[32]

Brown believed that 'the crimes of this guilty land will never be purged away but with blood'. Without an apocalyptic moment, American politics would bumble along from compromise to shabby compromise, and never find redemption. Yoshida dreamt too of radically and fundamentally changing every aspect of his society. His moment of revelation came when, back in 1854, he illegally attempted to board Commodore Matthew Perry's ship to discover the secrets of the West. After a spell in prison for his audacity, Yoshida opened his school in his home district of Choshu. Only a younger generation of samurai versed in the classics and traditional military skill could 'rectify' Japan. They also needed to learn from the West in order to defeat the West. In particular they needed to master modern inventions and weapons; technology married to tradition would make Japan invincible. When this happened, Japan would no longer be the quivering victim of foreign imperialists, but would dominate the world: 'we will seize Manchuria, thus coming face to face with Russia; regaining Korea, we will keep watch on China; taking the islands to the south, we will advance on India'.[33]

But instead of acting resolutely and revitalising the empire, Ii Naosuke's dictatorship combined servility to the West with repressive policies that inhibited Japan's capability to modernise and resist. Emperor Komei issued an order for the shogun to expel the barbarians by military force. The shogun enjoyed, after all, the title 'Barbarian-Subduing Generalissimo'. The new shogun was a pre-pubescent boy and the *bakufu*, under Ii, did nothing to fulfil this decree on his behalf. Yoshida's young warriors should therefore reject the *bakufu* and turn towards the emperor. Under the pines Yoshida converted dozens of young men to the higher cause with the intensity of his message and the imminence of the challenge; many of them would have a profound impact on the making of modern Japan. From Choshu, Yoshida established secret networks of political activists across the empire ready to begin a campaign of terrorism.

Both Yoshida and Brown wanted to purify and redeem their countries. Brown attempted to rouse the slave population in armed rebellion against their masters by capturing the US army's Armory and Arsenal at Harpers Ferry, Virginia, on 16 October. Poorly planned and executed, the whole thing became a fiasco. Rather than ignite a revolution, Brown and his small band of insurgents were trapped in Harpers Ferry and the old man was arrested. Similarly, Yoshida's first act of symbolic terrorism failed miserably and ended in his arrest.

In the ruins of their dreams and facing execution, both men achieved what had previously eluded them. Abraham Lincoln said later that John Brown's raid was 'absurd'. 'It was not a slave insurrection. It was an attempt by white men to get up a revolt among slaves, in which the slaves refused to participate.' True enough, the idea that a few fanatics could instigate a country-wide slave rebellion was absurd, just as a few high-minded and suicidal samurai could not see off the West. The unflinching and dignified ways in which Yoshida and Brown met their trials and deaths (the former was beheaded in Edo on 21 November at the age of twenty-nine; the latter was hanged in Charles Town, Virginia, a few days later on 2 December, aged fifty-nine) turned them from relatively impotent and little-known figures to men of monumental importance. Yoshida taught his followers that the secret of life was not to fear death. He bore his execution with a stoicism that inspired young Japanese to emulate him: 'To revere the emperor and expel the barbarian . . . may my death inspire at least one or two men of steadfast will to rise up and uphold this principle after my death.' John Brown wrote that although the raid was a disaster, 'by only hanging a few moments by the neck' he could turn defeat into victory. 'I am worth inconceivably more to hang, than for any other purpose.'[34]

Yoshida bequeathed his followers a plan to change Japan beyond recognition and an example of heroism and self-sacrifice. His execution could not arrest the creation of a powerful insurgent movement in Japan. Although Ii Naosuke appeared to hold Japan firmly in his grasp, beneath the surface of calm, forces were gathering that would shortly rip the country apart.

As Ii was approaching the gates of Edo Castle through the falling snow he was ambushed by a band of ronin samurai from Mito, the intellectual epicentre of anti-foreign sentiment and reverence for the emperor. The samurai overpowered the guards while one of their number pulled Ii from his palanquin and decapitated him. The assassin ran off with the head and then ritually disembowelled himself. The ronin left a statement declaring themselves to be the 'instruments of heaven' who had rid Japan of the instigator of the evil, the besmircher of the nation's honour. Their symbolic act inaugurated a wave of terrorism. Yoshida was dead, but his influence burned bright as Japan headed towards civil war.[35]

Although Brown failed in his ostensible aims, the Harpers Ferry Raid shook America to its core. At the moment he was hanged in Virginia, the telegraph prompted people across the Northern states to break out in collective and simultaneous mourning: black bunting was hung in numerous towns, bells tolled and prayer meetings were held. Thousands attended memorial meetings in New York, Boston and Philadelphia; books and lithographs flooded the market. For Henry David Thoreau, John Brown was comparable to Jesus Christ, a 'crucified hero', 'an angel of light'.

The outpouring of emotion and the canonisation of the terrorist confirmed, for the South, all it suspected about the homicidal intentions of abolitionists. The North, they said, with its endorsement of Brown, had declared war on the South. In late 1859 thousands of white Southerners braced themselves for Northern-led slave rebellions. In numerous cases, Northerners living and working in the Southern states were tarred and feathered or expelled.

It was clear that the Harpers Ferry Raid had driven a wedge into the heart of America. With a presidential election less than a year away many asked if the South could now possibly live under a government of what Southerners called the 'Black Republican Party'. The governor of Florida prayed for 'eternal separation from those whose wickedness and fanaticism forbid us to live with them in peace and safety'. One newspaper said that before Brown's act of terrorism the majority looked upon the possibility of the Union's disintegration as 'a madman's dream'.[36]

But not now. Across the South at the close of 1859 whites began to think not just that secession was inevitable, but that it was positively desirable. The governor of South Carolina recommended to his state legislature that a Southern confederacy should be established in the event of the election of a Republican president. If John Brown had wanted to bring America to a point of crisis that could not be resolved by anything but violence, his act of self-destruction had succeeded in doing just that.

Like Garibaldi, Brown was perfect for the age of the telegraph and photograph – the age of real-time news and romantic celebrities. That the year 1859 produced such popular heroes who tried to change the world by radical force is perhaps not as coincidental as it seems. Brown and Garibaldi, with their intriguing back-stories, forceful personalities, uncomplicated messages and explosive interventions, grabbed and riveted the attention of the new media and its expanding consumer base.

At the close of a decade of convulsive events and accelerated progress many expectations had not been fulfilled. National self-determination in Europe was still stifled by tyranny, and the political and economic power of enslavers in the United States had never been stronger. Self-motivated heroic individuals detonated bombs that promised to hustle things to their conclusion. That sense of millennial imminence was present in Japan too. The tsunami so long anticipated in Japan crashed down in 1859 with the transformation of the tiny fishing village of Yokohama into an international port. Traders from all countries came to Yokohama to see what profits could be made from the opening of Japan. The village was inundated with warehouses and the offices of Western businesses. Such feelings of epochal culmination, or the realisation of historical events, are ever-fertile breeding grounds for terrorists, freedom fighters, revolutionaries – call them what you will.

On the very last day of the 1850s *The Times* cast its eye back over the decade and likened it to 'a serial romance, in which each successive chapter is distinguished by some strange event or unexpected turn of fortune'.

The parcelling of time into decades is commonly regarded as a peculiarly modern phenomenon. We use the 1960s or 1990s as convenient shorthand to sum up a mood or a phase in history in ways that previous generations did not. The *Times*, however, was prepared to define the 1850s as a distinctive period of time.

The paper developed a theme that characterised the decade. Unceasing international turbulence and a storm of events had gone hand in glove with unprecedented global prosperity. Across the world millions had migrated and established thriving communities and great cities in North America, Australia, New Zealand and South Africa. The rapid extension of railways, telegraphs and shipping lanes over various parts of the world, the spread of free trade and the discovery of gold on a colossal scale had revolutionised the global economy. The Crimean War, the Indian Rebellion, wars in China, friction between Britain and the US, the financial crash of 1857 and the looming crisis in America had not retarded the progress of the world. 'It may be doubted whether greater accumulations of wealth have ever taken place in a period of ten years in any age.' Add to that the engineering and scientific breakthroughs of the decade and it amounted to a glorious heyday.

And for *The Times* the year 1859, one of world-historical significance, magnified the themes of the decade. Severe international crises and threats of war in Europe, America and China had not prevented 'a steady increase of material prosperity'. From a purely British perspective, *The Times* looked forward to the 1860s with optimism: 'Every branch of industry is flourishing as abundantly as at any former period, and the England of 1860 is richer, stronger, and better contented than the wealthy and prosperous England which in 1850 commanded the respect and envy of the world.'[37]

The paper reflected the complacency of the time. What *The Times* did not appreciate was how radically different the world of 1860 was from 1850. At the beginning of the decade the motors of historical progress appeared to be free trade, unrestrained communication and liberalism, with Britain as the main agent of change. But new forces were beginning to dominate. Ideals of nationalism and liberty were bursting to the fore. And people from China and Japan to Europe and America were coming to see direct, violent action as the midwife of radical transformation.

The year 1860 was dominated by Garibaldi's latest armed intervention. In May a rebellion against the despotic King of the Two Sicilies (Sicily and Naples) gave him the opportunity to cut the Gordian knot of the Italian question. Garibaldi electrified the world by leading a band of 1,000 Red Shirts that conquered first Sicily and then stormed north through Naples towards the Papal States.

'Garibaldi seems to have the knack of drawing everyone into his vortex,' said Mowbray Morris, managing editor of *The Times*. He should know because the paper's correspondent abandoned his duties of impartial reporting to become one of Garibaldi's brigade commanders. British army officers, including veterans of the Crimea and the Indian Rebellion, took leave at short notice and raced across Europe to join the Red Shirts. Hundreds of other Britons hungry for adventure, along with Americans and Europeans, did likewise. 'Anyone in search of violent emotions, cannot do better than set off at once for Palermo,' wrote Ferdinand Eber of *The Times*. 'However blasé he may be, or however milk-and-water his blood, I promise it will be stirred up. He will be carried away by the tide of popular feeling.'[38]

Eber gave a flavour of what was happening in southern Italy in his description of the general's reception in Palermo: 'It was one of those triumphs which seem to be almost too much for a man . . . the popular idol, Garibaldi, in his red flannel shirt, with a loose coloured handkerchief around his neck, and his worn wideawake, was walking on foot among those cheering, laughing, crying, mad thousands . . . The people threw themselves forward to kiss his hands, or at least to touch the hem of his garment, as if it contained the panacea for all their past and perhaps coming sufferings.'[39]

Little wonder, then, that Garibaldi was the hero of the world in 1860 and his activities followed everywhere. With his beard, red flannel shirt, neckerchief, weather-beaten hat and rifle he wore the unofficial uniform of men the world over, those hardscrabble emigrants who sought gold or carved out a farm in the wilderness. 'The course of civilization and human freedom,' declared *Harper's Weekly*, 'depends, for a time, upon the success of Garibaldi . . . if Garibaldi's invasion of southern Italy be crushed out, millions of people will conclude that God is on the side of despotism.'[40]

The guerrilla general said that he would proclaim the unification of Italy in Rome. That was to raise the stakes dangerously high. Napoleon III, as emperor of a Catholic country, was pledged to defend the papacy. Cavour now had to intervene to prevent another war. While

Garibaldi pushed north, the Piedmontese army invaded the Papal States and headed south towards Naples. Then Cavour produced one of his masterstrokes: the people of Naples would hold a plebiscite on the simple Yes or No question whether they wanted to join a unified Italy with Vittorio Emanuele as their king. Over 99 per cent of the electorate voted Yes. In effect, Piedmont had annexed southern Italy and Garibaldi's revolution was halted in its tracks. The unification of the whole of Italy, bar Rome and Venice, was now imminent.

On 24 October Prince Gong formally capitulated to Elgin's demands in Beijing. A few days later and thousands of miles away, American voters elected Abraham Lincoln as the sixteenth President of the United States. The day after that, 7 November, on the other side of the Atlantic, Laurence Oliphant was among the ecstatic, roaring crowds in the square in Naples looking up at a balcony of the palace. The two fathers of Italy emerged onto that balcony and the guerrilla chief Garibaldi proclaimed Vittorio Emanuele King of Italy.[41]

The conjunction of these unrelated events on three continents in the space of a fortnight highlighted the turbulence of the world in 1860. For *Harper's*, Garibaldi's great campaign was not solely about the fate of the Italian people. The general had kindled an unquenchable flame that would blaze around the world. 'The question is, whether men – living in Italy, or in Germany, or in France, or in Russia, or anywhere – have a right to their rational freedom or not. This is the point to be decided . . . Democracy, in the abstract, hangs in the balance.' The adventure in Italy dominated worldwide reporting; the ferment generated wider expectations of imminent change.[42]

'*Interpellatione, sempre interpellatione!* . . . what is the use of questions? What do they ever come to?' Within a few weeks of Lincoln's election as President – and three months before his inauguration – senators and congressmen from Southern states signed an address to their constituents that recalled Garibaldi's words to Laurence Oliphant earlier in the year: 'The argument is exhausted. All hope of relief in the Union, through the agency of committees, Congressional legislation, or constitutional amendments, is extinguished.'[43]

On 20 December South Carolina seceded from the Union. The differences of North and South, declared the Address of the People of South Carolina, had made them 'totally different peoples . . . All fraternity of feeling . . . is lost, or has been converted into hate; and we, of the South, are at last driven together by the stern destiny which controls the existence of nations'.[44]

A new nation was born. Over the course of January 1861, Mississippi, Florida, Alabama, Georgia and Louisiana (in that order) voted to secede from the Union; Texas followed on the first day of February. Senator Judah P. Benjamin of Louisiana wrote of the 'wild torrent of passion which is carrying everything before it . . . It is a revolution . . . of the most intense character . . . and it can no more be checked by human effort . . . than a prairie fire by a gardener's watering pot.' People in the seven breakaway states took to the streets and danced and sang.[45]

The Provisional Confederate States Congress met in Montgomery, Alabama, on 4 February. Within a few days the Confederate States of America was born as a fully fledged nation, with a constitution, flag, currency, legislature, provisional government and president. Bodies of militia in blue and red uniforms, brass bands playing 'Dixie' and a procession of carriages escorted the provisional president of the new nation, Jefferson Davis, from the Exchange Hotel in Montgomery to the Alabama Statehouse for his inauguration on 18 February. 'The man and the hour have met!' declared William L. Yancey as he introduced the president to the cheering crowd. 'The time for compromise is now passed,' Davis said in his inaugural speech. 'The South is determined to maintain her position and make all who oppose her smell Southern powder and feel Southern steel.'[46]

March 1861 witnessed another coincidence of radical events. On 3 March Tsar Alexander II issued his Emancipation Manifesto, freeing 23 million Russian serfs. The next day Abraham Lincoln was inaugurated as President of the United States and in Montgomery the Confederate flag was raised for the first time. On the 17th of that month the Kingdom of Italy was formally proclaimed by the new Italian parliament. Four days later in a speech in Savannah, Georgia, Alexander Stephens told a joyous, raucous crowd that 'we are passing through one of the greatest revolutions in the annals of the world'. The cornerstone of the government, he declared to echoing cheers, rests 'upon the great truth that the negro is not equal to the white man; that slavery, subordination to the superior race, is his natural and moral condition. This, our new government, is the first in the history of the world based upon this great physical, philosophical and moral truth.'[47]

William Howard Russell of *The Times* happened to be in Norfolk, Virginia, on Sunday 14 April when he was swept along by a crowd. He asked someone what the commotion was all about. 'Come along,' the man replied, 'the telegraph's in at the *Day Book*. The Yankees are whipped!'

News freshly arrived down the wire and pinned to the wall outside the *Day Book*'s offices told that Fort Sumter, a manmade granite island in Charleston Harbor, South Carolina, had just fallen to Confederate forces. 'At all the street corners men were discussing the news with every symptom of joy and gratification.' The next day, at Goldsboro, North Carolina, Russell witnessed mobs in the streets, the men with 'flushed faces, wild eyes, screaming mouths, hurrahing for "Jeff Davis" and the "Southern Confederacy"'. Their yells of rebellious joy drowned out 'discordant' bands playing 'Dixie Land'; above them fluttered Confederate flags. This was an outburst of emotion comparable to that seen so recently in Italy. 'Here was the true revolutionary furore in full sway. The men hectored, swore, cheered, and slapped each other on the back; the women, in their best, waved handkerchiefs and flung down garlands from the windows. All was noise, dust, and patriotism.'[48]

A day after Sumter fell, Lincoln issued a proclamation calling for 75,000 volunteers to enter national service and put down the Southern insurrection. Across the Northern states meetings and rallies were held in support of the Union; in New York 250,000 people took to the streets.

As in the South, there was an explosion of patriotism. Men everywhere reached for their rifles, had uniforms made and enlisted in militia regiments; hospitals were inundated with women begging to be trained as nurses. Be they American or European, thousands of volunteers were prepared – *eager* – to risk their lives in defence of their country and what Lincoln called the great republican 'experiment' begun on 4 July 1776. The election of Abraham Lincoln and the secession of the Southern states were no less revolutionary for Northerners as they were for the South. For fifty-three years out of the seventy-two that separated the inaugurations of George Washington and Abraham Lincoln, the White House had been occupied by presidents who either owned or had owned slaves. Up to 1850 at least half of the membership of the Senate at any one time came from states where slavery flourished. All American political parties had hammered compromises to permit the survival and territorial expansion of slavery until the foundation of the Republican Party in 1854 in the wake of the Kansas-Nebraska Act.

Now, in 1861, the grip of slaveocracy in the US had been prised open with the election of a Republican president. The party declared that the 'normal condition of all the territory of the United States is that of freedom'. The excitement sparked by the election of a Republican president was not so much to do with the thought that slavery would suddenly be abolished – not even Lincoln proposed that. It was the conviction that America was now free to enter the next glorious phase of the republican experiment. 'The great revolution has actually taken place,' marvelled Charles Francis Adams the day after the general election. 'The country has once and for all thrown off the domination of the Slaveholders.'[49]

'Thank God! we have a country to fight for,' said one American that summer, a country 'to live for, to pray for, to fight for, and if necessary to die for'. That sentiment could have been uttered word for word by someone on either side of the divide.* Federalists and Confederates alike felt they had nations to build; that they were the heirs of the great experiment in republican democracy. Each side now had the world to convince that theirs was the legitimate cause.[50]

* They were expressed, as it happens, by a Mississippian supporter of the Confederate States.

15

Blood, Iron, Cotton, Democracy

BOMBAY

This is an age of great and rapid events: their quick succession as remarkable as their importance . . . The restless and revolutionary spirits, distracted by the choice of materials, pause in the selection; not from satiety, but from sheer perplexity to decide where most mischief can be accomplished.

Benjamin Disraeli, House of Commons, 19 June 1861

The island of Caprera might not win first prize for its beauty. 'No picturesque fishing boat gives life to her waters – no pleasant little spot appears along her shores – no ruined forts crown her heights – but one mountain chain upon another raises its rugged masses in amphitheatric form before the wondering eye of the stranger. All that surrounds him here is severe and vast.'[1]

The small island sits off the northern tip of Sardinia. First-time visitors were struck by the great granite boulders that greeted them from the water; if you were seeking solitude you would surely find it here. The monotone 'barren mass of rock' that was the island was broken by a white speck. As arriving boats got nearer, the dot turned out to be a small whitewashed bungalow built in the style of a hacienda. Approach the farmhouse on foot and you would be greeted by numerous friendly dogs. It was an oasis in the aridity of the granite island: bubbling springs watered the well-maintained flower beds, cypresses, fruit trees and vegetables that grew on the steep slopes.[2]

Few people did not know of this picturesque, remote bungalow and its diligent hermit farmer. Immediately after proclaiming Vittorio Emanuele King of Italy, Giuseppe Garibaldi returned to Caprera and the hacienda he had built, bearing his rewards for unifying his country: a bag of seeds gleaned on his campaign. He might have retired insisting that he desired solitude, but the island was inundated with visitors. Their books and articles describing the rugged beauty of the island and the bucolic lifestyle of its resident became central to the legend of Garibaldi as the true hero of the nineteenth century – the simple farmer and 'the modern Cincinnatus' who eschewed power and status. All of them depicted Garibaldi spade or pickaxe in hand, toiling against the rocks of Caprera to carve out a patch of Eden. 'I now only wage war with stone,' he told his on-off lover Esperanza von Schwartz. In reality, the general was most likely to be found reading the heaps of letters that arrived from all over the world. This vast correspondence included indecent proposals from women, gushing fan mail from besotted men, gifts of books, songs composed recounting his deeds and invitations of all kinds – invitations to visit cities, to give his name to lost causes or to liberate all manner of places.[3]

In 1861 the world's press was itchy with speculation that one of the letters was a summons to America. Was Garibaldi, the icon of national unity, about to leave his island of self-imposed exile to reunify the United States? He told an American diplomat, 'If your war is for freedom, I am with you with 20,000 men.'[4]

The rumours that the 'Washington of Italy' was about to head to the US to take command of the Union army came almost exclusively from Caprera itself. But the idea entranced the Northern public and the government in Washington. American diplomats sounded Garibaldi out and it was clear that he was prepared to cross the Atlantic. These reports reached the State Department on 17 July. Four days later on the outskirts of Washington at the Battle of Bull Run – the first major engagement of the Civil War – the poorly trained and led volunteers of the Union army were routed. In a memorable and notorious piece of journalism William Howard Russell described being swept back by a torrent of retreating Union soldiers as he attempted to reach the front line. Wagons packed with men were 'grinding through a shouting, screaming mass of men on foot, who were literally yelling with rage at every halt, and shrieking out, "Here are the cavalry! Will you get on?"'

Writing up his account of the retreat the next morning in Washington, Russell looked out of the window and 'saw a steady stream of men covered with mud, soaked through with rain, who were pouring irregularly, without any semblance of order up Pennsylvania Avenue towards the Capitol'. The victorious Confederate army camped within sight of Washington.[5]

That day Lincoln asked for 500,000 more volunteers. Less than a week after the crushing and shocking defeat, the Union government opened up official communication with the 'distinguished Soldier of Freedom', Giuseppe Garibaldi. After consultation with Lincoln, Secretary of State William Seward wrote to Henry Sanford, the US Ambassador to Belgium and unofficial American spy chief in Europe, asking him to begin overtures to Garibaldi. Sanford was to tell the guerrilla general that 'the fall of the American Union' – a very real possibility now after Bull Run – 'would be a disastrous blow to the cause of Human Freedom equally here, in Europe, and throughout the world'.[6]

In short, Lincoln was looking for a figurehead of liberty and nationalism to galvanise the Union war effort at its lowest ebb. The presence of Garibaldi at the head of an army would inform the world that the cause of the North was the cause of the world.

Sanford arrived at Caprera in September and the two men talked as the sun set, no doubt enjoying the magnificent view of Corsica's blue-tinted mountains on the other side of the Straits of Bonifacio. There was one question that Garibaldi had asked before of American emissaries, and he asked it again: 'Tell me . . . if this agitation is regarding the emancipation of the Negroes or not?'[7]

No amount of prevarication could get round giving the general the simple, blunt truth. During his inaugural address Lincoln stated that he had 'no purpose, directly or indirectly, to interfere with the institution of slavery in the states where it exists. I believe I have no lawful right to do so, and I have no inclination to do so.' On 25 July, immediately after the shock of Bull Run and fearful that more slave states would join the Confederacy, Congress passed the Crittenden-Johnson

Resolution by thumping majorities in the House and the Senate. According to the Resolution, if states now in rebellion returned to the Union no action would be taken to abolish slavery.[8]

The general turned down the offer of leading a Union army. If, Garibaldi told Sanford, the Civil War was not about enfranchising slaves then it was merely a civil war 'like any civil war in which the world at large could have little interest or sympathy'.[9]

Over in England, Charles Darwin told an American correspondent that he prayed to God 'that the North would proclaim a crusade against slavery'. The feeling was shared by millions of Darwin's fellow Britons. Instinctive sympathy for the North was diluted by the feeling that the conflict in America was, as Garibaldi said to Sanford, 'like any civil war', a matter of union or disunion without any elevating ideals of freedom. And so, in the early months of the Civil War, the Confederates were gifted a great chance.[10]

Throughout the boom of the 1850s the slave states had become convinced not just of the critical importance they had to the American economy, but to the fortunes of global capitalism. The great crash of 1857 had, for them, exposed the fragility and illusionary nature of the North's and West's prosperity. Cotton exports had continued to increase even during the Panic and the subsequent recession. In Georgia the *Augusta Daily Constitutionalist* boasted that in the aftermath of the crash 'the most highly prosperous people now on earth are to be found in these very States [of the South]'. A paper in Indiana had to admit in 1859 that 'At present the South is beyond question, outstripping the North and East in the accumulation of wealth, if not in the increase of population.'[11]

Cheap slave-grown cotton fuelled the world's rampant economic expansion in the 1850s, Southerners held: take it away and growth would splutter to a halt. A stoppage in the flow of cotton, declared W. H. Chase of Florida, would result in revolution in Britain. 'The flow of cotton must not cease for a day,' he said. 'The great cotton zone of the world must never cease to be cultivated; the plough, and the hoe, and the cotton gin, must never cease to move.' Cotton bound the American South to Liverpool, Manchester and the million or so sailors, stevedores, hauliers, factory workers and other labourers who helped, directly or indirectly, turn raw cotton into spun cloth. Of the 1.4 billion pounds of raw cotton imported into Lancashire in 1860, 1.1 billion came from the American South. Similarly, slave cotton accounted for 90 per cent of French cotton imports (192 million pounds), 92 per cent of Russian (102 million) and 60 per cent of German (115 million). What would happen if the South refused to grow cotton, asked Senator James Henry Hammond of South Carolina back in 1858: 'I will not stop to depict what every one can imagine, but this is certain: England would topple headlong and carry the whole civilized world with her, save the South. No, you dare not make war on cotton. No power on earth dares make war upon it. Cotton *is* king.'[12]

But the South had more to offer the Old World than threats of imminent economic apocalypse. It held out an immeasurably greater prize.

And that prize was something that the British cherished above all else: free trade. In the same year that Britain made a giant leap to becoming a free trade nation with the repeal of the Corn Laws – 1846 – the United States Congress reciprocated by passing the Walker Tariff that lowered import and export duties. In 1857 the tariff was further reduced. The thread that has run through *Heyday* has been Britain's attempt to reshape the world through the power of free trade. Aside from opening up Asia to the supposedly rejuvenating winds of trade, nothing was more important to Britain's worldwide crusade than its partnership with the United States. More precisely, it was the partnership with Western grain producers and Southern cotton growers – two powerful groups that depended on unfettered access to the British market. Northern manufacturers, its iron producers and wool producers were opposed to free trade because the flow of cheap British goods undercut them and stifled industrial growth. But they had been frustrated in their efforts to raise duties during the '50s because the Westerners and Southerners together tipped the political balance in favour of free trade.

In early 1861, with the Southern senators absent, the Republicans at last passed the protective Morrill Tariff, aimed at British imports. By building a wall around one of the world's most important and dynamic economies, the Republicans at a stroke undermined the foundations of Britain's global system. For many people in Britain, free trade was a sacred thing; it alone promised the renovation of the world. 'There are two subjects on which we [the British] are unanimous and fanatical – personal freedom and free trade,' Richard Cobden wrote to Charles Sumner. The American measure was dubbed the 'Im-Morrill Tariff' and savagely attacked in the British press as a retrograde and reckless step. Protectionism, for the British, ranked alongside slavery as a roadblock to progress, and as an evil in the world that needed to be vanquished.[13]

Lincoln was not proposing to put an end to slavery. But Congress had put up a barrier to free trade. The Confederate States now offered the world – and Britain in particular – an enormous duty-free market. The 'great staples' of America – wheat, rice, sugar, tobacco and cotton – 'would flow freely through the rivers and railways to Southern ports,' promised Chase. The dreams of the 1850s – uninhibited trade – would be realised as the great trading nations of Europe shared in a stupendous bonanza. 'Their ships would come loaded with products of every nation, and delivering them at Norfolk, Charleston, Savannah, Fernandina, Pensacola, Mobile and New Orleans, would be reloaded with the rich products of the South and West.' Rather than go by rail to Chicago and New York the produce of Minnesota, Iowa, Missouri, Illinois, Kansas and other burgeoning states would flow down the Mississippi to New Orleans and out to the world market.[14]

The *New York Times* was alive to the danger in which the Morrill Tariff had placed the Union. Although, editorialised the paper, the British were – even by New England standards – ultra-abolitionists, 'in trade we look only at value and price. If the manufacturer at Manchester can send his goods into the Western States through New Orleans at a less cost than through New York, he is a fool for not availing himself of the advantage'. More worrying, however, for the *New York Times*, was the way in which the Morrill Tariff was wrecking foreign relations with the European powers. As *The Times* of London put it, by suddenly adopting protectionism the North was acting 'in a narrow, exclusive and unsocial spirit' that lost it 'the sympathy and regard of mankind'. The South, by contrast, with its promise of free trade 'extends the hand of good fellowship to all mankind, with the exception of its own bondsman'. Southerners agitating for help in Europe could not believe their luck and the folly of the Republicans.[15]

The Morrill Tariff seriously damaged relations between the North and Britain. Assessing the Civil War now became a matter not of principle but of commercial self-interest. Long habituated to regarding trade as the arbiter of human affairs, *The Times* said that trade would settle the matter. The British might not like that the United States was rent asunder, 'But the tendencies of trade are inexorable, and our manufactures will infallibly find their way to the best market with the regularity of a mechanical law . . . It may be that the Southern population will now become our best customers.' And good customers – even slave-owning customers – inevitably became political partners. Or as Palmerston put it pithily to a Union agent in London: 'We do not like slavery, but we want cotton, and we dislike very much your Morrill Tariff'.[16]

Sympathising with the South, however, was very different from doing anything to help. Britain and Europe had to be forced to recognise the sovereignty of the Confederate States, it was held in the South. And the only way that could happen was to send an electric shock through the industrialised world. To that end, the South embargoed the export of cotton to remind everyone how abjectly dependent they were on slave-grown produce. 'All these States and Empires are, and will be, under obligation to us,' wrote the influential Southern novelist and historian, W. Gilmore Simms. 'We have them in a net. *They know* the power of cotton! They must and will have it. Let us be tranquil and wait, and make no concessions. *We can exist and be independent, without selling a bale of cotton for years!*'[17]

With the fate of hundreds of thousands of Lancastrian mill workers and their dependants in its hands, not to say the country's economy, the British government would be forced to come cap in hand and recognise the Confederate States' full sovereignty, or so the thinking went. Not only that: Lincoln had placed the South under blockade to prevent commodity exports. In all likelihood the Union's navy would, according to Chase, be 'swept away by English fleets . . . hovering on the Southern coasts, to protect the free flow of commerce, and especially the free flow of cotton to English and French factories'.[18]

But it soon became clear that there was a young pretender to old King Cotton's all-mighty sceptre. By the early 1860s King Corn was pressing the old monarch hard. American wheat exports had been hard hit after the Crimean War ended, precipitating the Panic of 1857; but by 1860 demand had returned. In 1859 wheat exports had dipped to 16 million bushels of wheat (down from 27 million in 1856); by the first year of the Civil War they had soared to 54 million. Britain imported between 40 and 50 per cent of its foodstuffs from America. Lancashire's mills had to spin, but the people had to eat. The North was Britain's breadbasket and one of its biggest customers; hundreds of millions of pounds were invested in railroads and other enterprises. For its own sake, Britain had to appease both of America's two competing commodity kings. To the disgust of both sides, Queen Victoria issued a Proclamation of Neutrality.

Davis' South resented the fact it could not hold Britain to ransom over cotton. But neutrality was most offensive to Lincoln's North. By awarding legal rights of belligerency to both sides, Britain was not treating the conflict as an insurgency, with the North as the legitimate government putting down an illegal uprising, but as a contest between two nations. The Proclamation did not go as far as recognising the Confederacy as an independent state, but as far as Jefferson Davis was concerned it was a step in that direction.

Thanks to Lincoln's inaugural address and the Crittenden-Johnson Resolution, Confederate delegates in Britain could bypass the issue of slavery in their appeals for international recognition. Instead they could highlight the protective duties and the blockade. If the Union was not making the war an issue of freedom, then the Confederacy could. In London, William Yancey, the zealot who had introduced President Davis to the screaming crowds in Montgomery, told ministers and the press that the patriotic people of the South were fighting for independence. In a world where national self-determination had become so important, surely the American South could demand the same as Latin America or Greece in the 1820s and Italy, Hungary and Poland in the present day: liberation from a larger imperial power. As one Union agent in Europe noted, the Confederate emissaries were making headway, talking 'of free trade and cotton to the merchant and the manufacturer, and of the right of self-government to the liberal'.[19]

Free trade and constitutionalism: nothing could appeal to the British more. They gave the South some sort of glamour in their struggle. On the same day that Britain granted belligerent rights to the Confederacy, *The Times* called for the same rights to be extended to the Taiping insurgency. The paper's editorial hints at the link – long forgotten – between the bloody civil wars fought simultaneously in China and the United States. In the spring of '61 the British world system was being buffeted from two directions. The outbreak of the Civil War (along with the Morrill Tariff and the cotton embargo) coincided with the intensification of the Taiping Rebellion. For the British, the two events were inextricably bound together. China and the United States made up the eastern and western arms of Britain's global trading empire. Two-thirds of the tea that British merchants exported from China was sold in the States. That market collapsed as the war began, causing consternation from Shanghai and Canton to the City of London. With both of its primary trading partners disintegrating at exactly the same time, Britain faced economic shipwreck. But the global storm presented an opportunity as well.[20]

The southern insurgents of both China and the US might seem, in their different ways, like fanatics and freebooters. But, to a commercial mind, they possessed attractions. The Taipings controlled the 'great waterway of China' just as the Southerners of the US controlled access to the Mississippi Valley. Newspapers and members of parliament were beginning to argue that the British government should recognise the Heavenly Kingdom as a separate state and force the Qing dynasty to do likewise. 'With peace', commented *The Times*, 'we might walk right into the heart of China, and win her to civilization and intimacy by the strongest passion she has – her love of gain. At this moment the door stands wide open; our merchants are ready to enter it.' Like the entrepreneurs of the American cotton states, the Taipings could become valuable customers. In the early summer of 1861 the British were flirting with two major insurgencies on either side of the globe, weighing up the gains that could be made from exploiting them.[21]

Perhaps a partitioned China and a partitioned United States was in Britain's commercial interest after all. Indeed, many people in Britain were relishing the conflict. Politicians like Palmerston and his Foreign Secretary, Lord Russell, were delighted by the divisions in America. The latter in particular saw Washington's international isolation as payback for its close relations with Russia during the Crimean War. More delicious were the political ramifications. American democracy and republicanism had been held up as examples that the world should follow. That experiment had spectacularly blown up. As the author of 'Democracy on Trial' in the *Quarterly Review* put it, republican democracy 'has collapsed, as its predecessors had done, into a chaos of anarchy and bloodshed. The end has come quicker than it did to the democracies of Athens and Rome, slower than it did to the democracies of France and Spanish America; but the same event awaits them all.' If the world had 'aped the extravagances of America in the heyday of her folly' it would have been 'sucked into the same current' to be wrecked on 'the same fatal shore'. The Civil War was further evidence that British constitutionalism and liberty were the models the world should emulate.[22]

Despite their advantages, the Confederate agents were irritated that they were not making much progress in London. The power of cotton and free trade was not enough to overcome a powerful scruple. 'The anti-slavery sentiment is universal,' an exasperated Yancey reported from London. '*Uncle Tom's Cabin* has been read and believed.'[23]

But then something happened that threatened, or promised, to turn the situation on its head. In November, two Confederate envoys were in Cuba en route to Europe to renew the quest for international recognition. They boarded the British mail packet *Trent* so as to avoid the Union blockade and the ships searching the waters for them. On the 7th the *Trent* left Havana and was proceeding through the Bahama Channel when it was stopped by a Union steam frigate. The Confederate agents were arrested and taken to Boston where they were imprisoned.

The news, when it reached Britain, predictably sparked outrage. Secretary of State Seward received a letter from an American in London stating that 'The people are frantic with rage, and were the country polled I fear 999 out of 1,000 would declare for immediate war . . .

The Southern men in London . . . already see the South recognized by England and France in unison and cannot conceal their exultation.' Arresting a foreign ship, a mail ship at that, in time of peace was as blatant a violation of international law as could be imagined. What is more, it was the worst kind of insult to the British flag.[24]

Of the three occasions when Britain and the US had come close to war since 1851, the *Trent* affair was by far the most dangerous. Palmerston dispatched 10,000 troops to Canada and the Royal Navy was put on a war footing. Once again there was talk of capturing and burning US coastal cities. The British government delivered an ultimatum to the Federal government: release the prisoners or face war.

These were nervous times. The markets behaved as if war was imminent as the press thundered for vengeance. In America the mood was as bellicose. The captain of the US warship was lauded and the North rejoiced at the capture of two notorious secessionists. Lincoln's government faced enormous pressure from the public not to back down. At a ball William H. Russell overheard Seward saying 'We will wrap the whole world in flames.' Weeks passed as the Federal government delayed responding to the ultimatum. By Christmas it was clear that France would join Britain in upholding international law. The Union faced war against the world's two greatest naval powers as it was fighting the Confederacy.[25]

In the end, it was the Union that backed down after a month of war fever on both sides of the Atlantic. Lincoln apologised and the Confederate emissaries were released to continue their voyage to Britain.

The world sighed in relief as the two sides backed away from the precipice. All the world, that is, apart from the Confederacy. Had Britain and France fought the US on the high seas the Union's blockade of the South would have been destroyed. The hopes of the Confederacy, however, were not dimmed. At the beginning of 1862, Lancashire was facing a cotton famine, with thousands out of work and hundreds of thousands facing reduced wages and hunger. Britain and Europe were tired of the Civil War, tired of the damage being done to the global economy by the shortage of cotton, tired of the reports of bloody and inconclusive campaigns and battles. From July, with news of General Robert E. Lee's victories in Virginia and Maryland, there was a growing conviction that the European powers would have to intervene to put an end to the bloodshed and the misery caused to mill workers throughout the Continent. Already 150,000 people had died in the Civil War and the result was stalemate; in Lancashire at least 700,000 factory workers and their families were surviving on charity.

In Paris and London ministers were discussing a coalition that would bring an end to the Civil War. First, the Confederate States would at long last be recognised as an independent nation. Then Britain and France, perhaps in alliance with Russia, Prussia and Austria, would offer to mediate between the two 'nations'. If mediation was refused by the Union, as surely it would be, the great powers would intervene militarily to force the North to accept peace and the permanent division of the United States.

Aware of the hardening international mood, Abraham Lincoln knew that 'a distinct anti-slavery policy would remove the foreign danger'. During the summer of 1862 he committed himself to elevating the cause of the Union to a loftier plain, that of emancipation. He informed the Cabinet of his decision on 22 July, but for the moment the announcement was put off for fear of the political consequences during a summer of disaster for the Union.[26]

Then something extraordinary erupted in Europe. The instigator, of course, was Garibaldi. In August he led a band of Red Shirts once again from Sicily across the Straits of Messina and north through southern Italy. Their destination was Rome, which Garibaldi would liberate from the Pope and make Italy's capital. *Roma o Morte!* they chanted: Rome or death! If Garibaldi expected Vittorio Emanuele to join him, he was sorely mistaken. Napoleon III took very seriously his pledge to maintain the papacy's temporal majesty; he would go to war to prevent Rome becoming part of Italy. Eager to appease France and prevent a Garibaldian revolution, the Italian prime minister sent troops south to intercept the Red Shirts. The two forces came face to face at Aspromonte, which means 'sour mountain', on 29 August. *Viva Italia*, Garibaldi was shouting as two bullets smashed into his leg. He was arrested for treason.[27]

Once again, turmoil in Italy impacted on Europe. In France people growled in discontent at Napoleon III for propping up the pontiff. In

Belfast 70,000 Protestants marched in support of Garibaldi, sparking violent clashes with Catholics that lasted for five nights. Four days later, on 29 September, when a group called the Working Men's Garibaldian Committee began addressing a crowd in Hyde Park they were attacked by hundreds of Irish bricklayers and labourers. Between 10,000 and 20,000 Londoners who were in the park rushed to the melee to take on the Irish; the meeting broke down in violence, but the Committee vowed to return on the next Sunday.[28]

When the London working-class Garibaldians reconvened the following Sunday in Hyde Park there were over 100,000 of them. The excitement was generated by the appearance of a sensational letter in the British press two days before the meeting. What Garibaldi had to say in 'To the English Nation' had a direct bearing on the American Civil War.

Garibaldi began by praising the British for their help in securing Italian unification and for providing a refuge for those fleeing tyranny. 'Follow your path undisturbed, O unconquered nation!' he said, 'and be not backward in calling sister nations on the road to human progress.' The people of Britain should link arms with the people of France and bring to an end Napoleon III's 'dominion of the spirit of evil' that was holding back European freedom. Then he turned to the United States. 'Call the great American Republic,' he urged his British supporters. 'She is, after all, your daughter, risen from your bosom; and however she may go to work, she is struggling today for the abolition of slavery so generously proclaimed by you. Aid her to come out from the terrible struggle in which she is involved by the traffickers in human flesh. Help her, and then make her sit by your side, in the great assembly of nations, the final work of human reason . . . Rise, therefore, O Britannia! and lose no time. Rise with uplifted brow and point out to other nations the road to follow.'[29]

The American North, Garibaldi was saying, was dedicated to the abolition of slavery, even if that fact was not yet proclaimed. Britain should aid the United States in defeating the slave-owning rebels and proclaim a new era in the history of the world. On Sunday 5 October working-class Londoners turned out to show their support for Garibaldi. Once again the park descended into horrific violence. The much-boasted peace of London had not been disturbed like this for decades.

On that very day the Royal Mail steamship *Australasian* dropped off telegraphic messages at Queenstown (Cork) en route to Liverpool. She bore news that she had netted out of the water off Cape Race, Newfoundland, a week earlier; it was cabled immediately to newspapers across Britain and published in full on Monday 6 October.

It had happened at last. On 22 September President Abraham Lincoln issued a proclamation stating 'That, on the 1st day of January [1863] . . . all persons held as slaves within any state . . . the people whereof shall then be in rebellion against the United States, shall be then henceforward and forever free.'

Union victory at the Battle of Antietam in Maryland on 17 September gave him the opportunity to make such a risky promise. Now, surely, was the time for the civilised world to wheel in line behind the North. Not so. Lincoln's Preliminary Emancipation Proclamation made the case for intervention more urgent than ever for the European hawks. With his eye on the border states loyal to the Union but in which slavery existed, Lincoln was promising only that those slaves in *rebellious* states would be emancipated as of 1 January. Those who took an informed assessment of the American Constitution and the political constraints under which the president operated could see that the Proclamation would make slavery untenable everywhere in the event of a Northern victory. But the Indian Rebellion of 1857 still exerted a powerful influence over politicians in Britain. They believed that Lincoln had made a slave uprising and a horrific race war inevitable; the South would be drenched in blood.

The time had come to intervene, not for the sacred cause of freedom, but to force the North to accept the fact of a Confederate States of America.

On 7 October, the day after the news of emancipation hit the papers, William Gladstone gave a speech in Newcastle. The Chancellor of the Exchequer and heir apparent to Palmerston spoke of the 'mournful subject' of the situation in Lancashire, where, he said, half of all mills were closed after the price of cotton quadrupled and the people were living through a period of misery. That tale of woe led him naturally on to America. Britain, he said, never wanted to see the break-up of the Union, as some Americans alleged, but 'I can understand those who say that it is for the general interest of nations that no state should swell to the dimensions of a continent'. Britain had experienced the trauma of a break-up of a political union, Gladstone continued, and it had benefited in the long run; now it was the North's turn to accept the same fate.

With that opinion, the chancellor was putting the North in the position of the British Empire in 1775 and casting the Southerners as the heirs of the Founding Fathers. That might have been offensive enough to Northern American ears, but Gladstone was only just warming up. The North had to face the reality that the Confederacy could never be defeated. 'We may have our opinions about slavery; we may be for or against the South,' Gladstone told his cheering audience; 'but there is no doubt that Jefferson Davis and other leaders of the South have made an army; they are making, it appears, a navy; and they have made what is more than either – they have made a nation.'[30]

Making a nation. That was one of the key phrases of the 1860s. The breakaway South was to be compared to the glorious cause of Italy and the wider noble struggle for self-government in such places as Poland and Hungary. The bombshell of Gladstone's speech exploded not just in Newcastle but in American consulates across Europe that very night, thanks to the telegraph. His words were interpreted as a signal that the British government was about to recognise the Confederate States and intervene militarily to partition the old United States. 'We are now passing through the very crisis of our fate,' said Charles Francis Adams, the US Ambassador to London, with real apprehension. The fate of America seemed about to be settled by foreign invasion. Turmoil buffeted the financial markets as war reared its head.[31]

But even as Gladstone spoke, the political weather was changing. The Newcastle speech went down very badly; the public was dead set against supporting a slave republic. Charles Francis Adams dated the turning of the tide against military intervention to Garibaldi's defeat on the 'sour mountain'. Europe was on edge as a result, fearful of revolution spreading; eyes were diverted from America. Adams reported from London that Lord Russell had told him that Garibaldi's Roman adventure had derailed 'any idea of joint action of the European powers in our affairs'. The pro-Garibaldi riots in Hyde Park deeply impressed Adams. He wrote to US Secretary of State William Seward: 'I am inclined to believe that perhaps a majority of the poorer classes rather sympathise with us in our struggle, and it is only the aristocracy and the commercial body that are adverse.'[32]

A statue in Brazenose Street, Manchester, recalls the change of mood evident in 1862. On the plinth are extracts of a letter written in the name of the Working People of Manchester by beleaguered cotton workers during a meeting at the Free Trade Hall on the last day of the year. Inscribed also is the reply from its recipient, Abraham Lincoln, whose statue stands on the plinth.

Since the beginning of the Civil War many cotton towns in the north of England had flown the Confederate flag. After Lincoln's Proclamation of 22 September things began to change. The cotton workers told Lincoln that they were sure that the 'foul blot on civilisation and Christianity', slavery, would soon be wiped away and 'will cause the name of Abraham Lincoln to be honoured and revered by posterity'.

In his reply, which was to become famous across the world, Lincoln acknowledged and deplored the 'sufferings which the working people of Manchester and in all Europe are called to endure in this crisis'. Under these trying circumstances, continued Lincoln, 'I cannot but regard your decisive utterances on the question as an instance of sublime Christian heroism which has not been surpassed in any age or in any country. It is indeed an energetic and re-inspiring assurance of the inherent truth and of the ultimate and universal triumph of justice, humanity and freedom.'

The cotton workers of Lancashire were rewarded for their forbearance. The South believed it could manipulate the world by manipulating the price of cotton. It sold cotton bonds on the international money markets to finance its war effort. They were a good bet: with the price of cotton rising, lots of people bought them, including Gladstone. But the South's ability to profit by trading its bonds depended upon its ability to turn back on the supply of its staple. When the Union captured New Orleans it was deprived of that vital freedom. And then the

South lost its monopoly on supply. A few months after the working people of Manchester wrote to Lincoln, cotton began to reach the mills once more.

The Times saw the crisis in America as an opportunity. Marvelling at the interconnectedness of the world, the paper said that the Civil War 'is quickening the industry of millions in the ancient kingdoms of the East'. As soon as news of the secession of the Southern states reached Bombay from Montgomery 'it was obvious that the cotton markets of the world might be revolutionized . . . A new world is opened to Eastern commerce, and a new stimulus given to Eastern industry.' The paper urged the plutocrats of Manchester – the Cotton Lords – to provide the capital to reshape the world.[33]

'A trade like the cotton trade needs organization on a large scale,' the paper said. New cotton fields had to be created throughout the world and a means found to connect them to the global market. Restructuring the world of cotton required influxes of capital and often stringent measures of political and legal control. The American Civil War transformed large swathes of Egypt, Turkey, Algeria, Mexico, Peru, Brazil, Central Asia, China, West Africa and central India. In these cotton-growing regions, farmers and capitalists responded by sowing cotton seeds and raising loans on the money markets to invest in cotton gins and presses, railroads, canals and roads. Across the world the demand for new sources of cotton led to upheavals. Russian imperial expansion increased, attracted by the rich cotton-growing soils of Central Asia. Muhammad Sa'id, Viceroy of Egypt, converted 40 per cent of the lower Nile delta into cotton fields and began importing slaves from Sudan. Railways extended into western Anatolia to carry off the cotton crop. In West Africa slaves toiled to satisfy the demands of Manchester; in Brazil credit from the City of London encouraged farmers to uproot their subsistence crops and convert practically every inch of their land to the exclusive cultivation of white gold.[34]

Dalhousie's whirlwind modernisations in India began to pay dividends. Berar, which he had seized in 1853, became one of the most important cotton regions in the world. Work on the railways connecting Bombay to the cotton districts of Nagpur, Gujarat, Berar and Southern Maratha began in 1858. It had to overcome a formidable obstacle, the Western Ghats. The Ghats are not so much a mountain range as 'a precipitous line of cliffs, some 2,000 to 3,000 feet high, forming the vertical wall of the Deccan highland; this line of precipices being only broken by deep ravines, or rifts along its face, with occasionally projecting spurs running down to the Konkan [plain]'.[35]

Every mile completed on the Ghat inclines cost on average 1,667 lives and £70,000. At the peak of construction, 42,000 men laboured to remove 130 million cubic feet of hard basalt rock on *each* incline. In places the Ghats were vertical and devoid of footholds, so workers were lowered on ropes to drill and blast the way through the basalt. Many men were dashed to pieces in the ravines when the ropes snapped, or were killed in gunpowder accidents. Most of the casualties, however, came from cholera or malaria: the monsoon sent down the rain 'not in drops, but sheets of water'. Work stopped on most of the line during the monsoon, when 'every ravine became a roaring torrent bed', but not in the tunnels, where drilling continued non-stop all year round for all of the eight years.[36]

Conditions in the workers' camps were primitive and sanitation non-existent. At some times of the year the rainfall was prodigious; at others it was completely dry. An outbreak of disease could kill 25 per cent of the European officials overseeing the operation and wipe out numbers of workmen 'beyond accurate calculation' (the average number of deaths per mile is probably a gross underestimate). Outbreaks of rebellion and resistance on the Incline were frequent and violent. The Ghats were not the only obstacles to overcome. Massive iron bridges strong enough to resist the power of the monsoon floods had to be built across the great rivers of the Gangetic Plain. An average mile of Indian railway required a ship to bring 600 tons of material from Britain.[37]

This horrendous labour meant that by the early 1860s India was well placed to make up much of the shortfall in cotton in the global markets. By 1863 Indian cotton was reaching the industrial zones of northern England, France and New England, relieving the misery of the cotton famine. 'The effect upon the city of Bombay . . . was . . . electrical.' With railways bringing the white gold to Bombay the city experienced a vertiginous upward trajectory of population growth and economic output comparable to Chicago's in the 1850s.[38]

So sudden was the demand, so high the prices and so stupendous the profits that the city experienced the giddy rollercoaster ride associated with all spectacular bubbles: with money from Manchester sloshing around, crazy speculation was rife, as was conspicuous consumption. Hundreds of thousands poured into the city not only from the hinterland, but from all over Asia and the Middle East. 'Financial associations formed for various purposes sprung up like mushrooms; companies expanded with inflation as that of bubbles,' one witness said of these years. Another remembered that 'there was a new bank or "Financial" almost every day'.[39]

Bombay became the 'City of Gold', the Indian emblem of white-hot modernity. In the words of an American visitor, 'Cotton has built the splendid stores and warehouses, which are unequalled in any city in the East. Cotton has collected the hundred steamers and the thousands of native boats that are anchored in the harbour.'[40]

King Cotton's assumption of power in western India and elsewhere had been as sudden as his deposition in Dixie. Large parts of the world untouched by the boom of the previous decade were utterly transformed by the American Civil War as they were belted into an expanded 'cotton zone'. Increasing prosperity, infrastructure and participation in the networks of global trade and finance called into being other forms of economic activity. Less positively, the sudden expansion of the global cotton belt increased the demand for coerced labour and colonisation. It tied these places to the booms and busts of the world economy and to the caprice of nature. With so much money, labour and land invested in one commodity, they were dangerously vulnerable to depression in Europe or a local food crop failure. Either could spark dearth and famine almost overnight. The high price paid for cotton on international exchanges in Europe and the US meant that textile manufacturers in places like India, Egypt or Brazil could no longer afford their locally grown cotton; many went out of business in a process of de-industrialisation. Farmers all over the world borrowed money at high interest rates to enter the booming global market for the first time. All very well when prices were rising during the dislocation of the Civil War years – but being so enmeshed in debt left them exposed to fluctuating prices.

Little wonder that information became as precious a commodity in Bombay and Cairo as it was in New York and Liverpool; Julius Reuter extended his operations to feeding the latest commercial data to Indian speculators. People anxiously scanned newspapers for information fresh off the ship or cabled inland: the latest news of battles in America and market movements in Liverpool – anything that would tip prices one way or the other, making or unmaking their fortunes. With submarine telegraph cables in the Mediterranean by this time, and faster ships, messages could be exchanged between India and Britain in a mere fifteen days.

The Times had its eyes fixed on the ways in which conflict in America was reshaping the planet in an increasingly interconnected age. Throughout the 1850s, it said, Britain and the US had been yoked together. But they had become too dependent on each other. The textile giants of Manchester were addicted to cheap cotton, happily oblivious to the political storm that was gathering. In addition 'our workshops were employed upon American orders, and English capital found in the wants and products of the United States the most eligible sphere for its exercise'. The result had been a heyday of prosperity. But times had changed: 'As America, with her prohibitory tariffs, her smoking mounds of burning cotton, her impoverished people, and her inevitably approaching bankruptcy, ceases to be able to buy or sell, she loses her importance in the eyes of a commercial people.' But, continued the editors, if Britain had been utterly dependent on America in the '50s and was suffering a hangover from the boom in 1862, 'it should be a relief to us to discover that in this time of our need the good seed we sowed, and harrowed, and watered in the Far East is springing up and bearing fruit'.[41]

In other words, the Civil War and the Morrill Tariff was beneficial, in the judgement of *The Times*, because they forced the cotton lord to open up immeasurable tracts of land for cultivation, and the manufacturer to conquer new markets in India and China. The American Civil War was momentous for Asia. Not only did farmland give way to cotton plants and other cash crops, but British merchants shifted

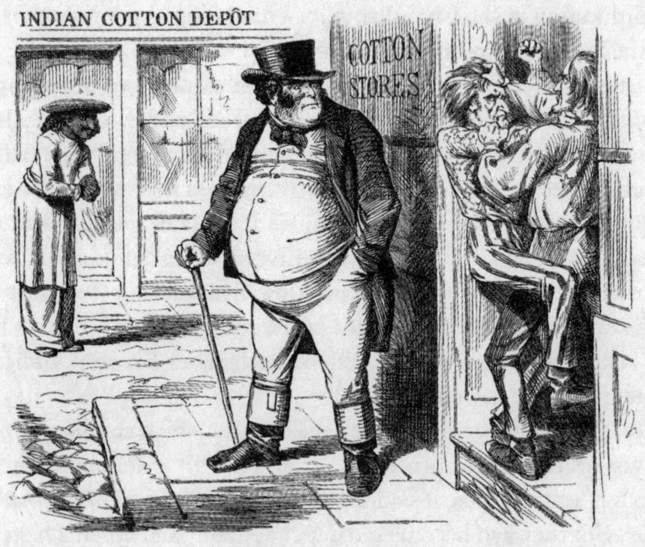

OVER THE WAY.

MR. BULL. "OH! IF YOU TWO LIKE FIGHTING BETTER THAN BUSINESS, I SHALL DEAL AT **THE OTHER SHOP.**"

John Bull discovers a new source for cotton in a Punch *cartoon*

attention from the West to the East, scenting in India and China lucrative new opportunities. The next phase of the heyday of the '50s would surely be enacted in Asia. British papers began to see the Treaty of Tientsin and Elgin's occupation of Beijing as providential. They marvelled at what they saw as the 'spontaneous growth of our commercial relations with China. Hankow, the very centre of the Empire, a city which but yesterday was vaguely discussed as if even its existence might be doubted, has already become a mart of British trade, having its "Price Current" and "State of the Markets" and its regular column in the *China Overland Trade Report*, where it ranks with Shanghai, Hongkong, and Canton'.[42]

In the race for the Golden Fleece of the enormous Asian markets, the British forgot the rush of enthusiasm for the Taipings that had briefly burst forth in 1860–1. Now the rebels were regarded as bloodthirsty fanatics. Their worst crime, seen through the commercial lens, was to threaten the enormous tea and silk markets at Shanghai and

Ningbo. The empire of tea could not be placed in jeopardy as the empire of cotton had been. *The Times* wrote that the 'dragon who threatens us and our golden apples should be killed by somebody'. The dragon was the Taiping movement; the 'golden apples' meant tea and trade; and the 'somebody' was the Qing dynasty. The British government began to arm the Chinese imperial forces and encouraged mercenaries to join them. 'We have almost as much material interest at stake in the battles between the Taepings [*sic*] and the Imperialists in China as we have in the contests between Unionists and Secessionists in America,' said *The Times*. The British press began to cheer on the Qing forces, seeing each imperial military victory as a victory for trade that vibrated all the way to 'the tall chimneys of Lancashire'.[43]

Millions of people in Asia, Africa and South America had American enslavers to thank or curse. And the South could only watch as others rushed to supply the factories of the industrialised world with their precious crop. Any hope that Europe would come to the aid of the Confederacy faded as ships from Egypt, Brazil, Turkey and India arrived in Liverpool. Much later William Seward visited the Taj Mahal and immediately afterwards a cotton-gin in Agra. 'From the tomb of the Mogul monarch of Old India, Akbar,' he said, 'we passed to the tomb of the pretended monarch of America, King Cotton.'[44]

The global ramifications of the Civil War are easy to trace in Bombay or Alexandria. Its effects on the balance of global political power are harder to discern. On 22 September 1862 two unrelated things happened that would have lasting consequences, even if their significance took a while to become apparent. Lincoln announced the first step in emancipation, and Otto von Bismarck became Minister President of Prussia.

Bismarck came to power and surveyed a European order that had been shattered by the Crimean War and Italian unification. The power that Russia, Austria and Britain once wielded over the Continent was at an end, and it was Prussia's chance to seize the moment and unify Germany under its leadership. A few days after he assumed power, Bismarck told the Budget Committee of the Prussian Landtag: 'Prussia must build up and preserve her strength for the favourable moment, which has already slipped by several times . . . The great questions of the day will not be resolved by speeches and majority decisions . . . but by blood and iron.'

In the eleven years that separate the Great Exhibition in Hyde Park and the Garibaldian riot in the same place, Britain had reached the pinnacle of its power in the world. By the early 1860s Britain was, without question, the world's premier superpower, its banker, money-lender, investor, insurer, warehouser, wholesaler, shipper, newsagent and engineer. During that time it had intervened, often with devastating effect, to open up the world for trade and progress. The massive expansion of the global economy in the 1850s gave Britain its sense of mission in leading the march of progress across the planet under the banners of free trade, unbridled communication and parliamentary government. As far as the British were concerned, their values had triumphed, while the rest of the world languished under slavery, tyranny, economic backwardness or the anarchy of democracy.

The kind of world envisaged depended upon unrestricted mutual exchange and an international order based on peace among the great nations. Under Bismarck, Prussia's imperial ambitions threatened this world system. The telegraph, national print cultures and railways did not diffuse power and foster cosmopolitanism, but concentrated it and gave birth to nationalism and imperialism. With its vast army at a state of constant readiness and its industry shielded behind a tariff barrier, Prussia became a military and industrial superpower on a colossal scale. Britain could only watch impotently from the sidelines as Bismarck built up a large, modern army and used it to wage war on Denmark, then Austria and then France. Nationalism had become the dominant force in Europe by the 1860s, and war was held up as a legitimate way of achieving its ends.

This was not the future of free trade and internationalism envisaged in 1851: this was a world of militarism, economic nationalism and brute struggle for existence among modern nation states. And if the future was being determined by blood and iron – hard power – the soft power of democracy was also becoming a major force. On that question too, Britain was out of kilter.

In his Newcastle speech, Gladstone said that the Confederacy was making a navy. In fact, it was British shipbuilders that were doing so. On 29 July the sleek and dangerous CSS *Alabama* snuck out of Liverpool, where she had been built in secret by John Laird Sons & Co., to begin a successful career hunting down and destroying the Union's merchant marine in the Atlantic. The powerful Confederate commerce raiders built on the Mersey shone a very poor light on Britain's claim for moral leadership of the world.

In the same way, the response of the press and politicians to Lincoln's Emancipation Proclamation began to seem self-serving and cynical. After 22 September, the Union could claim that it was fighting for human rights and that anyone who supported the Confederacy or called for intervention was helping to prop up the barbarous institution of slavery. 'Are Britain and France to throw their protection over the insurgents?' asked William Seward. 'Are they to enter . . . into this conflict, which . . . has . . . become a war between freedom and human bondage? Will they interfere to strike down the arm that so reluctantly but so effectually is raised at last to break the fetters of the slave, and seek to rivet anew the chains which he has sundered? . . . Is this to be the climax of the world's progress in the nineteenth century?'[45]

The British public knew the answer. People could see which way the tide of history was running. Throughout the 1850s it was held that free trade and free communications were the forces that would reshape the world. But the Civil War brought home an uncomfortable fact. Free trade did not blast away ancient crimes such as slavery. On the contrary, it appeared to rejuvenate them. The market did not discriminate between sugar and cotton grown by slave or free labour. Since the advent of full-throttle free trade, sugar-producing slave societies in Cuba and Brazil had flourished at the expense of the British West Indies, where the Afro-Caribbean population had been emancipated. In a world of laissez-faire economics, slave labour always priced free labour out of the market when it came to the intensive production of cotton, sugar, rice, tobacco, coffee and other tropical staples. As the pre-eminent trumpets of free trade, the *Economist* and *The Times*, constantly reminded readers, the market had settled on slave-grown cotton and it was contrary to the first law of economics 'to interfere with the results'.[46]

Devotees of free trade believed that when it came to slavery 'providence will, in its own way and its own time, work out a cure . . . because we believe improvement, progressive, though always slow and often interrupted, to be among the laws by which the earth is governed'. Those words were by one of the most respected British political economists, Nassau Senior, and they exemplify the problem of trusting to the 'invisible hand' of the free market to resolve human problems. Writing in 1855, Nassau Senior could not promise African-Americans toiling in the cotton fields of Louisiana that emancipation would come in the lifetimes even of their grandchildren.[47]

The Morrill Tariff flushed out Britain's instinctive sympathies. Did free trade trump all other considerations, particularly moral considerations? The answer, for many Britons, was yes, it did. 'Upon principle,' commented *The Times* discussing the free trade implications of the conflict, the South 'have . . . manifestly the best of it.' It is not a judgement upon which history looks very kindly: a far greater principle was at stake. The laissez-faire rhetoric of the South was like a siren song, and the issues contested in the Civil War were regarded through the hard, unsentimental prism of commercial self-interest. According to August Belmont, banker and Union agent in Europe, the British wilfully ignored the fact that the war was fought against 'a rebellious slave oligarchy', closing their ears to questions of principle and dwelling instead on commercial questions. The wife of a prominent Liberal politician told Belmont: 'I am sorry to say, we have been found wanting in the present emergency, and principles have to yield to interest'.[48]

But the 'inexorable tendencies of trade' and the 'mechanical law' of political economy could not free human beings from bondage. The Civil War showed that only physical force – not the invisible hand – was strong enough to shatter slavery. *The Times* and the *Economist* had long argued that the battle between free and slave labour would be settled on the battlefield of economics; how forlorn that nostrum seemed in the age of Gettysburg. One by one the assumptions that lay behind the progress and utopian dreams of the 1850s began to gurgle down the drain.

If people in Britain sniggered at the busted experiment of republican democracy when America fell apart in 1861, by the end of the next year it was becoming apparent that a new and more powerful example of democracy was being born in the US. Support for the Union began to build in Britain; the Civil War was regarded as a battle between good and evil that would define the modern world. For liberals and radicals the United States of Lincoln began to become a model of humanity, equality and civic virtue. The stirring example of republicanism and democracy in America after the Emancipation Proclamation and, the following year, the Gettysburg Address, gave a powerful boost to the flagging reform movement in Britain. Bismarck's Germany also offered salutary examples of state action at work, particularly in the field of public education.

'The initiative that today belongs to you might not be yours tomorrow,' Garibaldi warned the British public in his open letter in October. Ideals of nationalism, state action and democracy stood in stark contrast to impersonal and unemotional forces of laissez-faire and free trade. Britain, which had made the political weather in the 1850s and turned upside down the lives of hundreds of millions of people in China, India, Japan and Australasia, was less assured of herself in the following decade. Even the stirring tales of the suppression of the Indian Rebellion began to pall as the surge of emotions started to wear off, replaced with a nagging sense of shame. A correspondent for Ohio's *Cleveland Leader* travelling in India a few years later found that 'When all the civilized nations were shocked at the barbarity' of the reprisals meted out by the British, 'the perpetrators became cautious about mentioning the subject to strangers'. A number of books published in the 1860s began to point out the brutality of the British and the unblushing criminality of the looters and vandals. One of the first accounts of the Rebellion that offered some nuance, and even went so far as to hint that ruling India had corrupted the British, was William Howard Russell's superb and bestselling diary of his months as a war correspondent in India. Other writers debunked the outrageous rumours of atrocities committed by rebels and emphasised the darker side of the British. 'It is in India, when listening to a mess table conversation on the subject of looting,' commented Charles Dilke, 'that we begin to remember our descent from Scandinavian sea-king robbers. Centuries of education have not purified the blood.' Perhaps the Rebellion did not show the very best of the British national character after all; perhaps it showed the opposite.[49]

Elsewhere, the British seemed to have run up against their limits. As he watched the Summer Palace burn on the outskirts of Beijing in 1860, one British observer mused that it was an act of creative destruction. A new China would be born in the ashes of burnt fripperies and vain luxuries, one that conformed to the rule of international law and which could 'keep pace with the march of progress'. China was the test case for the utopian notion that free trade and technology could raise people out of poverty and backwardness. Prompted by the American Civil War, the British attempted to intervene in the Chinese civil war. As they zigzagged from support for the Qing dynasty to the Taipings and then back again, it was clear that commercial self-interest was driving their policy, not principle.[50]

Even after the occupation of Beijing at the cost to the British treasury of over £10 million, China was worse off than ever thanks to Western meddling. The great hopes that the vast population of China would become an inexhaustible market for British manufacturers were never fulfilled. Soon after British troops withdrew, Prince Gong assumed power in a palace coup. Gong was welcomed by the West because he apparently wanted to deal with foreign powers and modernise China. In Japan, modernisation meant rejuvenating every branch of industry and trade so as to compete on an equal footing with the West. Gong and his advisers, by contrast, saw modernisation as getting enough military technology to defeat the Taipings and, in time, throw out the intruding Western barbarians. He had no desire fundamentally to change China. This was particularly true of the pre-eminent symbol of modernity and economic expansion, the telegraph. 'We have examined the application for techniques for sending [information] via copper wires in China,' said Gong, 'and found that they are extremely inconvenient; therefore they can never be introduced.' When the French ambassador presented him with books on telegraphy, the prince politely returned them unread.[51]

The occupation of Beijing, then, was an important moment not just for China but for Britain: it represented both a triumph and a defeat.

British military skill was vindicated, but Britain's claim that it could rejuvenate humankind by means of the invisible hand of free trade was increasingly in doubt. Britain in the 1860s could only watch as the tide of history began to pull in the opposite direction. Slowly, but perceptibly, Britain's power began its long recession from the peak of the 1850s. Where it had been self-confident and energetic, it was now hesitant and increasingly isolationist.

At exactly the same time that Elgin bullied Gong into ratifying the Treaty of Tientsin, Garibaldi had proclaimed Vittorio Emanuele as King of Italy. Despite its boast that it arbitrated the affairs of Europe, Britain had been ineffectual in its bid to encourage unification and contain France. Its power in Europe was exposed as an illusion; Bismarck took note. In the same way, any attempt to throw its influence into the American Civil War was rebuffed with contempt.[52]

In truth, in trying to master the world Britain had become so overstretched, with troops on every land mass, that it could not influence affairs in its own backyard. At the same time as its political power was waning, its industry began to experience serious competition from the world's waxing industrial nations, Germany and the United States, countries that had rejected the British free trade gospel. The 1860s saw the beginning of a flow in global industrial, financial, military and ideological power, barely discernible at first, away from Britain. Where it had few serious rivals on the world stage in 1851, by the 1860s it had plenty.

Italian unification is known as *Risorgimento* – the Resurgence. The word defined the 1860s for many countries that to this day look back on that decade as their foundational moment. The decade belonged to Lincoln, Bismarck and Garibaldi with their brands of romantic nationalism, state activism and democracy. The assumptions of the previous decade that the world could be regenerated organically by trade, migration and technology had been replaced. In Britain, once so confident of its ability to change the world, the ideals espoused by the troika of world leaders began to shape politics and awaken the cry for political, social and educational reform. Garibaldi had it right in his open letter to the British: the initiative they possessed was easily squandered.[53]

The ruins of an abandoned telegraph station on the remote far-western tip of Ireland at a village called Crookhaven are testament to that change. Where once the news of the world flowed out of Britain, now it flowed in, and it shaped politics.

From 1862, interest in the American Civil War intensified as it became a struggle for freedom. People waited for the next instalment of an epic struggle with which they had come to identify so strongly. Hungry for the latest information, the London newspapers chartered steamers to wait off Roche's Point in County Cork to intercept the news carried on liners bound for Liverpool from New York or Boston. These ships would race the headlines back to Queenstown, where they would be telegraphed to Fleet Street. The daily papers were muscling in on Julius Reuter's act. He responded by building a private telegraph line from Cork to the remote village of Crookhaven, eighty miles further west. Now his steamer could get the news a few hours before any other news service.[54]

From New York the Associated Press telegraphed breaking news to Cape Race, where it was picked up from the sea by a mail steamer. When it reached its rendezvous point with Reuter's little steamer it let off a rocket. The clock started ticking. Reuter's men searched for the phosphorous flare that marked the tiny bobbing news canister. Safety-netted and brought on board, the clerks unscrewed the waterproof lids and rendered the news into Morse code as the ship bounced back over the waves to Crookhaven. As soon as they reached dry land they wired the news to Reuters HQ, which then distributed it to newspapers in Britain, the Continent, Russia and India. The Reuters men scooped the news from the ocean, and scooped their rivals by a few hours, perhaps minutes.

Epilogue
1873

Earth and sea . . . seemed to be at [Fogg's] service; steamers and trains at his beck and call; wind and steam . . . joined forces in furthering his voyage.
Jules Verne, *Around the World in Eighty Days* (1873)

Jules Verne's Phileas Fogg is a man of the modern world with his ruthless, monomaniacal drive to become the master of time. In *Around the World in Eighty Days*, Verne memorably describes Fogg's 'regular pace marking the seconds like the pendulum of an astronomic clock'. The world is there to be conquered, not relished. When he arrives in Bombay, for example, he does not even think to visit any of the port's wonders: 'not the Town Hall, the magnificent library, the forts, docks, cotton markets, bazaars, mosques, synagogues, Armenian churches, or the splendid pagoda of Malabar Hill with its twin polygon-shaped towers. He had no wish to see the masterpieces of the Island of Elephanta with its mysterious *hypogea* . . . not the Kanheri Grottoes on Salserre, those wonderful ruins of Buddhist architecture. No, nothing!'

Fogg's watchword is 'punctuality' and his Bible is *Bradshaw's Guide*, which gives him all the information he needs to know about the world: the inviolable arrival and departure times of railways and ocean liners. It is left to Passepartout, Fogg's more curious manservant, to cast his gaze on exotic sights and deviate from the rigid constraints of the timetable that dictates the race around the world, with comically disastrous results.

Jules Verne had a number of inspirations for his novel. One globetrotter who suggested himself as a model for Phileas Fogg was the eccentric American businessman George Francis Train, who claimed to have gone around the world in eighty days in 1870. Shuttling around the planet in record time was, for 'Citizen' Train, a tangible manifestation of his overarching business philosophy. By travelling 'under stress' he wanted to show how small the world really had become so that businessmen could 'reduce the long months and days consumed on voyages from country to country'. In his long and erratic business career he dedicated himself to making things run faster: he set up street tramways in British cities, financed railroads in America and invested in steam shipping lines. 'I have lived fast,' Train recalled. 'I have ever been an advocate of speed. I was born into a slow world, and I wished to oil the wheels and gear, so the machine would spin faster . . . I wished to add a stimulus, a spur, a goad – if necessary – that the slow, old world might go on more swiftly "and fetch the age of gold" with more leisure, more culture and more happiness. And so I put faster ships on the oceans, and faster means of travel on land.'[1]

Another inspiration for Verne was the British entrepreneur of tourism, Thomas Cook. Cook first came to prominence bringing hundreds of thousands of working-class visitors to the Great Exhibition in 1851. In the following decade he led tours to the most popular destinations for middle-class British sightseers: Italy, Switzerland and the United States. Jules Verne read a newspaper advertisement publicising Cook's latest venture as he was planning his novel: an announcement of annual tours around the world. Cook and his pioneering clients set off for their eleven-month 25,000-mile jaunt in the summer of 1872, a few weeks before the fictional Phileas Fogg began his race against time.

In many ways the technological and political breakthroughs that made an express world tour entirely by steam power possible at this time had their genesis in the heyday of the 1850s. The new generation of global tourists found the excursion safe, reliable and comfortable; telegraphs criss-crossed the planet, linking more or less instantaneously places that had twenty years before been separated by months. The vision conjured up at the time of the Great Exhibition of a world connected by electrical current and shrunk by railways, steamships and trade had been realised.

But gone was much of the utopianism and faith in unlimited progress that had defined the 1850s. In this Epilogue we jump forward a decade to look back at the mid-century heyday; only in doing so can the character of that period of change stand out in sharper relief. Economic expansion had continued in the intervening years. But the energy and passion that had sustained that remarkable surge of progress had dissipated.

The world of 1873 was darker and much less certain, a time of bitter competition between nations for resources and markets, of aggressive imperialism and economic nationalism. The advent of instantaneous communication and freer trade was supposed to have fostered internationalism, universal enlightenment and peace. The opposite seemed to be the case by the time Thomas Cook set out to encircle the earth. And by the time he returned to London in 1873 the world had been plunged into fresh turmoil.

Cook's real tour, and Phileas Fogg's fictional one, followed a path around the globe that bound together many of the regions that have been described in *Heyday*: Chicago, the prairies of Illinois, the Platte valley and the Great Plains; San Francisco, Yokohama, Shanghai and Hong Kong; central India and Suez. This line of communication and hops from port-city stepping-stone to stepping-stone represented what one world tourist called the 'Anglo-Saxon highway'. Traversing this great planetary circuit would have been virtually impossible before 1851 and arduous until four major breakthroughs occurred within a few months of each other. The first came into being on 10 May 1869, when the American Transcontinental Railroad was ceremonially completed at Promontory Summit, Utah, six years after ground was first broken in Sacramento. This great engineering feat was a dream of the 1850s. Like so many other things, it had been delayed by sectional divisions in the US and then by the Civil War.[2]

When Cook brought his clients from Liverpool to New York on the 'Atlantic Ferry' he could have hustled them to San Francisco by rail in a mere seven days had they not preferred very un-Phileas Fogg-like sojourns to tourist hotspots like Niagara and Chicago. Thomas Cook described the views as the train chuffed through Nebraska and Wyoming: 'Prairie fires on all sides, antelopes, wolves and Indians kept us in a state of almost constant excitement. The Sioux tribes were evidently on the move to Southern quarters, as they were mounted, in great force, on both sides of the line.' Then there were the Rockies, which 'disappointed us', and the astounding views of the Sierra Nevada, which Cook complained were mainly obscured by the fifty miles of snow sheds that protected the line from the elements and avalanches.[3]

But still, Cook evidently wanted to impress on would-be world tourists that long-distance travel had suddenly become luxurious. Many more people would read books about world tours than could ever hope to go on one. But, as in *Around the World in Eighty Days*, the chief attraction of such books was not so much the topographical detail as the miracle of modern communications and the phenomenon of speed. Danger, and adventure, could be bypassed or glimpsed with interest while you munched on the choicest morsels. 'Along the line of railway over the prairies, the Rocky Mountains, and at the summit of the Sierra Nevada,' Cook and his clients were treated with 'every variety of the best food, fruits etc'. The world in the 1870s was being seen and experienced in an entirely new, and to our eyes, modern way.[4]

After sightseeing in California, a voyage of just over three weeks brought tourists from San Francisco to Tokyo Bay* and the port of Yokohama. For Phileas Fogg, Japan was as essential as the American Pacific railroad in effecting a record-breaking speedy passage around the world because ships crossing the Pacific needed to take on coal there. The opening of Japan to foreign tourists was the second of the developments that made the express tour of the world possible. Although Matthew Perry had 'opened up' Japan back in 1854, Western visitors to Japan risked their lives treading on the soil of the forbidden empire. Modern Japan had been born in a bitter civil war in 1867/9 that resulted in the overthrow of the shogun and the restoration of the emperor to full power. The intense anti-foreign feelings of the

* Edo was renamed Tokyo in 1868.

last years of the Tokugawa shogunate gave way to a desire to learn as much and as quickly as possible from the West so that Japan could take its place in the first rank of world nations. Yokohama was expanding quickly to become one of the world's biggest and busiest ports.

The first tourists were entranced by Japan and the sense of newness and exoticism they experienced there. Most visitors in the early 1870s recorded deep pangs of regret on leaving. The empire, at this point, was a place of wonders and fascination for the West; few places on earth were more photographed during this period of rapid transition from ancient to modern. The Japan that emerged from its civil war, eager to learn from the West and modernise, was regarded as justifying Western intrusion into Asia. 'We look back upon the hills . . . terraced and cultivated in crops . . . and watch the fertile valleys that shelter an industrious, contented and happy people, as they fade from sight,' wrote one tourist. 'Japan is a country with a future.'[5]

Yokohama, an international port guarded by Western troops, was a stepping-stone on the 'Anglo-Saxon highway'; so too was Shanghai. From then on, until tourists reached the Mediterranean, in the words of an American traveller, 'almost all the way it is England, England, England'. According to Verne in *Eighty Days*, 'Hong Kong looks exactly like a busy town in Kent or Surrey transported through the globe to the Chinese locality.' The exception was the crowds of Chinese, Japanese, Jews, Indians, Parsees and Europeans, but this made it 'all very much like Bombay, Calcutta or Singapore . . . There is thus, so to speak, a trail of English towns all round the world.'[6]

When circumnavigating tourists disembarked their ship at the Hooghly River in Calcutta they boarded the first train they had seen since San Francisco. After almost two decades, during which the network had expanded thousands of miles through jungle, over rivers, up mountainsides and across the plains, Dalhousie's revolution culminated in March 1870 (ten months after the completion of the American Transcontinental Railroad) when the line linking Calcutta to Bombay was completed. It slashed journey times between West and East. On his world tour in 1871, US Congressman James Brooks spent the bare minimum time on the subcontinent: 'India "done" up!' he marvelled in the style of Phileas Fogg. 'One week in it! – from

Calcutta to Bombay, 1,420 miles, in sixty-two hours! Who can beat that!'[7]

Midway between the completion of the American Transcontinental Railroad and the Indian trans-subcontinental railway, the fourth of the breakthroughs that enabled worldwide travel, the Suez Canal, opened to the shipping of the world. As far as tourists were concerned, its completion did not speed things up because there had been a rail link connecting the Mediterranean and Red Sea since 1858. The Canal, however, gave cargo ships a short-cut to Asia, bypassing the 3,300-mile route via the Cape of Good Hope. Even before then, however, trade between West and East had been substantially changed. In 1866 the steam cargo ship *Agamemnon* returned from Foochow to Liverpool in just fifty-eight days, cutting thirty days off the world record set by a clipper. Until then, steamships could not compete with clippers; they had to give up too much hold space to store fuel and they made frequent stops to re-coal.

Agamemnon changed all that. The long-distance trader Alfred Holt of Liverpool equipped her with his new fuel-efficient compound steam engine. She could travel 8,500 miles without needing to take on coal. And not only was she considerably faster than the fastest clippers, but she could carry three times more freight than them. Long-range, steam-powered cargo ships and the Suez Canal opened up Asia even more to the trade of the world and a string of ports from Gibraltar to Yokohama boomed as a result. The lives of millions of peasant farmers were transformed because of the advent of the bulk carrier, the railway and the Suez Canal: freight prices dropped, encouraging the intensive farming of cash crops like cotton, rice, tobacco, coffee, rubber, jute, vegetable oils and so on for the world market. Distance had been vanquished; so too had time.

When he was in Penang on his world tour, Thomas Cook went to a telegraph office and spent £5 on a message two words long: 'All's Well'. Sent at 10 a.m. local time on 21 December 1872, it was read by its recipients in London at 7 a.m. GMT that same day.

The world Cook travelled in 1872–3 was girdled by submarine telegraph cables. He might have sent his telegram from anywhere on his trip – San Francisco, Yokohama, Shanghai, Hong Kong, Singapore,

India or Suez. Had he ventured to Australia he could have done the same. He chose Penang because he happened to be there at Christmas time and wanted to send his message in time for the holiday.

The breakthrough in intercontinental communication came after a hiatus caused by the Civil War. The dauntless Cyrus West Field had never given up on his dream of an Atlantic cable. By the end of the Civil War the science of deep-sea telegraphy had made giant strides and the leviathan *Great Eastern* provided the perfect cable-laying ship. In London the former cotton tycoon John Pender facilitated the amalgamation of the Gutta Percha Co., the maker of telegraph cable cores, with Glass, Elliot & Co., the maker and layer of cables. The new company, called the Telegraph Construction and Maintenance Co., Ltd (Telcon), made a cable that was technologically far in advance of its predecessor of 1856.

After the successful laying of the Atlantic Cable in 1866 the work of wiring the world accelerated. John Pender laid a series of submarine cables between Porthcurno in Cornwall and Bombay. That line was complete by 1870. With every technological breakthrough the process sped up. A year later the spider's web of cables extended over land and sea to Penang, Singapore, Hong Kong, Shanghai, Nagasaki and Vladivostok. Two years from then and the worldwide network embraced Adelaide, Rio de Janeiro, Havana and Panama City as well. By then London was in instantaneous communication with 20,000 towns and cities around the world over a network of 650,000 miles of overland and submarine cable.

Criss-crossed by shipping lanes, railways and telegraph wires, there were few places off the grid or sheltered from the blizzard of news. On his world tour in 1871 Congressman Brooks looked forward to the 'outer darkness' of the Pacific Ocean as he sailed between San Francisco and Yokohama. But his holiday 'beyond the reach of steam or telegraph' was ruined when in the loneliness of the ocean his ship encountered a Pacific Mail steamer carrying world news that had been telegraphed to Shanghai a few days earlier. Welcome to the modern world.[8]

From the window of a train or steamer, the tourist's view of the world was reassuringly positive. New cities, shiny railway lines, faster ships and mile upon mile of telegraph cable were testaments to the most explosive period in the world's history – two decades of breakneck growth and tumultuous change.

The well-heeled globetrotters revelled in the warm glow of modernity and enjoyed, like all tourists, superficial encounters with the lands they raced through and the people they met. The 'Anglo-Saxon highway' made this possible: the way was marked with Western railways, Western ships, Western concessions and Western colonial cities. Only by choice did one have to detour from the road and actually *see* the world. Thomas Cook took his tourists on a sightseeing excursion to the old city of Shanghai, but was nauseated: 'Narrow, filthy, and offensive streets, choked and almost choking bazaars, pestering and festering beggars in every shape of hideous deformity; sights, sounds, and smells all combine to cut short our promenade of the native city, to which no one paid a second visit, and the chief part of our stay at Shanghai was spent in the American, English, or French concessions.'[9]

There were many other places in close proximity to the global highway where the inhabitants had had their lives uprooted and turned over by the plough of progress. For Thomas Cook, the scene from the Transcontinental Railroad was picturesque, especially the view of the mounted Sioux warriors. Little thought did he give to the ways in which the railroad had pressed upon the already beleaguered inhabitants of the Great Plains as he gazed out of the window.

Throughout the 1860s war had become a feature of the Plains. The construction of the railroad threatened the Native Americans with final extinction. There were numerous Sioux and Cheyenne ready to resist. In 1867 attacks mounted, including an act of sabotage at Plum Creek, Nebraska, where Cheyenne braves derailed a locomotive. Throughout the summer, raids on railroad workers continued. Grenville Dodge, chief engineer of the Union Pacific, declared: 'We've got to clear the damn Indians out or give up building the Union Pacific Railroad. The government may make its choice.' Henry Morton Stanley (the man who later found the lost Livingstone in Africa) was with General William Sherman when he told a council of Indians at North Platte: 'you cannot stop the locomotive any more than you can stop the sun or moon, and you must submit . . . If our people in the east make up their minds to fight you they will come out as thick as a herd of buffalo, and if you continue fighting you will all be killed . . . We now offer you this, choose your own homes, and live like white men, and we will help you all you want.'[10]

In the following year more federal troops came west to protect the railroad and the telegraph against sabotage. Elsewhere in the world, peaceful interactions between indigenous peoples and incomers broke down. In Minnesota a vicious war erupted with the Dakota after years of treaty violations in 1862. It culminated in the hanging of thirty-eight Dakota, the largest mass execution in American history. New Zealand, which so resembled Minnesota in its early settler period, saw war between Europeans and Maori at the same time. Until then the Maori were formidable, with their adept use of firearms and their defensive fortifications, the *pa*. Westerners had also been dependent on them for foodstuffs, fuel and building materials. But with the European population doubling in size in the 1850s and again in the 1860s, the discovery of gold in Otago and the take-off of wool production, competition for land and resources became ever more intense and ruthless. The balance between Maori and Pakeha in terms of demographics and power tipped towards the newcomers. As in Minnesota, violence arose over conflicting notions of land ownership and definitions of sovereignty. When a minor Maori chief of the Te Ati Awa tribe sold 600 acres of land in the Taranaki district to the Crown, Wiremu Kingi, paramount chief of the Te Ati Awa, vetoed the sale. Settlers were hungry for land, to be sure, but the colonial governor was determined to affirm the legal authority of the Pakeha government, even over so trivial a land deal. Troops occupied the land. Equally determined to halt encroachments on Maori lands, the Te Ati Awa made its own stand. The incident sparked a series of wars between imperial troops and the Maori. British regular troops, veterans of the Crimea and India, armed with Enfield rifles, Colt revolvers and modern artillery and adopting a scorched-earth strategy, beat down determined resistance. By the end of the conflicts in excess of 4 million acres had been confiscated from the Maori.

The New Zealand experience was typical of a pattern emerging around the world. In North America and Australasia, in southern

Africa, Jamaica and South America the cascade of settlers in the 1850s resulted in numerous brutal wars on the frontiers in the succeeding decade. Once it was held that frontier lands represented the vanguard of progress: interactions of pioneers and indigenous people would carry the civilising progress into the 'wilderness'. In the 1860s, however, a new attitude ruled: civilisation had to be imposed by force. The uncompromising nature of this attitude reached its height in another region this book has ventured to: the Caucasus.

The mountains provided another popular destination for Western tourists. The wild lands defended for so long by Schamyl and his followers became the resort of sportsmen and mountain climbers in the 1860s. There was little trace in the awe-inspiring grandeur and silence of the Caucasus, however, of the proud peoples of the mountains or the wars they had fought. Visitors could travel through the once-lawless terrain for days and not see a living soul. After the final defeat of Schamyl and Circassia by the Russians as many as 2 million people were deported from the region in the brutal endgame of imperial conquest. They were crammed on ships and taken across the Black Sea to Turkey in appalling conditions. Those who survived the horrors of the voyage found themselves dumped in vast and insanitary refugee camps.[11]

In other places on the worldwide circuit, the impact of Western expansion was often harder to detect. In Bombay, for example, the cotton boom of the Civil War years transformed the city. Property prices went through the roof and the city was turned into a noisy building site. Anecdotes almost identical (and equally unbelievable) to those told during the Californian and Australian gold rushes entered circulation: once-impoverished peasant proprietors carted foodstuffs into boom-town Bombay on silver-rimmed cartwheels, their bullocks shod with golden shoes. 'No one ever drank anything but champagne in those days.'[12]

Indians enriched almost overnight contributed their good fortune to founding new schools and hospitals. 'Not even Chicago ever took a greater leap than did Bombay,' reported the correspondent of the *Cleveland Leader*. The citizens and government of the city resolved to give Bombay the architecture that befitted its new status, unsentimentally destroying ancient landmarks in the quest to build a modern world city. These grand new public structures boldly and brazenly lined up on the waterfront as if on parade, facing out to sea to impress upon Western visitors Bombay's new global importance.[13]

'The Bombay of today,' wrote a tourist from Ohio, 'is ruled by a potentate we once knew in America, "King Cotton", and his sway is here as absolute as it ever was in Charleston or New Orleans.' But he was a capricious monarch. The end of the Civil War in April 1865 sent cotton prices over the precipice at once in expectation of the re-entry of Southern plantations to the global market. Property prices in Bombay gurgled down the drain; banks failed; millionaires were reduced to penury; and businesses toppled like skittles. One citizen of Bombay surveyed the wreckage: 'Never had I witnessed in any place a ruin so widely distributed, nor such distress following so quickly on the heels of such prosperity.'[14]

By the time Thomas Cook and other world travellers arrived in the city it had recovered from its crash and possessed the sheen of modernity they were looking for. But the rollercoaster experience of Bombay dramatised a situation facing millions around the world. The global demand for commodities, combined with the telegraph, railways and the Suez Canal, tied peasants and city workers to the topsy-turvy world market, a bewildering and painful adjustment. In Berar, for instance, as an Asian newspaper said, 'A pressure unknown before was put upon the people to grow cotton' and abandon the diversified agriculture they had relied upon. Weavers and other workers once involved in local production of textiles had been forced to toil in the cotton fields because imported Manchester piece goods were underselling Indian cloths. Until Dalhousie took over this region in 1853 it remained far remote from the wider world, its economy village-based. Within a few years it had become one of the richest cotton regions on the planet.[15]

On a tour in 1867 the British cotton commissioner reported that Berar's inhabitants were perplexed that their lives had become bound up with the periodic 'great rise and the sudden fall in the price of cotton' on the world market. They found their existence from day to day determined by the 'throbbing pulse' of electricity sent down the wire from the Manchester Exchange. Because they were sidelined from the

Khamgaon cotton market, India, in 1870. The railway that connected the region to the global market is visible in the background.

instantaneous movements of information and capital along the transcontinental telegraph lines their horizons were still village-based. But they were inextricably bound to faraway forces of which they had little knowledge and absolutely no control. The precariousness of monoculture was made painfully clear in the later 1870s. Plummeting cotton prices combined with rising food prices plunged Berar into famine. The history of Berar exemplifies the experience of cash crop farmers the world over, be they growers of coffee, tea, jute, rubber or cotton.[16]

The tourists who blazed a trail across the planet in the early 1870s spoke the language of progress: they marvelled at the annihilation of time and distance and expressed wonder at the suddenness of it all. But gone were the expressions of radiant hope and faith in the collective march of the human race to perfectibility. The mood of 1851 was alive with expectation and optimism; that of 1873 cynical and dark with forebodings. The belief that non-Western societies could be regenerated merely by the force of modernisation and the contagion of example – so strong in the 1850s – was dead in the water; ideals of multiracial co-operation broke down in a series of small wars and innumerable reprisals in the succeeding decade as the logic of imperialism, commercial penetration and mass migration played out. Scrape beneath the wonder of globetrotting tourists and you find a time riven with discord, not only between the West and the rest of the world, but within the West itself.

Reverse the direction of tourism and yet another picture emerges. I mean this in a literal and a figurative sense. At the same time that Thomas Cook and many others took the 'Anglo-Saxon highway' across America, the Pacific and Asia, a group of high-ranking Japanese emissaries led by Iwakura Tomomi took the same route in the opposite direction on a world tour to the US, Britain, Russia and Continental Europe and back home via Egypt and South-East Asia.

A San Francisco newspaper greeted the ambassadors as representatives of 'the most progressive nation on the globe'. Over the course of two packed years they pursued an exhaustive itinerary of factories, shipyards, telegraph stations, museums, art galleries, parliaments, universities and law courts in every major Western country. What they saw and learnt about the industrial power of the West provided invaluable information about how to modernise Japan in the wake of the imperial restoration. But one meeting stands out above all others. On 15 March 1873 the highest-ranking members of the Iwakura Mission were invited to the Berlin palace of Otto von Bismarck.[17]

Over dinner the Iron Chancellor addressed his guests. Japan, he said, was like the Prussia of his youth, weak and the victim of more powerful nations. International law, he went on, was an illusion. The law of the world was the law of the strongest; the powerful pretended to respect treaties, but when it suited them they disregarded rights and trampled on the weak. Prussia, he said, had strengthened itself by the patriotic will of its people. Its wars against Denmark (1864), Austria (1866) and France (1870) had resulted in the unification of Germany under Prussian leadership. But war, said Bismarck, was not a weapon of national aggrandisement; it was a necessary form of defence against the real predatory states, Britain and France, countries that hypocritically hid behind a paper-thin veneer of courtesy and respect for international law. In that regard, Germany and Japan had a lot in common.[18]

Like tourists in Asia and America, Iwakura Tomomi and his companions were naturally attracted towards the wonders of modern progress; their exhausting tours of industrial plants and inspections of modern machines could fill volumes. But their meeting with Bismarck in the spring of 1873 revealed that behind this bombastic progress lay unrest and uncertainty. At about the same time, Charles Kingsley was telling a lecture hall that Darwin's theory of evolution was showing the importance of hereditary powers 'from the lowest plant to the highest animal'. In the same way, science was proving 'how the more favoured race . . . exterminates the less favoured, or at least expels it, and forces it, under penalty of death, to adapt itself to new circumstances; and, in a word, that competition between every race and every individual of that race, and rewarding according to deserts, is . . . an universal law of living things'.[19]

A decade on from the explosion of Darwin's theory, the 'survival of the fittest' was being used to justify violence committed against people who stood in the pathways of empire: adapt or become extinct. Richard Burton wrote of the destruction of Australian Aborigines and New Zealand Maori that the Anglo-Saxon race 'found it necessary to wipe out a people that could not be civilized – a fair instance of the natural selection of the species'. And the same supposed natural law of competition was applied to Western countries as well. During the boom of the 1850s the most conspicuous beneficiaries were Britain and her dependencies and the United States, which had the biggest global impact because of their startling geographical extension. But many European countries exceeded this growth, with their industrial capacity expanding at breakneck velocity. For Belgium, Austria, France, Germany and Piedmont the 1850s was a transformative decade. While Britain's total steam power increased from 1.3 million horsepower to 4 million between 1850 and 1870, that of Germany surged from 260,000 HP to 2.5 million. For these ambitious nations, expanding industrial power had to be harnessed to, and directed by, the state. As Napoleon III told his people: 'we have immense tracts of uncultivated lands to clear, roads to open, ports to create, rivers to make navigable, canals to finish, our railway network to complete. These are the conquests I am contemplating and all of you . . . you are my soldiers.' And as Bismarck made clear in his 'blood and iron' speech, industrial growth and nationalism were inextricably linked.[20]

The modern state was built in the crucible of the changes of the 1850s and beyond. Rather than diffuse power, these technologies concentrated it in the centralised state; and rather than foster internationalism they intensified national identities. Telegraphs and railways bound large countries together and gave rise to the modern form of media that created a common culture and fostered mass nationalism. This assertive patriotism coincided with rapid industrial expansion, which was parlayed into military power.

When Napoleon III went to war with Austria in 1859 he utilised his nation's industrial capacity and its network of railways and telegraphs. The North in the American Civil War did this to an even greater degree. Mass-produced firearms gave the countries that possessed them incomparable advantages; telegraphs directed the movement of armies numbering over 300,000 to the battlefield by railway. Germany's rapid advance ensured victories over the two superpowers that had long resisted unification: Austria in 1866 and France in 1870–1. With more advanced railways, rifles and artillery guns produced by the Krupps steel works, the German forces possessed overwhelming advantages. As Bismarck lectured the Iwakura Mission, relatively weak nations such as Prussia (or Piedmont or Japan) could become very strong very quickly. By honing the tools of modernity they would ensure their survival by out-competing their rivals in a ruthless, dog-eat-dog world.

In a speech in the House of Commons, John Stuart Mill looked back at the utopianism of the 1850s: 'The world was fresh from the recent triumphs of Free Trade,' he said, ' – fresh from the Great Exhibition of 1851, which was to unite all nations, and inaugurate the universal substitution of commerce for war.' But even though there had been long periods of peace, an 'unexampled' multiplication of international commerce and the advance of free trade, Mill continued, 'we have since had opportunities of learning a sadder wisdom'. The hope of the 1850s had given way to pessimism and uncertainty. Technologies that promised peace were harnessed as the means of waging a new form of high-tech, industrialised war. As Mill said, 'inventive genius' and the 'lights of modern science' were 'bringing forth every year more and more terrific engines for blasting hosts of human beings into atoms'. The great powers of the Continent had armed against each other in a career of 'mad rivalry'. The Europe that the Japanese travellers encountered in 1872–3 was divided, fearful and increasingly militarised. And this battle of the survival of the fittest was being exported outside Europe as well.[21]

Back in 1851 Britain was not only the most advanced industrial state, it was the only country whose commercial and military tentacles reached into all the corners of the globe. By the early 1870s it had lost its industrial primacy to the United States and Germany. And not only that; the rapid progress of other countries during the boom of the 1850s made them ambitious to extend their power and trade around the world. If the 1860s saw bitter struggle between settlers and indigenous people, the same was true between Western powers in the wider world, and for much the same reason: economic expansion intensified the competition for resources. Many European rulers worried that domestic growth was about to reach its limits; they had to look abroad, to Africa, the Pacific and Asia, to sustain their profits. A similar pressure affected all those who had profited in the golden years of the mid-century: expand or perish.

The values that Britain had espoused at the zenith of its power in the 1850s – those of liberalism, free trade and internationalism – were increasingly irrelevant by the 1870s. Those who had predicted that formal empires would melt away in a world of free trade were sorely mistaken. Britain had sought to extend its informal control over the world in the 1850s, disavowing direct annexation except in certain circumstances. As Palmerston once said comparing traditional imperialism with British free-trade imperialism, 'we have achieved triumphs, we have made aggressions, but we have made them of a very different kind. The capital and skill of Englishmen are spread over the whole surface of the globe.' This invisible or informal power had made Britain the most powerful – and feared – country on earth without the need for direct imperial expansion.[22]

But with the appearance of acquisitive rivals all over the planet – Russia in Central Asia; Germany, Belgium, Italy and France in Africa, Asia and the Pacific – it was pulled into the scramble for territory. The swaggering self-assurance of the middle part of the century gave way to fear and anxiety about the future now that Britain had so evidently passed the peak of its industrial, commercial and political power. Intimations of decline, not strength and confidence, made the British an imperial leviathan. John Ruskin put it bluntly in 1870: 'This is what England must either do, or perish: she must found colonies as fast and as far as she is able.'[23]

When Britain made demands of China or Japan or Siam in the 1850s, it did so in the name of free trade. 'Her Majesty's Government have no desire to obtain any exclusive advantages for British trade in China, but are only desirous to share with all other nations any benefits which they may acquire in the first instance specifically for British commerce.' For a while it seemed that Britain's idealistic free-trade crusade was reshaping the planet in the 1850s. It had been pressed on China, Japan, Persia, Turkey, Egypt, Siam and India. Mexico, Brazil and Argentina were eager adherents to the British liberal model. And across Europe, countries had reduced their tariffs after sustained British lobbying. But liberalism's triumph was fleeting. Participants in the 'new imperialism' of the 1870s were more interested in securing exclusive zones of economic activity in Africa and China and seizing land as a way of pre-empting rivals in the hunt for lucrative new markets and resources. The world was being carved up, not joined together.[24]

With so much discord and antipathy manifest in the world, no one could say in 1873 that economic, scientific and technological advance led to enlightenment, reason, peace and freedom. The final nail in the coffin of the notion of exponential human progress came in September of that year. On the 18th – Black Thursday – the collapse of the financial house Jay Cooke & Co., which had over-invested in the Northern Pacific Railroad, sparked a series of bank failures and the temporary closure of the New York Stock Exchange. The panic pulsed along the Transatlantic Telegraph to Europe. This crash was of a magnitude several times greater than that of 1857. Within twelve months companies controlling 21,000 miles of American railroad were bankrupt, half the blast furnaces in the world's iron-producing regions had closed, the price of cotton halved, shares were in free fall and unemployment soaring. The world economy had recovered quickly after 1857; this time the recession lasted into the next decade, longer in some places.

The Long Depression had implications beyond the economy. Hard-pressed European countries in need of markets, cheap labour and exploitable resources were hungrier than ever for imperial acquisitions. At the same time free trade came under sustained attack. The United States' model of economic nationalism – protecting domestic industry by tariffs – seemed a better bet than Britain's open-door policies. European countries retreated into the stockade of protectionism. The party was over, and it was paid for with the mother of all hangovers.

Little wonder that Western people who were young and born in the 1850s (or who had just missed it) looked back on that decade as a golden age sandwiched between periods of economic stagnation and violence. Benjamin Disraeli rhapsodised: 'It is a privilege to live in this age of rapid and brilliant events. What an error to consider it an Utilitarian age! It is one of infinite Romance.'[25]

With a longer perspective we might judge the heyday of the 1850s to have been much darker. In the wake of accelerated economic expansion came death, dislocation and exploitation; the palls of smoke that obscured skies over Lucknow or Beijing smell more acrid to us now than they did to cocksure contemporaries. The West's triumphalism, its acts of creative destruction and celebration of masculinity and racial superiority put the spirit of the age violently at odds with our own values. If you are Chinese, African-American, Maori, South African, Indian or Native American, the implications of the events of the 1850s and 1860s are being played out to this day.

For good and ill, the incredible boom that occurred between, roughly speaking, 1850 and 1857 provided one of the most concentrated bursts of global change that has been seen in history. More than just a cyclical upswing in the global economy, the confidence and millennial hopes it generated resulted in a whirlwind of energy, innovation, creativity and risk-taking that hastened, and conditioned, what we regard as modernity.

A lot of the great breakthroughs of the century – particularly instantaneous communication, steam-powered ships and long-distance railways – were made in this frenzied and brief space of time; much of what followed was refinement and extension. Impressive as they are in an engineering or economic sense, they profoundly affected the way people related to each other and experienced the world. From a plantation worker in Berar to a reader of a newspaper in Dresden, people were bound to each other by new, invisible and often unsettling networks of information exchange. Electric communication over long distance, fast steamships and transcontinental railways sped up and shrank the world. But they also contributed to a sense that humankind had made a decisive break with the past and stood at a precipice in time; epochal change was imminent and people acted to realise it, sometimes with violence. It was no coincidence that the age of technology was crowded with fast-moving events: the Taiping Rebellion, the Indian Rebellion, the unifications of Italy and Germany, the American Civil War, the abolition of slavery and serfdom, and the clash of the West and Asia. Economic boom and technological innovation converged to accelerate explosive political changes all over the world.

Communications revolution, globalisation, the remaking of the news industry, military interventions in Asia, spectacular debt-fuelled booms, catastrophic busts, the ripping-apart of settled patterns of work, the waning of one superpower and the waxing of others, the popularity of beards: the parallels with our own time are striking. But perhaps seeing it as a parallel is not quite right. People in the 1850s experienced something that we are living through, and indeed what every generation since has: the continual drawing-together of the human race and the annihilation of time and distance by the power of technology. If the generation of the 1850s saw the emergence of a planetary telegraphic (distant-writing) system, their children saw the beginning of telephonic (distant-sound) communication by cable and then, internationally, by wireless. Succeeding generations experienced a rapid succession of developments: in their turn distant electrical vision – television and video-telephony – then the internetworking of billions of computers. The great leap forward was the commercialisation and mass domestication of these communication devices, then their portability.

What made them so radical was the successive freeing of information from the constraints of the physical world. First came gutta-percha-insulated copper cable to transmit the written word over limitless distance; latterly, it has been fibre-optic cable, satellites and the ability to share data in milliseconds and store it in virtually infinite quantities at minimal cost. In our world of 'the perpetual now' or 'perpetual next' it is easy to forget that these things have a history; that previous generations experienced the shock of change too, then absorbed it into their daily lives with similarly remarkable speed and complacency. And in our age of ubiquitous social media, we perhaps overlook the fact that telecommunication and high-speed transport have been continually reshaping human relationships since the middle of the nineteenth century.

If the 1850s seems eerily familiar it is because it marked the emergence of modernity, with all its recurring possibilities and problems. But the history of technology teaches us that what is astoundingly new today is wearingly old-fashioned tomorrow; technological progress makes us forget its own origins. It is worth remembering, however, that the modern age began with a ship carrying an experimental cable across the English Channel in 1851. And it is worth restoring to prominence the vanished material that made the telecommunication revolution possible: gutta-percha.

The wired world: this image of telegraph wires forming a canopy over Lower Manhattan later in the century illustrates the ubiquity of instantaneous communication

CHRONOLOGY OF EVENTS

1851

11 January	Taiping Rebellion begins
6 February	Black Thursday, Victoria, Australia
10 February	Illinois Central Railroad chartered
12 February	Edward Hargraves claims to have found gold in Australia
1 May	Great Exhibition opens
1 July	Colony of Victoria founded
23 July	Treaty of Traverse des Sioux
5 August	Treaty of Mendota
22 August	The *America* wins the Royal Yacht Squadron's race. The trophy was renamed the America's Cup
17 September	Treaty of Fort Laramie
18 September	*New York Times* founded
29 September	Channel telegraph cable successfully laid
October	Julius Reuter founds news agency in London
18 October	Great Exhibition closes
13 November	Channel telegraph opens to the public
2 December	Coup in Paris

1852

12 June	Taiping rebels enter Hunan
2 November	Franklin Pierce (Democrat) elected US President
24 November	Commodore Matthew Perry of the US navy leaves Norfolk, Virginia, to open relations with Japan
11 November	Palace of Westminster opened
2 December	Louis Napoléon becomes Emperor of the French

1853

4 March	Franklin Pierce inaugurated fourteenth President of the United States
20 March	The Taipings seize Nanking
16 April	First railway in Asia opened at Bombay
8 July	Matthew Perry arrives in Edo Bay
27 July	Tokugawa Iesada becomes Shogun of Japan
	Treaty of Fort Atkinson
12 August	New Zealand becomes self-governing
4 October	Russia and the Ottoman Empire go to war
3 November	William Walker captures La Paz
30 November	Russian fleet destroys that of the Ottoman Empire at Sinop

1854

13 February	William Walker defeated by Mexican forces
28 February	Republican Party founded
3 March	First telegraph line opened in Australia
27 March	Britain declares war on Russia
28 March	France declares war on Russia
31 March	Perry signs the Convention of Kanagawa with Japan
30 May	The Kansas-Nebraska Act becomes law
19 August	Grattan Massacre
20 September	Crimean War: Battle of Alma
9–11 October	Ostend Manifesto drafted by Buchanan, Soulé and Mason
25 October	Crimean War: Battle of Balaclava
5 November	Crimean War: Battle of Inkerman
3 December	Miners' rebellion at Ballarat, Victoria, defeated at the Eureka Stockade
23 December	First of the Ansei Great Earthquakes and tsunamis strikes Japan
24 December	Second Ansei earthquake and tsunami

1855

27 January	Panama Railroad opens
29 January	Lord Aberdeen's ministry falls in Britain over mismanagement of the Crimean War
5 February	Palmerston becomes British Prime Minister
2 March	Alexander II succeeds his father, Nicholas I, as Tsar of all the Russias
30 March	Missourian Border Ruffians cross into Kansas to elect a pro-slavery legislature
3 May	William Walker leaves San Francisco to intervene in Nicaraguan civil war
15 June	Stamp duty abolished in Britain
16 July	New South Wales granted self-government
2 September	Battle of Ash Hollow
19 September	Crimean War: Sevastopol falls to French and British
13 October	Walker captures Granada
11 November	Edo (Tokyo) devastated by the third Ansei earthquake
21 November	Fighting breaks out in Kansas

1856

7 February	The British East India Company annexes Oudh
31 March	Treaty of Paris brings Crimean War to an end
21 May	Lawrence, Kansas, sacked by Border Ruffians
22 May	Senator Charles Sumner caned in the hall of the US Senate by Congressman Preston Brooks
24 May	Potawatomie Massacre
12 July	Walker inaugurated president of Nicaragua
8 October	Chinese authorities seize the *Arrow* on the Pearl River
29 October	British troops enter Canton
1 November	Anglo-Persian War begins
4 November	James Buchanan (Democrat) elected fifteenth President of the United States
14 December	Walker's forces burn Granada and retreat to Lake Nicaragua

1857

3 March	Fall of Palmerston's ministry in Britain following a vote of censure regarding the war with China
24 April	Palmerston wins British general election, the so-called 'Chinese Election'
1 May	Walker surrenders to the US navy
10 May	Meerut: 3rd Light Cavalry of the Bengal army mutinies, beginning the Indian Rebellion
11 May	Indian Rebellion: Delhi captured by rebels
3–14 June	Indian Rebellion: sequence of mutinies in Oudh, North-West Provinces, central India, Rajputana and the Punjab
6 June	Indian Rebellion: siege of Cawnpore begins
27 June	Indian Rebellion: massacre of Cawnpore garrison at the Satichaura Ghat
17 July	Indian Rebellion: Havelock retakes Cawnpore
24 August	Ohio Life Insurance and Trust Company collapses, precipitating the Panic of 1857
28 August	Matrimonial Causes Act makes divorce easier in Britain
20 September	Indian Rebellion: British forces recapture Delhi
23 September	Indian Rebellion: Nicholson dies of his wounds
24 November	Indian Rebellion: Havelock dies in Lucknow

1858

1 January	British and French capture Canton
21 March	Indian Rebellion: Lucknow retaken by British
11 May	Minnesota admitted as the thirty-second state of the United States
13–17 June	Britain, France, the United States and Russia sign treaties with China at Tientsin
July	Gold rush to Colorado
29 July	The United States and Japan sign the Treaty of Amity and Commerce
2 August	The East India Company transfers its Indian territories to the British Crown

5 August	Atlantic Cable successfully laid
14 August	Tokugawa Iemochi becomes fourteenth Shogun of Japan
16 August	President Buchanan sends telegram to Queen Victoria

1859

9 March	Piedmont mobilises against Austria
25 April	Ground broken for the Suez Canal
29 April	Austrian troops cross Ticino River into Piedmont
26 May	The Cacciatori delle Alipi under Giuseppe Garibaldi defeat the Austrians at Varese
4 June	Battle of Magenta
15 June	Pig War begins on San Juan
24 June	Battle of Solferino
24/5 June	British and French forces repelled at the Taku Forts, northern China
11 July	Preliminaries of the Treaty of Villafranca signed
25 August	Imam Schamyl surrenders to Prince Baryatinsky
28 August	Solar storm begins
16 October	John Brown raids Harpers Ferry Armory
12 November	Yoshida Shoin executed
24 November	Charles Darwin's *On the Origin of Species* published
	La Gloire launched
2 December	John Brown executed

1860

17 March	First Taranaki War, New Zealand
4 April	Uprising in Palermo
6 May	Garibaldi leads the Expedition of the Thousand to Sicily
2 July	Vladivostok founded
7 September	Garibaldi captures Naples
12 September	William Walker executed
18 October	Summer Palace burnt
	British and French troops enter Beijing

24 October	Treaty of Tientsin ratified by Prince Gong in Beijing
26 October	Garibaldi recognises Vittorio Emanuele as King of Italy
6 November	Abraham Lincoln (Republican) elected sixteenth President of the United States
20 December	South Carolina secedes from the Union

1861

9–11 January	Mississippi, Florida and Alabama secede from the Union
19 January	Georgia secedes from the Union
26 January	Louisiana secedes from the Union
1 February	Texas secedes from the Union
4 February	The Provisional Confederate States Congress formed in Montgomery, Alabama
8 February	Confederate States of America established
9 February	Jefferson Davis elected provisional President of the Confederacy
3 March	Serfdom abolished in Russia
4 March	President Abraham Lincoln takes office
17 March	Unification of Italy proclaimed
21 March	Vice President Alexander Stephens of the Confederacy gives his Cornerstone Speech
13 April	Fort Sumter surrenders to Confederate forces
17 April	Virginia secedes from the Union
6/7 May	Arkansas and Tennessee secede from the Union
13 May	American Civil War: Queen Victoria issues Proclamation of Neutrality
20 May	North Carolina secedes from the Union
21 July	First Battle of Bull Run
25 July	Crittenden-Johnson Resolution passed by Congress

1862

25 April	Union forces capture New Orleans
17 August	Dakota War begins in Minnesota

17 September	Battle of Antietam
22 September	Abraham Lincoln makes preliminary announcement of his Emancipation Proclamation
	Otto von Bismarck appointed Minister President of Prussia
3 October	Garibaldi's 'Letter to the English Nation' published
5 October	Riots in Hyde Park, London
	News of the Emancipation Proclamation reaches Europe

ACKNOWLEDGEMENTS

Much like the subject matter of *Heyday*, this book has been a transatlantic affair. I was lucky to benefit from the support, insights and enthusiasm of two brilliant editors, Bea Hemming (in London) and Lara Heimert (in New York). I would also like to thank Roger Labrie for his help in shaping the book. Clare Conville, as ever, has been the most wonderful agent and friend. The book was sped into production with the expert guidance of Cindy Buck, Holly Harley, Linden Lawson, Matthew Marland, Leah Stecher and Melissa Veronesi. The maps were expertly prepared by John Gilkes. Dr Miriam Berry did some excellent scouting.

I am extremely grateful to the London Library Trust for providing financial assistance in the form of a Carlyle Membership. This book would not have been written without the Library's generosity.

Thanks and love to Marney, Ariane, Conrad and Claire.

NOTES

PREFACE

1 Herman Merivale, *Lectures on Colonization and Colonies. Delivered before the University of Oxford in 1839, 1840 and 1841* (2 vols, London, 1841), vol. I, p. 134
2 *Illustrated London News* [*ILN*], 7/6/1851

INTRODUCTION

1 Jesup W. Scott, 'The Great West', *De Bow's Review*, vol. 15, no. 1 (July 1853), p. 51
2 William T. Brannt, *India Rubber, Gutta-Percha, and Balata* (Philadelphia, 1900), pp. 230ff.
3 'A Visit to the Gutta-Percha Works', *The Illustrated Exhibitor and Magazine of Art*, vol. I (London, 1852), pp. 18ff.
4 For a list of uses of gutta-percha see *Allen's Indian Mail*, 4/2/1851, p. 87
5 *Times*, 15/11/1851
6 Charles Briggs and Augustus Maverick, *The Story of the Telegraph* (New York, 1858), p. 12
7 Thomas Hardy, 'The Fiddler of the Reels', in *The Fiddler of the Reels and Other Stories, 1888-1900* (London, 2003 edn), p. 191; G. M. Young, *Victorian England: Portrait of an Age* (Oxford, 1936), p. 77
8 Karl Marx and Friedrich Engels, *Collected Works* (Moscow, 1983), vol. 39, p. 70
9 *Times*, 22/3/1850, p. 5
10 Dionysius Lardner, *The Great Exhibition and London in 1851* (London, 1852), pp. 84, 107; *ILN*, 3/5/1851
11 'Effects of the Discoveries of Gold', *British Quarterly Review*, vol. XVII (May 1853), p. 546
12 George F. Train, *Young America Abroad in Europe, Asia and Australia* (London, 1857), p. 2

13 For the scale and scope of the nineteenth-century settler revolution the key text is James Belich, *Replenishing the Earth: The Settler Revolution and the Rise of the Angloworld* (Oxford, 2009)
14 William Howitt, *Land, Labour, Gold* (2 vols, London, 1858), vol. I, p. 27
15 *Times*, 22/3/1850, p. 5
16 For the global importance of slave-grown cotton in sustaining the boom of the 1850s see two recent and important works: Sven Beckert, *Empire of Cotton: A New History of Global Capitalism* (London, 2014) and Edward E. Baptist, *The Half Has Never Been Told: Slavery and the Making of American Capitalism* (New York, 2014)
17 *Tenth Annual Report of the Aborigines' Protection Society* (London, 1847), p. 9
18 John Tallis, *Tallis's History and Description of the Crystal Palace* (London, 1852), vol. I, p. 149
19 For the use of technology as a tool of imperialism see the following works by Daniel R. Headrick: *The Tentacles of Progress: Technology Transfer in the Age of Imperialism, 1850–1940* (Oxford, 1988); *The Invisible Weapon: Telecommunications and International Politics, 1851–1945* (Oxford, 1991); *Power Over Peoples: Technology, Environments, and Western Imperialism, 1400 to the Present* (Princeton, NJ, 2010)
20 Karl Marx and Friedrich Engels, *Manifesto of the Communist Party* in *Collected Works* (Moscow, 1983), vol. 1, p. 16
21 David Gillard, *The Struggle for Asia, 1828–1914: A Study in British and Russian Imperialism* (London, 1977), pp. 112–13

CHAPTER 1: 1851: ANNUS MIRABILIS

1 Charlotte Brontë, *The Letters of Charlotte Brontë* (ed. Margaret Smith, Oxford, 2000), vol. II, p. 630. For modern accounts of the Great Exhibition see: Jeffrey Auerbach, *The Great Exhibition of 1851: A Nation on Display* (New Haven, CT, 1999); Michael Leapman, *The World for a Shilling: How the Great Exhibition Shaped a Nation* (London, 2001); Jeffrey Auerbach and Peter Hoffenberg (eds), *Britain, the Empire and the World at the Great Exhibition* (Aldershot, 2008); Hermione Hobhouse, *The Crystal Palace and the Great Exhibition* (London, 2002)
2 Tallis, vol. I, p. iii; *ILN*, 10 May 1851; *Times*, 1/5/1851
3 Tallis, vol. I, p. 207
4 Ibid., p. 199
5 Ibid., pp. 159–60

6 Queen Victoria's diary entry for 9/6/1851, http://www.queenvictoriasjournals.org
7 Greeley, *Glances*, p. 31; *Times*, 1/5/1851
8 Tallis, vol. III, pp. 69–70; *Times*, 1/5/1851
9 Tallis, vol. I, p. 196
10 *The North American Review*, vol. LXXV, no. 157 (October 1852), p. 358
11 Fredrika Bremer, *England in the Autumn of 1851; or, sketches of a tour of England* (Boulogne, 1853), pp. 1ff.
12 Ibid., pp. 4–6
13 House of Commons, 27/2/1846, col. 242
14 T. B. Macaulay, *Speeches* (2 vols, London, 1853), p. 71
15 David Kynaston, *The City of London: A World on its Own, 1815–1890* (London, 1994), pp. 196–7
16 Auerbach, *Great Exhibition*, pp. 61ff.
17 *Times*, 13/10/1851
18 Greeley, *Glances*, p. 31
19 *The North American Review*, vol. LXXV, no. 157 (October 1852), p. 361
20 *Times*, 15/5/1851; Senate Committee on Military Affairs, *Report . . . as to the relative efficiency of the repeating pistols invented by Samuel Colt* (31st Congress, 2nd session, Rep. Con. No 257, 1850), p. 2
21 *Times*, 27/5/1851, 2/9/1851
22 Russell Fries, 'British Response to the American System: The case of the small-arms industry after 1850', *Technology and Culture*, vol. 16, no. 3 (1975)
23 *Economist*, 18/3/1854, p. 281, quoted in D. C. M. Platt, *Foreign Finance in Continental Europe and the USA, 1815–1870* (London, 1984), p. 165
24 George Russell, *The Narrative of George Russell* (Oxford, 1935), pp. 293–4
25 Kynaston, pp. 178, 181
26 Ibid., p. 167
27 For British free trade see: Anthony Howe, *Free Trade and Liberal England, 1846–1946* (Oxford, 1997); Bernard Semmel, *The Rise of Free Trade Imperialism* (Cambridge, 1970); D. C. M. Platt, *Finance, Trade and Politics in British Foreign Policy* (Oxford, 1968); P. J. Cain, *Economic Foundations of British Overseas Expansion, 1815–1914* (London, 1980); D. C. M. Platt, *Business Imperialism, 1840–1930* (Oxford, 1970); Albert H. Imlah, *Economic Elements in the Pax Britannica: Studies in British Foreign Trade in the Nineteenth Century* (Cambridge, MA, 1958); John Gallagher and Ronald Robinson, 'The Imperialism of Free Trade', *Economic History Review*, 2nd series, vol. VI (1953); Oliver MacDonaugh, 'The Anti-Imperialism of Free Trade', *Economic History Review*, vol. XIV (1962); C. McLean, 'Finance and Informal Empire', *Economic History*

Review, vol. XXIX (1978); Peter Cain, 'Capitalism, War and Internationalism in the thought of Cobden', *British Journal of International Studies*, vol. V (1979); P. K. O'Brien and G. A. Pigman, 'Free Trade, British Hegemony and the International Economic Order', *Review of International Studies*, vol. 18, no. 2 (1992)
28 *Times*, 10/11/1856
29 Tallis, vol. II, p. 17
30 *Times*, 29/10/1851
31 Tallis, vol. I, pp. 55 and 102; vol. II, p. 94
32 Ibid., vol. I. pp. 92ff., vol. III, pp. 75–6; *Times*, 25/6/1851
33 *Times*, 18/10/1849
34 *Edinburgh Review*, vol. xciv, no. cxcii, July–Oct 1851, pp. 596–7
35 Ronald Hyam, *Britain's Imperial Century, 1815–1914: A Study of Empire and Expansion* (London, 1976), p. 54
36 Greeley, *Glances*, pp. 88–9
37 *American Telegraph*, 26/7/1851, p. 2
38 Hyam, p. 49
39 Dostoyevsky, *Notes from the Underground*, Part I, chapter VII; *New York Daily Tribune*, 26/1/1856
40 Tallis, vol. II, p. 94
41 *ILN*, 7/6/1851

CHAPTER 2: THE HAIRYSTOCRACY

1 D. T. Coulton, 'Gold Discoveries', *Quarterly Review*, vol. XCI, no. 182 (September 1852), p. 504
2 John Hunter Kerr, *Glimpses of Life in Victoria* (Edinburgh, 1872), p. 95
3 Kerr, p. 101; George Butler Earp, *The Gold Colonies of Australia* (London, 1852), p. 173
4 Kerr, pp. 111–12, 119ff.
5 W. E. Adcock, *The Gold Rushes of the Fifties* (Melbourne, 1912), pp. 32, 73
6 George Butler Earp, *What We Did In Australia* (London, 1853), p. 144; William Craig, *My Adventures on the Australian Goldfields* (London, 1903), pp. 15, 195ff.
7 John Capper, *The Emigrant's Guide to Australia* (Liverpool, 2nd edn, n.d.) p. iv; Geoffrey Serle, *The Golden Age: A History of the Colony of Victoria, 1851–1861* (Melbourne, 1963), p. 38
8 Mrs Charles Clacy, *A Lady's Visit to the Gold Diggings of Australia in 1852–52* (London, 1853), pp. 29–30; Kerr, p. 128
9 Clacy, p. 9

10 Serle, p. 31; Adcock, p. 66
11 Earp, *Gold Colonies*, p. 175; Earp, *What We Did*, p. 113
12 Capper, p. 148
13 Howitt, p. 24
14 Kerr, p. 130; Howitt, pp. 4–5, 34–5
15 Howitt, pp. 8ff; Serle, p. 124, Clacy, p. 17
16 Howitt, pp. 28–9
17 Charles Stretton, *Memoirs of a Chequered Life* (3 vols, London, 1862), vol. II, pp. 38, 43–4, 77; Craig, pp. 44–6; Howitt, pp. 259–60
18 Howitt, p. 18; Stretton, pp. 52ff.
19 Clacy, pp. 283–4; Capper, pp. 235ff.
20 Clacy, p. 276; Howitt, pp. 248–9
21 Howitt, p. 80
22 Howitt, pp. 118, 175; Clacy, pp. 79–80
23 Howitt, p. 124; Clacy, p. 183
24 Henry Brown, *Victoria As I Found It* (London, 1862), pp. 300–1; Earp, *What We Did*, pp. 141–2
25 Howitt, pp. 239ff.
26 Brown, *Victoria*, pp. 352ff.
27 Charles Ferguson, *Experiences of a Forty-niner During Thirty-four Years' Residence in California and Australia* (Cleveland, OH, 1888), pp. 242, 249
28 Howitt, p. 126
29 Ibid., p. 137; Tallis, vol. I, p. 54
30 Craig, p. 43
31 Howitt, pp. 137ff.
32 Brown, *Victoria*, pp. 321ff.; Kerr, p. 210; Serle, pp. 216ff.
33 Kerr, p. 274
34 Howitt, pp. 267ff.
35 *Times*, 3/1/1855
36 Howitt, pp. 220ff.
37 Serle, p. 228
38 Susan H. Farnsworth, *The Evolution of British Imperial Policy* (New York, 1992), p. 259
39 *Argus* (Melbourne), 10/5/1867; Earp, *What We Did*, p. 10
40 Brown, *Victoria*, p. 273; Serle, p. 220

CHAPTER 3: BONANZA

1 Marx to Joseph Weydemeyer, in Karl Marx and Friedrich Engels, *Collected*

Works, vol. 39, p. 70

2 Basil Lubbock, *The Colonial Clippers* (Glasgow, 1921), pp. 29ff.

3 Charles Hursthouse, *New Zealand, or Zealandia, the Britain of the South* (2 vols, London, 1857), vol. I, p. 625

4 Craufurd D. W. Goodwin, *The Image of Australia: British Perception of the Australian Economy* (Durham, NC, 1974), p. 42

5 *Neue Rheinische Zeitung Revue*, May–October 1850

6 Lubbock, pp. 23ff., 26ff., 85–6

7 Daniel Cornford, 'Labor and Capital in the Gold Rush', in James L. Rawls and Richard J. Orsi (eds.), *A Golden State: Mining and Economic Development in Gold Rush California* (Berkeley, CA, 1999), p. 82

8 Ralph J. Roske, 'The World Impact of the California Gold Rush, 1849–1857', *Arizona and the West*, vol. 5, no. 3 (Autumn 1963); Gerald D. Nash, 'A Veritable Revolution: The Global Economic Significance of the California Gold Rush', *California History*, vol. 77, no. 4 (Winter 1998/1999); A. C. W. Bethel, 'The Golden Skein', in Rawls and Orsi (eds), p. 254

9 George F. Train, *Young America*, vol. I, p. 2; Serle, p. 121

10 Howitt, pp. 25, 205ff.

11 Train, *Young America*, pp. 387, 471; Robert B. Minturn, *From New York to Delhi, by way of Rio de Janeiro, Australia and China* (New York, 1858), p. 24; Kerr, p. 162

12 Clacy, p. 240; Lubbock, pp. 34–5

13 Richard McKay, *Donald McKay and His Famous Sailing Ships* (New York, 1928), p. 129

14 Léon Faucher, *Remarks on the Production of the Precious Metals, and on the Demonetization of Gold* (2nd edn, London, 1853), p. 97

15 Nathaniel Hawthorne, *Passages from the English Notebooks* (Boston, MA, 1870), p. 10

16 *Age*, 16/8/1855, p. 4

17 Hawthorne, pp. 28–9

18 *Household Words*, 17/7/1852, pp. 405 ff.

19 Hawthorne, p. 14

20 Kerr, p. 347

21 Serle, p. 321

22 *Boston Daily Atlas*, 21/9/1852

23 *Daily Alta California*, 1/4/1851; *ILN*, 22/5/1852; cf. *Times*, 8/4/1852

24 *Neue Rheinische Zeitung Revue*, May–Oct 1850, https://www.marxists.org/archive/marx/works/1850/11/01.htm

25 J. D. Borthwick, *Three Years in California* (Edinburgh, 1857), p. 11

26 Ibid., pp. 10ff.
27 Frank Marryat, *Mountain and Molehills; or, recollections of a burnt journal* (New York, 1855), pp. 24–5
28 Ibid., p. 26
29 Bethel, p. 254
30 Kynaston, p. 167
31 *The British Quarterly Review*, 1/5/1853, pp. 551, 555, 562
32 *Fraser's Magazine*, vol. XLIV, no. CCLX (August 1851), p. 227
33 *Times* 12/11/1856, p. 6
34 Henry M. Field, *The Story of the Atlantic Telegraph* (New York, 1892), p. 13
35 Matthew Maury, 'Submarine Telegraph Across the Atlantic', *De Bow's Review*, vol. 16, no. 6 (June 1854), pp. 626ff.
36 Isabella Field Judson, *Cyrus Field: His life and work* (NY, 1896), pp. 61–2
37 Judson, p. 68
39 *Times*, 14/11/1856, p. 12
39 *Times*, 12/11/1856, p. 6
40 *Times*, 12/11/1856

CHAPTER 4: ON THE ROAD

1 Sandra L. Myres (ed.), *Ho for California: Women's Overland Diaries* (San Marino, CA, 1980), p. 93
2 Elliott West, *The Contested Plains: Indians, goldseekers, and the rush to Colorado* (Lawrence, KS, 1998), p. 193
3 Ferguson, pp. 48ff.
4 Bayard Taylor, *El Dorado; or, adventures in the path of empire* (New York, 1854), p. 281
5 'Travellers' Books for 1851', *Fraser's Magazine*, vol. XLIV, no. CCLX (August 1851), p. 227; William Kelly, *Across the Rocky Mountains, from New York to California* (London, 1852), p. vi
6 Ferguson, p. 66
7 Horace Greeley, *An Overland Journey, from New York to San Francisco* (New York, 1860), p. 272
8 Myres, p. 177
9 For the overland trail see John D. Unruh, *The Plains Across: Emigrants, Wagon Trains and the American West* (Champaign, IL, 1979)
10 Kenneth L. Holmes, *Covered Wagon Women: Diaries and Letters from the Western Trails, 1850* (Lincoln, NE, 1996), p. 85; Unruh, p. 85
11 Unruh, pp. 391ff.

12 Richard Burton, *The City of the Saints and Across the Rocky Mountains to California* (New York, 1862), pp. 53, 141
13 West, pp. 208–28
14 Greeley, pp. 47–8.
15 Burton, pp. 6, 13, 18–19, 64, 70, 148, 276, 321
16 Ibid., p. 42
17 Ferguson, p. 21
18 West, pp. 88ff., 192ff.
19 Unruh, p. 170
20 James Gamble, 'When the Telegraph Came to California', *The Californian* (1881) http://www.telegraph-history.org/transcontinental-telegraph/index.html; Burton, p. 9
21 Myres, pp. 106ff.
22 Loretta Fowler, *Arapahoe Politics, 1851–1978: Symbols of Crises of Authority* (Lincoln, NE, 1986), p. 32
23 Leo E. Oliver, 'Fort Atkinson and the Santa Fe Trail, 1850–1854', *Kansas History: A journal of the Central Plains*, vol. XL, no. 2 (Summer 1974), pp. 212–33; West, p. 200
24 Burton, p. 32; Oliver, 'Fort Atkinson'; West, pp. 278ff.
25 Thomas Twiss, 'Letter of Thomas S. Twiss, Indian Agent at Deer Creek, U.S. Indian Agency on the Upper Platte', *Annals of Wyoming*, vol. 17, no. 2 (July 1945), pp. 148ff.
26 Burton, pp. 32, 46; West, p. 201
27 George Caitlin, *Caitlin's Notes of Eight Years' Travels and Residence in Europe* (New York, 1848), p. 62
28 West, pp. 256–7
29 Ibid., pp. 1ff.
30 J. B. Peires, 'The Central Beliefs of the Xhosa Cattle-Killing', *The Journal of African History*, vol. 28, no. 1 (1987), pp. 43ff.
31 Twiss, pp. 148ff.
32 Burton, p. 3
33 Walter G. Sharrow, 'William Henry Seward and the Basis for American Empire, 1850–1860', *Pacific Historical Review*, vol. 26, no. 3 (August 1967), p. 339

CHAPTER 5: STAR OF EMPIRE

1 Hursthouse, vol. I, pp. 588–9
2 Anthony Trollope, *The New Zealander* (London, 1995), pp. 3ff.; T. B.

Macaulay, 'The Ecclesiastical and Political History of the Popes', *Edinburgh Review*, vol. LXXII (October 1840), p. 228

3 Earp, *What We Did in Australia*, pp. 26–7

4 Jesup Scott, 'Westward the Star of Empire', *De Bow's Review*, August 1859

5 Micajah Tarver, 'The Growth of Cities in the United States', *Western Journal of Agriculture, Manufactures, Mechanic Arts, Internal Improvement, Commerce and General Literature*, March 1851; Jesup Scott, 'The Great West', *De Bow's Review*, July 1853; Charles N. Glaab, 'Visions of Metropolis: William Gilpin and Theories of City Growth in the American West', *Wisconsin Magazine of History* (Autumn 1961)

6 E. Sandford Seymour, *Sketches of Minnesota: The New England of the West* (New York, 1850), p. 34

7 Hursthouse, vol. I, pp. 97–8; Sarah Tucker, *The Southern Cross and the Southern Crown* (London, 1855), p. 2; J. Donnelly, 'Minnesota and the North West', *De Bow's Review*, July 1856

8 Hursthouse, vol. I, pp. 102, 656

9 J. Donnelly, 'Minnesota and the North West', *De Bow's Review*, July 1856; 'Towns and Statistics of Minnesota', *Western Journal*, May 1851; Carlton C. Qualey, 'A New El Dorado: guides to Minnesota, 1850s–1880s', *Minnesota History*, vol. 42, no. 6 (Summer 1971), pp. 218, 220–21

10 J. W. Bond, *Minnesota and its Resources* (New York, 1853), p. 24; G. H. Scholefield, *New Zealand in Evolution* (New York, 1909), p. 39

11 T. Cholmondeley, *Ultima Thule* (London, 1854), p. 4; Hursthouse, vol. I, pp. 192, 637

12 Hursthouse, vol. I, p. 637

13 James M. Woolworth, *Nebraska in 1857* (New York, 1857), pp. 16–17; cf. 'Emigration', *Western Journal*, January 1849

14 David Hamer, *New Towns in the New World: Images and Perceptions of the Nineteenth-century Urban Frontier* (New York, 1990)

15 Laurence Oliphant, *Minnesota and the Far West* (Edinburgh, 1855), p. 131; Hamer, p. 103

16 Hamer, pp. 98ff., 113ff.

17 Oliphant, *Minnesota*, pp. 130ff.

18 Hursthouse, vol. I, p. 359

19 Hamer, p. 48

20 Oliphant, *Minnesota*, pp. 159–60

21 Ibid., p. 165

22 Paul. W. Gates, *The Illinois Central Railroad and its Colonization Work* (New York, 1968), pp. 68ff., 73

23 Gates, *passim*

24 William Cronon, *Nature's Metropolis: Chicago and the Great West* (New York, 1991), pp. 104ff.
25 Ibid., pp. 112, 113–14
26 T. B. Macaulay, *Speeches* (2 vols, 1853), p. 71
27 *Times* 24/10/1855
28 Oliphant, *Minnesota*, pp. 241, 244–7, 284
29 C. C. Andrews, *Minnesota and Dakotah* (Washington, DC, 1857), p. 82
30 Mary Lethert Wingerd, *North Country: The Making of Minnesota* (Minneapolis, MN, 2010), pp. 247–8
31 Oliphant, *Minnesota*, p. 252
32 Wingerd, pp. 209–10
33 Ibid., pp. 211, 248
34 Oliphant, *Minnesota*, p. 281
35 Ibid., pp. 281–2; John Fletcher Williams, *A History of the City of St Paul* (St Paul, MN, 1876), p. 377
36 Oliphant, *Minnesota*, p. 257
37 Seymour, p. 95
38 Scholefield, p. 32
39 James Belich, *Making Peoples: A History of the New Zealanders* (Auckland, 1996), pp. 212ff.
40 Tucker, pp. 1–2; Belich, pp. 226ff.

CHAPTER 6: THE HASHISH OF THE WEST

1 Williams, *History*, p. 379
2 *New York Herald*, 28/5/1856
3 Frederick Law Olmsted, *The Cotton Kingdom* (2 vols., New York, 1862), vol. I, pp. 277–8
4 Edward E. Baptist, *The Half Has Never Been Told*, pp. 343ff.
5 'Interests of the Slave and Free States and the Union', *De Bow's Review*, vol. 19, no. 4 (Oct 1855), p. 378
6 *Economist*, 17/12/1853. Cf. *Economist*, 15/9/1860; *Edinburgh Review*, vol. 101 (April 1855), pp. 151-76; *Times*, 24/11/1857
7 J. C. N. [Josiah Nott], 'The Future of the South', *De Bow's Review*, vol. 10, no. 2 (1851), pp. 132, 137, 142
8 Merivale, *Lectures*, vol. I, p. 295
9 Bonnie Martin, 'Slavery's Invisible Engine: mortgaging human property', *Journal of Southern History*, vol. 76, no. 4 (November 2010), pp. 840–1; *Times*, 19/12/1857

10 Baptist, pp. 352ff; Beckert, p. 110
11 J. C. N, 'The Future of the South', p. 132
12 'What the South is Now Thinking and Saying About the Course of the North', *De Bow's Review*, vol. 19, no. 2 (August 1855), pp. 139–40
13 'What the South is Now Thinking', pp. 139–40
14 David M. Potter, *The Impending Crisis: America before the Civil War, 1848–1861* (NY, 1976), p. 205
15 Thomas Gladstone, *The Englishman in Kansas: or, squatter life and border warfare* (NY, 1857), pp. 40–2
16 Williams, *History*, pp. 379ff.
17 Ibid., p. 380; *New York Daily Tribune*, 22/3/1856
18 Wingerd, p. 248; Jocelyn Wills, *Boosters, Hustlers, and Speculators: Entrepreneurial Culture and the Rise of Minneapolis and St Paul* (St Paul, MN, 2005), p. 87
19 'The Political Crisis in the United States', *Edinburgh Review*, vol. CIV, no. 212 (October 1856), pp. 561–2
20 Beckert, p. 133; James A. Mann, *The Cotton Trade of Great Britain: its rise, progress and present extent* (London, 1860), pp. 55–6, 85
21 *Westminster Review*, vol. 52 (October 1849–January 1850), pp. 213–14; *Manchester Guardian*, 20/6/1857
22 Mann, pp. 43, 83–6 , 87; Arthur Redford, *Manchester Merchants and Foreign Trade* (Manchester, 1934), p. 227; Arthur W. Silver, *Manchester Men and Indian Cotton, 1847–1872* (Manchester, 1966), pp. 85ff.; Beckert, pp. 124ff.; *New York Tribune*, 7/4/1856; *Morning Advertiser* quoted in the *Caledonian Mercury*, 6/3/1856; *Report of the Special Committee of the House of Representatives of South Carolina, on so much of the message of His Excellency Gov. Jas. H. Adams, as relates to slavery and the slave trade* (Columbia, SC, 1857), p. 47
23 'Future Supply of Cotton', *Manchester Times*, 14/3/1857; J. T. Danson, 'On the Existing Connection Between American Slavery and the British Cotton Manufacture', *Quarterly Journal of the Statistical Society of London*, vol. XX (March 1857), p. 7
24 *Report of the Special Committee*; *The American Cotton Planter*, vol. III, no. 1 (January 1859), pp. 11ff., vol. III, no. 4 (April 1859), pp. 105ff.; *New York Herald*, 28/5/1856, 11/6/1856, 6/12/1856
25 *New York Herald*, 28/5/1856; James Henry Hammond, *Selections from the Letters and Speeches* (New York, 1866), pp. 311–12, 316

CHAPTER 7: THE RAMPARTS OF FREEDOM

1 Fyodor Tyutchev, 'A Russian Geography' in, F. Jude (trans.),*The Complete*

Poems of Tyutchev in an English Translation (Durham, 2000), p. 137

2 Laurence Oliphant, *Episodes in a Life of Adventure: or moss from a rolling stone* (Edinburgh, 1887), p. 93; Oliphant, *Patriots and Filibusters, or, incidents of political and exploratory travel* (Edinburgh, 1860), pp. 10–11

3 John Milton Mackie, *Life of Schamyl; and narrative of the Circassian war of independence against Russia* (Boston, MA, 1856), p. 295

4 Mackie, pp. 29ff.; Louis Moser, *The Caucasus and its People* (London, 1856), pp. 29ff.

5 Mackie, p. 7

6 George Alexander Lensen, *The Russian Push Toward Japan: Russo-Japanese Relations, 1697–1875* (New York, 1971), pp. 300–1

7 John F. Baddeley, *The Russian Conquest of the Caucasus* (London, 1908), pp. 385–410

8 Ibid., p. 355

9 Peter Hopkirk, *The Great Game: On Secret Service in High Asia* (London, 1990), p. 155

10 Paul Biriukov [Pavel Ivanovich Biryukov] (ed.), *Leo Tolstoy: His Life and Work* (London, 1906), p. 132

11 Lesley Blanch, *The Sabres of Paradise* (London, 1960), pp. 293ff.

12 Tolstoy, diary entry, 6/1/1852

13 Orlando Figes, *Crimea: The Last Crusade* (London, 2010), p. 105

14 Figes, p. 157

15 *Times*, 29/10/1851

16 Sir John McNeill, *Progress and Present Position of Russia in the East* (London, 1838), pp. 142–5

17 *New York Herald*, 1/9/1856; *ILN*, 7/6/1851. On the supposed epochal confrontation between free trade and protectionism, see Richard Cobden, *What Next – and Next?* (1856), in F. W. Chesson (ed.), *The Political Writings of Richard Cobden* (2 vols, London, 1903), vol. II, pp. 461ff.

18 Karl Marx, 'The Real Issue in Turkey', *New York Daily Tribune*, 12/4/1853; Trevor Royle, *Crimea: The Great Crimean War, 1854–1856* (London, 1999), pp. 44–5

19 'Schamyl and the War in the Caucasus', *North American Review*, vol. 81, no. 169 (October 1855), pp. 389ff.; *Daily News*, 9/7/1855; *Morning Post*, 27/7/1855

20 Mackie, preface, pp. 221ff.; *ILN*, 26/10/1844, 29/8/1845; *Manchester Times*, 17/3/1852; *Morning Post*, 10/11/1854, 31/1/1855; 'Schamyl: the prophet-warrior of the Cauasus', *Westminster Review*, 1854, pp. 480ff.; 'Schamyl and the Caucasus', *Preston Guardian*, 19/11/1853; 'The Asiatic Theatre of the

War', *Jackson's Oxford Journal*, 22/7/1854; 'The Caucasus and the Circassians', *Belfast News-Letter*, 25/10/1854

21 Lensen, p. 301

22 Laurence Oliphant, *Trans-Caucasian Campaign of the Turkish Army Under Omer Pasha: a personal narrative* (Edinburgh, 1856), p. 225; Royle, *Crimea*, p. 45

23 Olive Anderson, *A Liberal State at War: English Politics and Economics during the Crimean War* (London, 1967), pp. 1–4.

24 Anderson, p. 72

25 *Times*, 31/10/1854, 1/3/1855

26 *Times*, 17/11/1855; *Edinburgh Review*, April 1856

27 Figes, pp. 194–5

28 Oliphant, *Trans-Caucasian Campaign*, p. 232; Blanch, pp. 257–8. Examples of books and articles on Schamyl and the Caucasus published during the Crimean War are: *Lloyds Weekly*, 30/7/1854; *Morning Post*, 17/11/1854 and 31/1/1855; *Daily News*, 7/11/1854. E. Spencer, *Turkey, Russia, the Black Sea and Circassia* (London, 1855), Mackie, *Life of Schamyl* (1856), Moser, *The Caucasus* (1856), Kenneth Mackenzie (ed.), *Schamyl and Circassia. Chiefly from materials collected by Dr Friedrich Wagner* (London, 1854), John Reynell Morrell, *Russia and England, their strength and weakness* (New York, 1854), Baron August von Haxthausen, *The Tribes of the Caucasus, with an account of Schamyl and the Murids* (London, 1855), Ivan Golovin, *The Caucasus* (London, 1854). For portraits of Schamyl see *Times* classified advertising 22/9/1854, sales by auction 18/12/1854. For the plays *Schamyl: The Warrior Prophet* and *Schamyl: The Circassian Chief* see *ILN* 11/11/1854; *Times* classified advertising 28/8/1854. For the racehorse see *Times* sporting intelligence 28/10/1854, 3/11/1854, 5/1/1855, 6/1/1855. For the dance see *Times* classified advertising 28/10/1854, 4/11/1854. For the stock market see *Times* 16/9/1854.

29 *Morning Post*, 31/1/1855

30 *York Herald and General Advertiser*, 7/10/1854

31 *Times*, 10/11/1855

32 Norman Luxenburg, 'England and the Caucasus During the Crimean War', *Jahrbücher für Geschichte Osteuropas*, December 1968; *Times* 15/9/1854

33 Blanch, p. 365

34 Dibir M. Mahomedov, 'Shamil's Last Testament', *Central Asian Survey*, vol. 21, no. 3 (2002), pp. 241–2

CHAPTER 8: EL PRESIDENTE

1 James M. McPherson, *Battle Cry Freedom: The Civil War Era* (Oxford,

1988), pp. 115–16

2 J. P. Parry, *The Politics of Patriotism: English Liberalism, National Identity and Europe, 1830–1886* (Cambridge, 2006), p. 238

3 *Times*, 2/2/1856; E. D. Steele, *Palmerston and Liberalism* (Cambridge, 1991), p. 58

4 *Edinburgh Review*, July 1856, p. 298

5 *Sacramento Daily Union*, 26/4/1870; Frédéric de Gaillardet, *Sketches of Early Texas and Louisiana* (Austin, TX, 1966), pp. 144ff.

6 For Sanders, his dinner party and the Young America movement see M. E. Curti, 'Young America', *The American History Review*, vol. 32, no. 1 (October 1926), pp. 48ff.

7 Potter, p. 190

8 McPherson, *Battle Cry*, p. 106

9 Robert F. Durden, 'J. D. B. de Bow: Convulsions of a Slavery Expansionist', *Journal of Southern History*, vol. 17, no. 4 (November 1951)

10 C. Stanley Urban, 'The Africanization of Cuba Scare, 1853–1855', *The Hispanic American Historical Review*, vol. 37, no. 1 (February 1957)

11 McPherson, *Battle Cry*, p. 109

12 *Times*, 24/3/1855; Royle, *Crimea*, pp. 86, 316ff.

13 *Times*, 28/11/1855

14 Oliphant, *Minnesota*, pp. 274ff.; Urban, p. 34.

15 Albert A. Woldman, *Lincoln and the Russians* (Cleveland, OH, 1952), p. 11

16 Hyam, p. 65

17 Frank A. Golder, 'Russian-American Relations During the Crimea War', *The American Historical Review*, vol. 31, no. 3 (April 1926), p. 464

18 *Edinburgh Review*, July 1856, p. 271

19 J. B. Conacher, 'British Policy in the Anglo-American Enlistment Crisis of 1855–1856', *Proceedings of the American Philosophical Society*, vol. 136, no. 4 (December 1992)

20 *Pennsylvanian*, 28/9/1855; Claude M. Fuess, *The Life of Caleb Cushing* (New York, 1923), vol. II, p. 172

21 *Times*, 10/11/1855; *Edinburgh Review*, July 1856, p. 267

22 *Age*, 17/4/1856, p. 2; Conacher, pp. 548–9, 553–4

23 Conacher, p. 550

24 Richard W. Van Alstyne (ed.), 'Anglo-American Relations, 1853–1857', *American Historical Review*, vol. 42, no. 3 (April 1937), pp. 495–7

25 Conacher, p. 534

26 *Morning Post*, 6/2, 12/2, 18/2/1856; *Brighton Herald*, 27/10/1855

27 Conacher, p. 568

28 *Morning Post*, 12/2, 12/9/1856; *Times* 2/2/1856; Don H. Doyle, *The Cause of All Nations: An International History of the American Civil War* (New York, 2015), p. 97
29 William G. Beasley, *Great Britain and the Opening of Japan, 1834–1858* (London, 1951), pp. 157–8; Train, *Young America*, p. 328
30 *New York Daily Tribune*, 21/4/1856
31 *Morning Post*, 6/2/1856; *Daily Telegraph*, 29/10/1855; *Boston Chronicle* quoted in *Times*, 13/3/1856
32 *Hansard*, 16 June 1856
33 Oliphant, *Patriots*, pp. 172–3
34 Ibid., p. 179
35 Ibid., p. 181
36 'Nicaragua and the Filibusters', *De Bow's Review*, vol. 20, no. 6 (June 1856), p. 673
37 Oliphant, *Patriots*, p. 193; 'Nicaragua and the Filibusters', *De Bow's Review*, vol. 20, no. 6 (June 1856), p. 673; 'The Experience of Samuel Absalom, Filibuster', *The Atlantic Monthly*, vol. IV, no. 26 (December 1859)
38 C. W. Doubleday, *Reminiscences of the Filibuster War in Nicaragua* (New York, 1886), pp. 92–3
39 *New York Herald*, 10/4/1856; Tennessee State Library and Archives: 'Grey-Eyed Man of Destiny':
www.tna.gov/tsla/exhibits/walker/index.htm
40 'Nicaragua', *De Bow's Review*, vol. 22, no. 1 (1 January 1857); *New York Daily Tribune*, 17/11/1856, 5/12/1856
41 'The Experience of Samuel Absalom, Filibuster', *The Atlantic Monthly*, vol. IV, no. 26 (December 1859); *New York Daily Tribune*, 6/5/1856
42 *New York Daily Tribune*, 23/10/1856
43 *Morning Post*, 5/12/1856
44 *New York Daily Tribune*, 6/6/1857; C. A. Bridges, 'The Knights of the Golden Circle: a filibustering fantasy', *Southwestern Historical Quarterly*, vol. XLIV (1941); 'Acquisition of Mexico – Filibustering', *De Bow's Review*, vol. 25, no. 6 (December 1858)
45 'Late Southern Convention at Montgomery', *De Bow's Review*, vol. 24, no. 6 (June 1858), pp. 603–4; 'Speech of Mr [Leonidas W.] Spratt', *De Bow's Review*, vol. 27, no. 2 (April 1859), pp. 210–11
46 Van Alstyne (ed.), pp. 499–500; Edward A. Pollard, 'The Central American Question', *De Bow's Review*, vol. 27, no. 5 (November 1859), pp. 550–661; *New York Daily Tribune*, 22/12/1856; *Morning Post*, 22/11/1856
47 'The Walker Expedition of 1856', *De Bow's Review*, vol. 24, no. 2 (February

1858), pp. 150–1; *New York Daily Tribune*, 14/4/1856

48 *Times*, 8/4/1852; Walter LaFeber, *The New Cambridge History of American Foreign Relations: The American Search for Opportunity, volume two, 1865–1913* (Cambridge, 2013), p. 8

CHAPTER 9: TSUNAMI

1 *United States Magazine and Democratic Review*, vol. 30 (April 1852), p. 332

2 M. William Steele, *Alternative Narratives in Modern Japanese History* (London, 2003), p. 15

3 Gregory Smits, 'Shaking Up Japan: Edo society and the 1855 catfish picture prints', *Journal of Social History*, vol. 39, no. 4 (Summer 2006)

4 William G. Beasley, *Japan Encounters the Barbarian: Japanese Travellers in America and Europe* (London, 1995), p. 30

5 Francis L. Hawks (ed.), *Narrative of the Expedition of an American Squadron to the China Seas and Japan* (Washington, DC, 1856), pp. 255–6

6 Smits, p. 1065

7 Ernest Mason Satow, *Japan 1853–1864, or, Genji Yume Monogatari* (Tokyo, 1905), p. 6; Beasley, *Japan Encounters*, p. 39

8 Beasley, *Japan Encounters*, pp. 39–40

9 Ibid., p. 44

10 Steele, *Alternative Narratives,* chapter 1, *passim*

11 Marius B. Jansen, *Sakamoto Ryoma and the Meiji Restoration* (Princeton, NJ, 1961), p. 83

12 Ibid., pp. 3ff

13 Ibid., pp. 77ff

14 Hawks, p. 17

15 *Times*, 8/4/1852

16 Hawks, p. 357

17 Richard Hildreth, *Japan: as it was and is* (New York, 1855), p. 522

18 Chushichi Tsuzuki, *The Pursuit of Power in Modern Japan, 1825–1995* (Oxford, 2000), pp. 40–41; Marius B. Jensen, *The Making of Modern Japan* (Cambridge, MA, 2000), p. 291

19 Hisao Furukawa, 'Meiji Japan's Encounter with Modernization', *Southeast Asian Studies*, vol. 33, no. 3 (December 1995), p. 507

20 Takehiko Hasimoto, 'Japanese Clocks and the Origin of Punctuality in Modern Japan', in *Historical Essays on Japanese Technology* (Tokyo, 2009), pp. 17ff.

CHAPTER 10: THE CIVILISING MISSION

1 Frederic E. Wakeman, *Strangers at the Gate* (Berkeley, CA, 1966), p. 76
2 Robert Fortune, *A Journey to the Tea Countries of China* (London, 1852), p. 3
3 Train, *Young America*, p. 76; Albert Smith, *To China and Back* (London, 1859), p. 39
4 J. D'Ewes, *China, Australia, and the Pacific Islands in the years 1855–56* (London, 1857), p. 240
5 E. J. Eitel, *Europe in China: the history of Hong Kong from the beginning to the year 1882* (Hong Kong, 1895), p. 571
6 David Todd, 'John Bowring and the Global Dissemination of Free Trade', *The Historical Journal*, vol. 51, no. 2 (June 2008), p. 379
7 John Bowring, *The Influence of Knowledge on Domestic and Social Happiness* (London, 1842), pp. 3–5; Todd, pp. 388–9
8 Eitel, p. 570
9 Train, *Young America*, p. 71
10 House of Commons, 3/3/1857, col. 1800
11 House of Commons, 26/2/1857, col. 1409
12 Jonathan Spence, *God's Chinese Son: The Taiping Heavenly Kingdom of Hong Xiuquan* (London, 1996), p. 134
13 Ibid., p. 137
14 Franz Michael, *The Taiping Rebellion: History and Documents* (3 vols, Seattle, WA, 1971), vol. III, p. 799
15 George Wingrove Cooke, *China: being* The Times *Special Correspondence from China in the years 1857–58* (London, 1858), pp. 67ff.
16 Ibid., pp. 68–9
17 Christopher Munn, 'Colonialism "in a Chinese atmosphere": the Caldwell affair and the perils of collaboration in early colonial Hong Kong', in Robert Bickers and Christian Henriot, *New Frontiers: Imperialism's New Communities in East Asia, 1842–1953* (Manchester, 2000), pp. 10ff.
18 Sir John Bowring, *Autobiographical Recollections* (London, 1877), p. 218
19 Todd, pp. 393–5
20 J. Y. Wong, *Deadly Dreams: Opium, Imperialism and the Arrow War* (Cambridge, 1998), p. 85.
21 Gerald S. Graham, *The China Station: War and Diplomacy, 1830–1860* (Oxford, 1978), p. 287; *Argus*, 3/10/1855
22 J. D. Grainger, *The First Pacific War: Britain and Russia, 1854–1856* (Woodbridge, 2008), p. 63

23 Graham, pp. 290ff.
24 Mark Bassin, 'The Russian Geographical Society, the "Amur Epoch", and the Great Siberian Expedition 1855–1863', *Annals of the Association of American Geographers*, vol. 73, no. 2 (June 1983), p. 246
25 Grainger, pp. 53ff., 173
26 Norman Saul, 'An American's Siberian Dream', *Russian Review*, vol. 37, no. 4 (October 1978), pp. 405–20
27 Graham, p. 293
28 Bassin, p. 243
29 David O. Allen, *India, Ancient and Modern* (Boston, MS, 1856), pp. 353–4
30 Mikhail Volodarsky, 'Persia and the Great Powers, 1856–1869', *Middle Eastern Studies*, vol. 19, no. 1 (January 1983)
31 *Times*, 29/12/1859, p. 8; Hopkirk, p. 289
32 Graham, p. 290
33 Ibid., p. 298
34 Stanley Lane-Poole, *The Life of Sir Harry Parkes* (London, 1894), pp. 245, 262
35 Ibid., p. 262
36 Ibid., p. 245
37 House of Lords, 24/2/1857, col. 1220
38 Theodore Walrond (ed.), *Letters and Journals of James, Eighth Earl of Elgin* (London, 1872), pp. 209, 213

CHAPTER 11: RETRIBUTION

1 E. D. Steele, p. 204
2 Walrond, p. 250
3 P. J. O. Taylor, *Chronicles of the Mutiny* (Delhi, 1992), pp. 23–4; Laurence Oliphant, *Narrative of the Earl of Elgin's Mission to China and Japan* (New York, 1860), pp. 26–7
4 Taylor, p. 24; George MacMunn, 'Mees Dolly: an untold tragedy of 57', *Cornhill Magazine*, 63 (July–December 1927), pp. 327–31
5 Elizabeth Muter, *Travels and Adventures of an Officer's Wife in India, China and New Zealand* (2 vols, London, 1864), vol. I, pp. 2ff.
6 Jane Robinson, *Angels of Albion: Women of the Indian Mutiny* (London, 1996), p. 33
7 John Kaye, *A History of the Sepoy War in India, 1857–1858* (3 vols, London, 1864), vol. I, p. 595

8 Edward Vibart, *The Sepoy Mutiny as seen by a subaltern* (London, 1898), pp. 252ff.; 'How the Telegraph Saved India', *Daily News*, 29/9/1897

9 John Kaye and G. B. Malleson, *Kaye and Malleson's History of the Indian Mutiny* (3 vols, London, 1892) , vol. II, p. x.

10 Deep Kanta Lahiri Choudhury, *Telegraphic Imperialism: Crisis and Panic in the Indian Empire* (Basingstoke, 2010), p. 40; Kaye and Malleson, vol. II, p. 193n, vol. VI, p. 71

11 Manindra Nath Das, *Studies in the Economic and Social Development of Modern India: 1848–56* (Calcutta, 1959), pp. 102–3

12 John Clark Marsham, *The History of India: from the earliest period to the close of Lord Dalhousie's administration* (3 vols, London, 1867), vol. III, p. 441

13 *Railway Times*, 15/1/1853; *ILN*, 4/6/1853

14 J. G. A. Baird, *Private Letters of the Marquis of Dalhousie* (Edinburgh, 1910), p. 284; Edwin Arnold, *The Marquis of Dalhousie's Administration of British India* (2 vols, London, 1865), vol. II, p. 164

15 Baird, pp. 169, 348

16 William Howard Russell, *My Diary in India, in the year 1858–9* (London, 1860), vol. I, pp. 253–4; cf. Minturn, p. 166

17 Ferdinand Mount, *The Tears of the Rajas: Mutiny, Money and Marriage in India, 1805–1905* (London, 2015), pp. 284ff.

18 Joan Leopold, 'British Applications of the Aryan Theory of Race to India, 1850–1870', *EHR*, vol. 89, no. 352 (July 1974), pp. 584–5 fn 4. For other criticism of the word 'nigger' see *Times* 20/10/1858; G. O. Trevelyan, *Cawnpore* (London, 1865), p. 36; Russell, *My Diary*, vol. I, p. 194; Robert Montgomery Martin, *The Indian Empire: its history, topography, government, finance, commerce and staple products* (3 vols, London, 1858), vol. II, pp. 11, 123

19 Kaye, vol. I, pp. 190–93

20 Ibid., p. 472

21 Kaye and Malleson, vol. I, p. 345

22 Sita Ram Pande, *From Sepoy to Subedar* (ed. James Lunt, London, 1970), pp. 24–5

23 Pande, pp. 26, 73, 173

24 J. A. B. Palmer, *The Mutiny Outbreak in Meerut in 1857* (Cambridge, 1966), p. 32

25 *New York Daily Tribune*, 15/7/1857; 'The Beginning of the End', *United States Democratic Review*, vol. 40, no. 5 (November 1857); *New York Herald*, 31/7/1857, 27/9/1857

26 Saul David, *The Indian Mutiny* (London, 2002), p. 282; Lawrence James, *Raj: The Making and Unmaking of British India* (London, 1997), pp. 258–277

27 Vibart, pp. 265–6
28 Ibid., p. 252; *Daily News*, 29/9/1897
29 F. D. Goldsmid, *Telegraph and Travel: a narrative of the formation and development of telegraphic communication between England and India* (London, 1874), pp. 32, 37–8
30 Kaye, *History*, vol. I, pp. 439, 442, 451, 548, 602, 612, 614, 615
31 Kaye, *History*, vol. II, pp. 121, 151
32 Goldsmid, pp. 38ff.
33 Ibid., pp. 32ff.; Lord Roberts of Kandahar, *Forty-One Years in India* (2 vols, New York, 1898), vol. I, pp. 115, 141 297, 405, 456
34 Kaye and Malleson, vol. II, p. 302; David, pp. 230ff.
35 William Dalrymple, *The Last Mughal: The fall of Delhi, 1857* (London, 2006), p. 315
36 David, pp. 258–9
37 Dalrymple, pp. 385–6
38 Anne Taylor, *Laurence Oliphant, 1829–1888* (Oxford, 1982), p. 50
39 Walrond, p. 199
40 Ibid., p. 212
41 Cooke, *China*, p. 315; Oliphant, *Narrative*, p. 97; Walrond, p. 214.
42 Walrond, p. 215
43 Cooke, *China*, p. 339
44 Walrond, pp. 251, 253
45 Ibid., p. 254
46 Masataka Banno, *China and the West, 1858–1861: The Origins of the Tsungli Yamen* (Cambridge, MS, 1964), p. 25
47 Ibid., p. 26
48 Sherard Osborn, *A Cruise in Japanese Waters* (Edinburgh, 1859), p. 127; Jansen, *Making of Modern Japan*, p. 283
49 Walrond, pp. 260, 261, 263, 265, 268, 282
50 *Times* 2/11/1858
51 *Times*, 4/2/1859
52 Jansen, *Making of Modern Japan*, pp. 280ff.
53 Walrond, p. 274
54 Russell, *My Diary*, vol. I, pp. 218ff.
55 Ibid., pp. 327ff.
56 Aditi Vatsa, 'When telegraph saved the empire', *The Indian Express*, 19/11/2012; H. C. Fanshawe, *Delhi Past and Present* (London, 1902), pp. 18, 331; Louis Tracy, *The Red Year: a story of the Indian Mutiny* (New York, 1907), p. 258

CHAPTER 12: EMPIRE OF NEWS

1 *ILN*, 26/9/1857
2 *ILN*, 24/8/1861
3 *Genesee County Herald*, 26/6/1858. For American perceptions of the global importance of the Rebellion see, for example, the *New York Herald*, 27/9/1857
4 *Times* 25/8/1858
5 *The History of* The Times*: The Tradition Established, 1841–1884* (London, 1951), p. 87
6 G. A. Cranfield, *The Press and Society: From Caxton to Northcliffe* (London, 1978), p. 207
7 Much of the following discussion draws on Graham Dawson, *Soldier Heroes: British Adventure, Empire and the Imagining of Masculinities* (London, 1994) and Christopher Herbert, *War of No Pity: The Indian Mutiny and Victorian Trauma* (Princeton, NJ, 2008), pp. 22ff.
8 Dawson, pp. 87ff., 94ff; Herbert, pp. 22ff.
9 Dawson, pp. 98–9
10 *Times*, 17/9/1857; Charles Ball, *The History of the Indian Mutiny* (London, 1858), vol. I, p. 75; Alexander Duff, *The Indian Rebellion: its causes and results* (New York, 1858), pp. 24, 63
11 Charles Dickens, *Letters from Charles Dickens to Angela Burdett-Coutts, 1841–1865* (London, 1953), p. 350; cf. letter to the editor, *Times*, 8/8/1857
12 Russell, *My Diary*, vol. I, pp. 2, 92, 117
13 Ball, vol. I, pp. 340ff., 379n, 380
14 Dawson, p. 78
15 John Charles Pollock, *Way to Glory: The Life of Havelock of Lucknow* (London, 1957), p. 153
16 William Brock, *A Biographical Sketch of Sir Henry Havelock* (London, 1864), pp. 133–4; *Times*, 13/1/1858
17 Roberts, *Forty-One Years*, vol. I, p. 60; for Nicholson's extraordinary career see Charles Allen, *Soldier Sahibs: The Men Who Made the North-West Frontier* (London, 2000)
18 R. G. Wilberforce, *An Unrecorded Chapter of the Indian Mutiny* (London, 1894), pp. 215–16
19 Kaye and Malleson, vol. III, pp. 301–2; Dalrymple, p. 307; Allen, p. 221
20 Anon., *The Siege of Delhi. By an officer who served there* (Edinburgh, 1861), p. 224

21 'Why Shave?', *Household Words*, 15/8/1853, vol. VII, pp. 560–63
22 'David', *The Beard! Why do we cut it off? An analysis of the controversy concerning it, and an outline of its history* (London, 1854), p. 7
23 *ILN*, 17/2/1855
24 *Times*, 14/11/1857
25 *Times*, 19/11/1857
26 David, *The Beard!*, pp. 6–7
27 *Times*, 14/11/1857
28 Ibid.
29 John Tosh, *Manliness and Masculinities in Nineteenth-century Britain* (Harlow, 2005), p. 94
30 Mrs [Sarah] Ellis, *The Women of England: their social duties and domestic habits* (London, 1839), pp. 12, 16, 22, 28
31 Ibid., pp. 50ff.
32 Ibid., pp. 16, 21, 39
33 Ibid., pp. 39, 50ff.
34 Robinson, *Angels*, pp. 131ff.; Dawson, p. 97; Herbert, pp. 164ff.
35 R. M. Coopland, *A Lady's Escape from Gwalior and Life in the Fort of Agra* (London, 1859), p. 116
36 Kaye, *History*, vol. I, p. xii; Ball, *History*, vol. I, p. 81
37 *Times*, 14/11/1857
38 Parry, pp. 69ff.
39 *ILN*, 26/9/1857
40 Parry, p. 10
41 *Times*, 14/11/1857
42 *Times*, 20/9/1858
43 House of Commons, 15/6/1858, col. 2113
44 *ILN*, 26/6/1858
45 Henry Adams, *The Education of Henry Adams* (Boston, MS, 1948), p. 59

CHAPTER 13: MASTER OF TIME

1 Briggs and Maverick, p. 21
2 'The Ocean Telegraph', *Frank Leslie's Illustrated Newspaper*, 21/8/1858; William Howard Russell, *The Atlantic Telegraph* (London, 1866), p. 26
3 '1858 New York Celebration', http://atlantic-cable.com/1858NY/index.htm
4 James L. Huston, 'Western Grains and the Panic of 1857', *Agricultural History*, vol. 57, no. 1 (January 1983), p. 16

5 Oliphant, *Episodes*, p. 74
6 Kynaston, p. 193
7 Wills, *Boosters*, p. 93; Hamer, pp. 157–8
8 Williams, *History of the City of St Paul*, p. 385; *Times*, 6/8/1858
9 'Memorabilia, Ephemera, and Promotional Material', http://atlantic-cable.com/Souvenirs/index.htm
10 *New York Times*, 27/8/1858, p. 1; Huston, p. 19
11 Briggs and Maverick, pp. 12, 22
12 *Times*, 25/8/1858; Briggs and Maverick, pp. 14, 20–21
13 Henry M. Field, *The Story of the Atlantic Telegraph* (New York, 1892), p. 220
14 Ibid., p. 218
15 *Times*, 25/8/1858
16 For a digest of world opinion on the Rebellion see the *New York Herald*, 27/9/1857
17 Donald Read, *The Power of News: The History of Reuters, 1849–1989* (Oxford, 1992), pp. 21–2
18 Ibid., pp. 26, 32–3
19 *Times*, 11/1/1859
20 *Times*, 14/1/1859; 21/1/1859
21 *Nashville Union and American*, 29/1/1859, p. 3; *New York Daily Tribune*, 29/1/1859, p. 54; Huston, p. 19
22 Read, p. 25; *Times*, 8/2/1859
23 Read, p. 26

CHAPTER 14: BEST OF TIMES, WORST OF TIMES

1 *New York Times*, 30/8/1859; 5/9/1859; Prescott, *History, Theory and Practice of the Electric Telegraph* (Boston, MA, 1866), pp. 321ff.; 'The Late Aurora Borealis and the Telegraph', *The Journal of Education for Upper Canada*, vol. XII, no. 9 (September 1859), p. 132; J. L. Green and S. Boardsen, 'Eyewitness Reports on the Great Auroral Storm of 1859', *Advances in Space Research*, vol. 38, no. 2 (2006), pp. 145–53; M. A. Shea and D. F. Smart, 'Compendium of the Eight Articles on the "Carrington Event" Attributed to or Written by Elias Loomis in the *American Journal of Science*, 1859–1861', ibid., pp. 313–85; *Times*, 19/9/1859; *ILN*, 24/9/1859; *Empire* (Sydney), 19/11/1859; *Moreton Bay Courier* 7/9/1859; *Argus*, 1/9/1859, 3/9/1859; *Bendigo Advertiser*, 30/10/1859; *Sydney Morning Herald*, 10/9/1859
2 Alvar Ellegard, *Darwin and the General Reader: The Reception of Darwin's*

Theory of Evolution in the British Periodical Press, 1859–1872 (Chicago, IL, 1958), pp. 43–4

3 Ibid., p. 35

4 *Westminster Review*, 17, 1860, p. 166

5 John K. Fairbank, *Trade and Diplomacy on the China coast: The Opening of the Treaty Ports, 1842–1854* (Cambridge, MS, 1953), vol. I, p. 173

6 Christian Wolmar, *Engines of War: How Wars were Won and Lost on the Railways* (London, 2010), p. 31

7 Imlah, p. 17

8 Lucy Riall, *Garibaldi: Invention of a Hero* (New Haven, CT, 2007), pp. 128ff., 185ff.

9 Howard R. Marraro, *American Opinion on the Unification of Italy, 1846–1861* (New York, 1932), p. 234

10 Parry, pp. 223, 232

11 *ILN*, 21/5/1859, p. 486

12 Mary J. Philips-Matz, *Verdi: A Biography* (Oxford, 1993), p. 394

13 *Times*, 25/1/1859, p. 5

14 *New York Daily Tribune*, 9/12/1859, p. 6

15 *Times*, 27/9/1859

16 Banno, pp. 110ff.; *New York Daily Tribune*, 15/11/1859, p. 4

17 *Times*, 12/9/1859

18 Ibid.

19 Graham, p. 383

20 G. J. Wolseley, *Narrative of the War with China in 1860* (London, 1862), p. 189

21 Robert Swinhoe, *Narrative of the North China Campaign of 1860* (London, 1861), p. 301

22 Ibid., p. 306; Wolseley, p. 226

23 Wolseley, pp. 273–4

24 Walrond, p. 366; Swinhoe, p. 330

25 Wolseley, p. 280

26 Ibid., p. 279

27 Wm. Theodore de Bary, Carol Gluck and Arthur E. Tiedemann (eds), *Sources of Japanese Tradition: volume 2, 1600 to 2000* (New York, 2005), p. 654; Stephen B. Oates, *To Purge This Land with Blood: A Biography of John Brown* (New York, 1970), p. 61; Oliphant, *Episodes*, p. 169

28 Riall, p. 172; *Times*, 26/7/1859, p. 9

29 *Weekly Arizonian*, 27/10/1859, p. 1

30 Ibid.; *Times*, 14/11/1859

31 For Garibaldi as the pre-eminent celebrity of his age, and the power this celebrity gave him, see Riall's enthralling account, especially chapters 6 and 7; *Times*, 14/11/1859

32 *Harper's Weekly*, 9/6/1860, p. 354; Oates, p. 272; de Bary (et al.), pp. 654, 656

33 Oates, p. 351; Beasley, *Japan Encounters*, p. 43

34 *The Campaign of 1860: comprising the speeches of Abraham Lincoln . . .* (etc.) (Albany, NY, 1860), p. 8; de Bary (et al.), p. 657; Oates, pp. 335, 338

35 James Murdoch, *History of Modern Japan* (3 vols, London, 1925), vol. III, p. 702

36 Potter, p. 384

37 *Times*, 31/12/1859, pp. 6–7

38 *The History of* The Times, p. 290; *Times*, 13/6/1860

39 *Times*, 13/6/1860, p. 12

40 *Harper's Weekly*, 9/6/1860, p. 354

41 Oliphant, *Episodes*, pp. 183–4

42 *Harper's Weekly*, 9/6/1860, p. 354

43 McPherson, *Battle Cry*, p. 254

44 Edward McPherson, *The Political History of the United States of America During the Great Rebellion* (Washington, DC, 1865), p. 15

45 McPherson, *Battle Cry*, p. 237

46 Ibid., p. 259

47 Henry Cleveland, *Alexander H. Stephens, in Public and Private: with letters and speeches* (Philadelphia, PA, 1866), pp. 718, 721

48 William Howard Russell, *My Diary North and South* (Boston, MS, 1863), pp. 84, 92, 98

49 Eric Foner, *Free Soil, Free Labour, Free Men: The Ideology of the Republican Party before the Civil War* (Oxford, 1995), p. 223

50 E. Merton Coulter, *The Confederate States of America, 1861–1865* (Baton Rouge, LA, 1950), p. 57

CHAPTER 15: BLOOD, IRON, COTTON, DEMOCRACY

1 Elpis Melena [Esperanza von Schwartz], *Recollections of General Garibaldi* (London, 1861), p. 224

2 *ILN*, 9/2/1861, p. 113

3 Melena, p. 223

4 *New York Daily Tribune*, 13/8/1861

5 Russell, *My Diary North and South*, pp. 453, 467

6 Doyle, *Cause*, pp. 19ff.
7 Ibid., p. 20
8 Abraham Lincoln, 'First Inaugural Address, Monday March 4 1861', http://www.bartleby.com/124/pres31.html
9 Doyle, *Cause*, p. 24
10 Frederick Burkhardt (ed.), *The Correspondence of Charles Darwin: volume 9, 1861* (Cambridge, 1994), p. 163
11 Huston, pp. 23–4
12 W. H. Chase, 'The Secession of the Cotton States: its status, advantages, and its power', *De Bow's Review*, vol. 30, no. 1 (January 1861), pp. 94–5; John Henry Hammond, *Selections from the Letters and Speeches*, pp. 316–17.
13 Marc-William Palen, 'The Great Civil War Lie', *New York Times*, 5/6/2013
14 Chase, 'Secession', pp. 93–4
15 *New York Times*, 'The Great Question', 30/3/1861; *Times*, 12/3 and 5/5/1861; Martin Crawford, *The Anglo-American Crisis of the Mid-Nineteenth Century:* The Times *and America, 1850–1862* (Athens, GA, 1987), pp. 93ff.
16 *Times*, 8/3/1861; August Belmont, *A Few Letters and Speeches of the Late Civil War* (New York, 1870), pp. 63–4
17 W. Gilmore Simms, 'Our Commissioners to Europe', *De Bow's Review*, vol. 31, no. 4 (October–November 1861), p. 415
18 Chase, 'Secession', p. 94
19 Frederic Bancroft (ed.), *Speeches, Correspondence and Political Papers* of Carl Schurz (6 vols, New York, 1913), vol. I, p. 187
20 *Times*, 13/5/1861; Stephen R. Platt, *Autumn in the Heavenly Kingdom: China, the West and the epic story of the Taiping Civil War* (London, 2012), pp. 231ff.
21 *Times*, 31/5/1861
22 'Democracy on Trial', *Quarterly Review*, July 1861, pp. 256, 283
23 Malcolm C. McMillan, *The Alabama Confederate Reader* (Tuscaloosa, AL, 1963), p. 103
24 United States War Department, *The War of the Rebellion*, vols 1–8, serial numbers 114–121 (Washington, DC, 1897), p. 1107
25 Russell, *My Diary North and South*, p. 587
26 Amanda Foreman, *A World on Fire: An Epic History of Two Nations Divided* (London, 2010), pp. 236–7
27 Doyle, *Cause*, pp. 225ff.
28 Sheridan Gilley, 'The Garibaldi Riots of 1862', *Historical Journal*, vol. 16, no. 4 (1973), pp. 697–732
29 *Times*, 3/10/1862, p. 7

30 *Times*, 9/10/1862, pp. 7–8
31 Doyle, p. 235
32 Ibid., 232
33 *Times*, 24/1 and 8/4/1861
34 *Times*, 14/6/1862; Beckert, pp. 256ff.
35 G. W. MacGeorge, *Ways and Works in India* (London, 1894), p. 349
36 Ian J. Kerr, *Engines of Change: The Railroads that Made India* (Westport, CT, 2007), pp. 40ff.; Richard Temple, *Men and Events of My Time in India* (London, 1882), pp. 255–6
37 Kerr, pp. 46ff.
38 Temple, p. 269
39 Ibid., pp. 268–70; G. E. Tindall, *City of Gold: The Biography of Bombay* (London, 1982), p. 223
40 William Perry Fogg, *Round the World: letters from Japan, China, India and Egypt* (Cleveland, OH, 1872), p. 198
41 *Times*, 8/4/1861, 14/6/1862; *Morning Post*, 26/11/1862
42 *Times*, 14/6/1862
43 *Times*, 14/6 and 12/12/1862
44 Beckert, p. 264
45 George E. Baker (ed.), *The Works of William H. Seward* (5 vols, Boston, MA, 1884), vol. V, p. 363
46 Seymour Drescher, *The Mighty Experiment: Free Labour Versus Slavery in British Emancipation* (Oxford, 2002); *Economist*, 21/5/1853, 17/12/1853, 10/10/1857; *Times*, 19/12/1857, 14/6/1862
47 [Nassau Senior], 'Slavery in the United States', *Edinburgh Review*, vol. 101, no. 206 (April 1855), p. 331
48 *Times*, 8/3/1861; Belmont, p. 64
49 Fogg, p. 162; Charles Dilke, *Greater Britain: a record of travel in English-speaking countries during 1866 and 1867* (London, 1869), pp. 426ff. See also Herbert, *War of No Pity*, pp. 164ff.
50 Platt, pp. 335ff., 359
51 Swinhoe, pp. 330–31; Erik Baark, *Lightning Wires: The Telegraph and China's Technological Modernisation, 1860–1890* (Westport, CT, 1997), p. 69
52 Derek Beales, *England and Italy, 1859–60* (Edinburgh, 1861), pp. 166ff.
53 Parry, pp. 242ff.
54 Read, *Power*, pp. 34–5; Steven Roberts, 'Bridging the Gap – News Telegraphs, 1863–1870', http://atlantic-cable.com/Article/NewsTelegraphs/index.htm; John Entwisle, 'Fourscore and Seven Years Ago', http://blog.thomsonreuters.com/index.php/fourscore-and-seven-years-ago/

EPILOGUE: 1873

1 George F. Train, *My Life in Many States and in Foreign Lands* (New York, 1902), pp. 339–40
2 Dilke, p. 196
3 Thomas Cook, *Letters from the Sea and from Foreign Lands, Descriptive of a Tour Round the World* (London, 1873), p. 13
4 Ibid., pp. 14–15, 17, 22
5 Fogg, p. 80
6 James Brooks, *A Seven Months' Run Up, and Down, and Around the World* (New York, 1872), p. 283
7 Brooks, p. 314
8 Ibid., pp. 9, 50–51
9 Cook, p. 31
10 Stephen E. Ambrose, *Nothing Like it in the World: The Men who Built the Transcontinental Railroad, 1863–1869* (New York, 2000), p. 266
11 'The Circassian Exodus', *Times*, 7/2/1860, p. 6; Charles King, *The Ghost of Freedom: A History of the Caucasus* (Oxford, 2008), pp. 94ff., 123ff.
12 Tindall, p. 223
13 Fogg, p. 198
14 Ibid., p. 198; Temple, p. 273
15 Beckert, p. 297
16 Ibid., pp. 294–7, 302–3, 336–7
17 Ian Nish, 'Introduction', in Ian Nish (ed.), *The Iwakura Mission in America and Europe: A New Assessment* (Richmond, 1998), p. 8
18 Ulrich Wattenberg, 'Germany: . . . An Encounter Between Two Emerging Countries' in ibid., p. 76
19 Charles Kingsley, 'The Natural Theology of the Future', *Scientific Lectures and Essays* (London, 1893), p. 324
20 Burton, p. 476n; Robert Tombs, *France, 1814–1914* (Abingdon, 2014), p. 399
21 House of Commons, 5/8/1867, cols 876–884
22 E.D. Steele, p. 37
23 Parry, pp. 273–4; R. Faber, *The Vision and the Need: Late Victorian Imperialist Aims* (London, 1966), p. 54
24 A. J. Sargent, *Anglo-Chinese Commerce and Diplomacy* (Oxford, 1907), p. 109
25 W. G. Wiebe (et al.), *Benjamin Disraeli. Letters, volume 3: 1860–1864* (Toronto, 2009), p. 233

INDEX